LONG TERM DURABILITY OF STRUCTURAL MATERIALS

Elsevier Science Internet Homepage

http://www.elsevier.nl (Europe)
http://www.elsevier.com (America)
http://www.elsevier.co.jp (Asia)

Consult the Elsevier homepage for full catalogue information on all books, journals and electronic products and services.

Elsevier Titles of Related Interest

CHAN & CHUI
Non-Linear Static and Cyclic Analysis of Steel frames with Semi-Rigid Connections.
ISBN: 008-042998-X

CHAN & TENG
ICASS '99, Advances in Steel Structures. (2 Volume Set)
ISBN: 008-043015-5

DUBINA
SDSS '99 – Stability and Ductility of Steel Structures.
ISBN: 008-043016-3

FRANGOPOL, COROTIS & RACKWITZ
Reliability and Optimization of Structural Systems.
ISBN: 008-042826-6

FUKUMOTO
Structural Stability Design.
ISBN: 008-042263-2

GODOY
Thin-Walled Structures with Structural Imperfections: Analysis and Behavior.
ISBN: 008-042266-7

KELLY & ZWEBEN
Comprehensive Composite Materials.
ISBN: 008-042993-9

KO & XU
Advances in Structural Dynamics (ASD 2000).
(2 Volume Set)
ISBN: 008-043792-3

LUNDQUIST *ET AL.*
Life Cycle Engineering of Plastics.
ISBN: 008-043886-5

MAKELAINEN & HASSINEN
Light-Weight Steel and Aluminium Structures (ICSAS'99).
ISBN: 008-043014-7

OHTSUBO & SUMI
Proceedings of the 14th International Ship and Offshore Structures Congress. (3 Volume Set)
ISBN: 008-043602-1

OWENS
Steel in Construction. (CD-ROM Proceedings with Printed Abstracts Volume, 268 papers)
ISBN: 008-042997-1

USAMI & ITOH
Stability and Ductility of Steel Structures.
ISBN: 008-043320-0

VASILIEV & MOROZOV
Mechanics and Analysis of Composite Materials.
ISBN: 008-042702-2

WANG, REDDY & LEE
Shear Deformable Beams and Plates.
ISBN: 008-043784-2

ZARAS, KOWAL-MICHALSKA & RHODES
Thin-Walled Structures – Advances and Developments. Proceedings of the 3rd International Conference.
ISBN: 008-043955-1

ZHAO & GRZEBIETA
Structural Failure and Plasticity (IMPLAST 2000).
ISBN: 008-043875-X

ZINGONI
Structural Engineering, Mechanics and Computation (SEMC 2001). (2 Volume Set)
ISBN: 008-043948-9

Related Journals

Free specimen copy gladly sent on request. Elsevier Science Ltd, The Boulevard, Langford Lane, Kidlington, Oxford, OX5 1GB, UK

Applied Ocean Research
Advances in Engineering Software
CAD
Coastal Engineering
Composite Structures
Computers and Structures
Construction and Building Materials
Engineering Failure Analysis
Engineering Fracture Mechanics
Engineering Structures
Finite Elements in Analysis and Design
International Journal of Solids and Structures
Journal of Constructional Steel Research
Marine Structures
NDT & E International
Ocean Engineering
Structural Safety
Thin-Walled Structures

To Contact the Publisher

Elsevier Science welcomes enquiries concerning publishing proposals: books, journal special issues, conference proceedings, etc. All formats and media can be considered. Should you have a publishing proposal you wish to discuss, please contact, without obligation, the publisher responsible for Elsevier's civil and structural engineering publishing programme:

Ian Salusbury
Senior Publishing Editor
Elsevier Science Ltd
The Boulevard, Langford Lane
Kidlington, Oxford
OX5 1GB, UK

Phone: +44 1865 843425
Fax: +44 1865 843920
E.mail: i.salusbury@elsevier.co.uk

General enquiries, including placing orders, should be directed to Elsevier's Regional Sales Offices – please access the Elsevier homepage for full contact details (homepage details at the top of this page).

LONG TERM DURABILITY OF STRUCTURAL MATERIALS

DURABILITY 2000

Proceedings of the Durability Workshop,
Berkeley, California, 26–27 October, 2000

Edited by

P.J.M. Monteiro
Department of Civil and Environmental Engineering,
University of California at Berkeley, USA

K.P. Chong
Mechanics and Materials Program,
National Science Foundation, USA

J. Larsen-Basse
Surface Engineering and Material Design Program,
National Science Foundation, USA

K. Komvopoulos
Department of Mechanical Engineering,
University of California at Berkeley, USA

2001
ELSEVIER
AMSTERDAM – LONDON – NEW YORK – OXFORD – PARIS – SHANNON - TOKYO

ELSEVIER SCIENCE Ltd
The Boulevard, Langford Lane
Kidlington, Oxford OX5 1GB, UK

First edition 2001

British Library Cataloguing in Publication Data
Durability 2000 (2000 : Berkeley, California)
Long term durability of structural materials : proceedings of the Durability Workshop, Berkeley, California, 26-27 October, 2000
1.Building materials - Service life - Congresses
I.Title II.Monteiro, Paulo J. M.
691

ISBN 0080438903

Library of Congress Cataloging in Publication Data
A catalog record from the Library of Congress has been applied for.

ISBN: 0-08-043890-3

Transferred to digital printing 2005
Printed and bound by Antony Rowe Ltd, Eastbourne

FOREWORD

This volume addresses the basic mechanisms of long term deterioration of engineering materials and the development of short term laboratory or in-situ tests which will allow prediction of the extent of this long term deterioration. All machines, structures and facilities deteriorate over a period of years or decades and eventually wear out, break down, or become unproductive or unsafe. Replacement costs to counter the deterioration are enormous. For instance, the US Federal Highway Administration (FHWA) estimates that in the US there is an annual accruing "bridge deficiency" of $ 2.3 billion (in 1980 dollars) and that the total expenditure for bridge repair and replacement from 1981 through 2000 was $ 102.6 billion. Similar costs can be documented in many other sectors, from transportation and manufacturing to electronics and MEMS. Understanding how to design components and structures for optimal life performance is therefore very important and becomes essential when new materials or new application areas are considered. Of high importance is also the understanding of how to determine the optimal time to replace existing components or structures and development of techniques to prolong their useful life. Once this understanding has been reached and it has become possible to closely determine the life cycle of the most critical components of a structure or system, then it becomes feasible to develop design methodology and/or inspection, monitoring, and replacement strategies which allow significant extension of the life of the complete system. This will result in significant savings to society. In order to reach this desirable goal it is necessary to closely understand deterioration mechanisms at different scales and to have available short term tests that can be used to reliably predict long term deterioration, durability and performance, as discussed in this volume.

Objectives of the workshop

This book contains the proceedings of a workshop held at Berkeley, CA in October 2000. It brought together engineers and scientists, who have received research grants from the National Science Foundation under the 1998 initiative "Long Term Durability of Materials and Structures: Modeling and Accelerated Techniques" (NSF 98-42). The purpose was to share results from the study of long-term durability of materials and structures. The major objective was to develop new methods for accelerated short-term laboratory or in-situ tests, which allow accurate, reliable prediction of long-term performance of materials, machines and structures. To achieve this goal it was important to understand the fundamental nature of the deterioration and damage processes in materials and to develop innovative ways to model the behavior of these processes as they affect the life and long-term performance of components, machines and structures. The researchers discussed their approach to include size effects in scaling up from laboratory specimens to actual structures. Accelerated testing and durability modeling techniques developed were validated by comparing their results with performance under actual operating conditions. The main mechanisms of the deterioration discussed included environmental effects and/or exposure to loads, speeds and other operating conditions that are not fully anticipated in the original design. A broad range of deterioration damage, such as fatigue, overload, ultraviolet damage, corrosion, and wear was presented. A broad range of materials of interest was also discussed, including the full spectrum of construction materials, metals, ceramics, polymers, composites, and coatings. Emphasis was placed on scale-dependence and history of fabrication on resulting mechanical behavior of materials from the macroscale to the microscale.

In summary, the objective of this workshop was to establish a holistic discussion of deterioration mechanisms relevant to structural and construction materials. Topics included the physics and chemistry of the deterioration mechanism, develop new equipment to determine the degree of distress caused by the deterioration and test the new methodology in field conditions. We hope that the results of the workshop can lead to improved durability, life cycle performance, safety, reduced maintenance and lower cost which in turn should lead to superior machines and structures.

Paulo J.M. Monteiro
Ken P. Chong
Jorn Larsen-Basse
Kyriakos Komvopoulos

Editors

WORKSHOP ATTENDEES

Javier Balma
Civil & Environmental Engineering
University of Kansas
2008 Learned Hall
Lawrence KS 66045
785-864-3826
jbalma@ukans.edu

Zdenek P. Bazant, PhD, SE
Walter P. Murphy Professor of Civil
Engineering & Materials Science
Northwestern University
Evanston IL 60208-3109
847-491-4025/848-491-3351
z-bazant@northwestern.edu

Raimondo Betti, PhD
Civil Engineering & Mechanics
Columbia University
610 Mudd Bldg
New York NY 10027
212-854-6388
betti@civil.columbia.edu

Zednek Bittnar
Chair, Dept of Structural Mechanics
Czech Technical University
Prague 6, Czeck Republic
++420-2-2435-3869
bittnar@fsv.cvut.cz

Scott Case, PhD
Materials Response Group
Engineering Science & Mechanics
121 -C Patton Hall, MC 0219
Virginia Tech University
Blackburg VA 24061
540-231 -3140
scase@vt.edu

Ken P. Chong, PhD, PE
Director, Mechanics & Materials Program
CMS/Engineering Directorate
National Science Foundation
4201 Wilson Blvd, Suite 545
Arlington VA 22230
703-292-7008
kchong@nsf.gov

Richard M. Christensen, PhD
Research Professor
Aeronautics & Astronautics
Stanford University
Durand Bldg, Rm. 387A
Stanford CA 94305-40035
christensen@stanford.edu

Julio F. Davalos, PhD
C.W. Benedum Distinguished Teaching
Professor, Civil & Environmental
Engineering
College Engineering & Mineral Resources
ESB, Rm 611, Evansdale Drive
West Virginia University
Morgantown WV 26506-6103
304-292-3031, x.2632
davalos@cemr.wvu.edu

Grace Hsuan, PhD
475 Kedron Avenue
Folsom PA 19033
215-895-2785
ghsuan@cbis.ece.Drexel.edu

Y.C. Jerry Jean, PhD
Chemistry and Physics
Chair, Dept of Chemistry
University of Missouri-Kansas City
5009 RockNill Road
Kansas City MO 64110
816-235-2295
jeany@umkc.edu

Christopher H.M. Jenkins, PhD, PE
Director, Compliant Structures Laboratory
Mechanical Engineering
501 E. Saint Joseph Street
Rapid City SD 57701
605-394-2406
cjenkins@taz.sdsmt.edu

William Jordan, PhD, PE
Chair, Mechanical Engineering Program
Louisiana Tech University
Ruston LA 71272
318-257-4304
jordan@coes.latech.edu

Jacob Jorne, PhD
Chemical Engineering
University of Rochester
Rochester NY 14627
716-275-4584
jorne@che.Rochester.edu

Akira Kuraishi
Graduate Student
Aeronautics & Astronautics
Stanford University
Durand Bldg Rm. 006D
Stanford CA 94305-4035
650-723-3524
akirak@leland.stanford.edu

Kyriakos Komvopoulos, PhD
Mechanical Engineering
5143 Etcheverry Hall
University of California, Berkeley
Berkeley CA 94720-1740
510-642-2563
kyriakos@me.berkeley.edu

Jorn Larsen-Basse, PhD
Director, Surface Engineering & Material
Design Program
Civil & Mechanical Systems Division
National Science Foundation
4201 Wilson Blvd, Rm 545
Arlington VA 22230
703-292-7016
Jlarsenb@nsf.gov

Dr. Victor C. Li, FASCE, FASME
Professor & Director, ACE-MRL
Civil & Environmental Engineering
University of Michigan
2326 G.G. Brown Bldg
Ann Arbor Ml 48109-2125
734-764-3368
vcli@engin.um ich.edu

Richard A. Livingston, PhD
Senior Physical Scientist
Office of Infrastructure R&D, HRDI
Federal HighwayAdministration
6300 Georgetown Pike
McLean VA 22101
202-493-3063
dick.livingston@igate.fhwa.dot.gov

Hongbing Lu, PhD
School of Mechanical & Aerospace
Engineering
218 Engineering North
Okalahoma State University
Stillwater OK 74078
hongbin@master.ceat.okstste.edu

Wes Lum, Chief, Office of Infrastructure
Research, MS-42
California Department of Transportation
New Technology And Research Program
1101 R Street
Sacramento CA 95814
916-324-2713
wes.lum@dot.ca.com

Sankaran Mahadevan, PhD
Director of Graduate Studies
Dept of Civil & Environmental
Engineering
Box 6077, Station B
Vanderbilt University
Nashville TN 37235
615-322-3040
sankaran.mahadevan@vanderbilt.edu

Gerald H. Meier, PhD
Materials Science & Engineering
848 Benedum Hall
University of Pittsburgh
Pittsburgh PA 15261
412-624-9720
ghmeier@engrng.pitt.edu

Yasushi Miyano, PhD
Materials System Research Laboratory
Kanazawa Institute of Technology
3-1 Yatsukaho, Matto,
Ishikawa 924-0838, Japan
miyano@neptune.kanazawa-it.ac.jp

Paulo J. Monteiro, PhD
Civil & Environmental Engineering
725 Davis Hall
University of California, Berkeley
510-643-8251
monteiro@ce.Berkeley.edu

Doug Parks
Division of Materials Engineering &
Testing Services
5900 Folsom Blvd
Sacramento CA 95819-4612
916-227-7007
doug_parks@dot.ca.gov

Ramana M. Pidaparti
Mechanical Engineering
Purdue School of Engineering &
Technology, IUPUI
723 W. Michigan Street
Indianapolis IN 46202-5132
317-274-6796
ramana@engr.iupci.edu

Arron Rambach
California DOT (Caltrans)
5900 Folsom Blvd
Sacramento CA 95819
916-227-7236
arronJam bach@dot.ca.gov

Robert A. Reis
Division of Materials Engineering &
Testing Services
5900 Folsom Blvd
Sacramento CA 95819
916-227-7287
robJeis@dot.ca.gov

Alberto A. Sagues, PhD, PE
Distinguished University Professor
Civil & Environmental Engineering
University of South Florida ENB-118
4202 E. Fowler Avenue
Tampa FL 33620-5350
813-974-5819
sagues@eng.usf.edu

Tom Sandreczki
Dept of Chemistry
University of Missouri-KC
5009 Rockhill Road
Kansas City MO 64110
sandreczkit@umkc.edu

Jian-Ku Shang, PhD
Materials Science & Engineering
University of Illinois at Urbana-Champaign
1304 West Green Street
Urban IL 6180
217-333-9268
j-shang@uiuc.edu

C.T. Sun
School of Aeronautics & Astronautics
Purdue University
West Lafayette IN 47907-1282
765-494-5130
sun@ecn.purdue.edu

Michael Tolin
(Caltrans) Division of Materials &
Testing Services
5900 Folsom Blvd
Sacramento CA 95819
916-227-5297
michael_tolin@dot.ca.gov

Stephen W. Tsai, Professor/Research
Durand Building, Rm 381
Dept of Aeronautics & Astronautics
Stanford University
Stanford CA 94305-4035
650-725-3305
stsai@structure.stanford.edu

Clifton Vining
Dept of Mechanical Engineering
Louisiana Tech University
Ruston LA 71272

Aleksandra Vinogradov, PhD
Mechanical & Industrial Engineering
Montana State University
220 Roberts Hall
Bozeman MT 59717
406-994-6284
vinograd@me.montana.edu

Paul Virmani
Federal Highway Administration
6300 Georgetown Pike
McLean VA 22101
202-493-3052
paul.virmani@fhwa.dot.gov

Max Yen, PhD
Director, Materials Technology Center
Southern Illinois University
Carbondale IL 62901-6603
618-536-7525
jsulivan@siu.edu

Yunping Xi, PhD
Civil, Environmental & Architectural Engineering
Campus Box 428
University of Colorado
Boulder CO 80309
303-492-8991
xiy@bechtel.colorado.edu

WORKSHOP ATTENDEES

CONTENTS

Polymeric and Composite Materials

Test Methods

Appendix

Introduction

INITIATIVE ON LONG TERM DURABILITY OF MATERIALS AND STRUCTURES

J. Larsen-Basse and K.P. Chong
National Science Foundation
Arlington, VA 22230, USA

ABSTRACT

Fundamental research in durability of materials and structures have shown great potential for enhancing the functionality, serviceability and increased life span of our civil and mechanical infrastructure systems and as a result, could contribute significantly to the improvement of every nation's productivity, environment and quality of life. This initiative is aimed at developing innovative short-term laboratory or in-situ tests, which allow accurate, reliable prediction of long-term performance of materials, machines and structures. It is especially needed for new materials since such data are non-existing. The intelligent renewal of aging and deteriorating civil and mechanical infrastructure systems includes efficient and innovative use of high performance composite materials for structural and material systems. In this paper the NSF initiative on durability modeling and accelerated tests, as well as research needs are presented.

KEYWORDS

Durability, accelerated tests, modeling, designer materials, life-cycle performance.

INTRODUCTION

Demands for better-performing, longer-lasting, safer, more economical, and more environmentally friendly structures and machines are constantly pushing the envelope of technological capabilities and engineering practice. As a result there are relentless moves towards close tolerances and use of realistic life-cycle design, condition-based maintenance, and performance-based design. In this environment, the engineering designer is faced with the problem of finding useful and relevant materials property data for use in the design of machines and structures which are expected to provide top performance for an extended period of time. He or she will typically have access to "hard" data, i.e., repeatable and reproducible results from short term laboratory tests, such as simple hardness, fatigue, uniaxial yield and fracture tests; and even results from somewhat more complex standard tests, such as fracture toughness or short-term salt spray or ultraviolet chamber exposure tests. The designer usually also has some much more "soft", qualitative information on how a material has performed in the past under a given combination of time, temperature, mechanical load, environment, etc. in the same or similar applications. The skilled designer will usually be able to draw from his or her experience of connections between these sets of "hard" and "soft" information for one particular material to make educated estimates of how a different material, which gives somewhat similar results in short term tests, will perform under not-too-different sets of long-term complex mechanical and environmental loading conditions.

This approach has served reasonably well in the past but has several shortcomings relative to the new demands on tight design for performance. For example:

- it depends heavily on the experience and background of the designer,

- it does not deal well with synergistic interactions of several different types of loading, for example cyclic stress and long-term corrosion. For creep of metals and viscoelastic materials, use of the Larson-Miller parameter and similar approaches serve as semi-empirical ways to add the effects of temperature and time, but similar parameters are lacking for most other situations,
- it does not readily allow for comparison of different classes of materials, for which the short-term test results vary substantially, for example steel and polymer-matrix composites. This inhibits or delays the adoption of new materials for many critical applications, and
- it does not provide adequate information to allow direct design-for-performance in specific, long-term applications. For example, wear testing by the popular pin-on-disk apparatus may be a simple way to discriminate between different materials for, say, hip implant use, but it is only marginally relevant to a material's performance in the actual service and it does not permit any useful life-cycle design.

With this general background in mind a number of NSF program directors held informal discussions over a period of time. Some of the questions considered were the following:

- Have recent advances in the fields of modeling, computation, understanding of basic materials properties, sensing, control, probability analysis, etc., reached the stage where we really can do better than outlined by the problem set above, where we can begin to predict long-term performance from short-term tests by quantitative approaches? And where we can confidently operate with lower safety margins or safety factors and closer prediction of life to failure or time until maintenance is necessary?
- Do we understand the different processes well enough to be able to closely predict their long-term synergistic interactions, such as the combined effects of stress, corrosion and temperature variations?
- Is there some basic generic approach, which has general applicability in diverse cases, maybe, including model-based simulation and uncertainty and probability considerations?
- Do we have some quantitative or semi-quantitative ways of dealing with new materials, new combinations of external loadings and environmental effects, or changes in these factors during the life of a machine or structure?
- Are there any new short-term tests or NDE techniques that need to be developed to provide some of the necessary information in a useful manner? And
- What new research should we try to stimulate in order to expedite development in this field?
- What are the long-term field data available and how do them compare with the research results of proposed methods?

Durability of new materials involves the synthesis, laboratory and field testing, accelerated tests and modeling, etc. Fig. 1 illustrates the size effects and mechanics (Boresi and Chong, 2000) involved.

Materials		**Structures/machines**	**Infrastructure**
Submicro-level	***meso-level***	***macro-level***	***systems integration***
Molecular scale	**Microns**	**Meters**	**Up to km scale**
~micro-mechanics	**~meso-mechanics**	**~beams**	**~bridge systems**
~nanotechnology	**~interfacial**	**~columns**	**~airplanes**
designer materials		***smart structures***	***high performance systems***

Fig. 1. Scales in Materials and Structures

The discussions initially resulted in the support of a workshop focused on problems in the infrastructure materials area, funded by NSF and organized by the Board on Infrastructure and the Constructed Environment under the National Advisory Board of the National Research Council of the Academy (NRC 1999). From the report of this workshop and additional discussions an NSF research investment initiative was subsequently developed. It led to the research discussed at this meeting.

THE NRC WORKSHOP

The workshop was held at the National Academy of Sciences on August 24 and 25 of 1998. It defined its role as ". a reconnaissance-level assessment of models and methods that are being used, or potentially could be used, to determine the long-term performance of infrastructure materials and components." (NRC 1999) The objectives were (NRC 1999):

- "define the objectives for infrastructure-based research that would use accelerated testing and computational simulations to determine life-cycle performance
- assess the state of the knowledge base to identify gaps and overlaps in research activities
- establish outcome-oriented metrics for setting research priorities
- identify promising lines of research and collaborations"

The participants agreed that a "root cause" of the deterioration and failure of any system is related to materials but that accelerated-testing methods, while they may potentially be used to rank the performance of materials in real-world systems, they are not at present sufficiently reliable to make system-life predictions.

The workshop proposed that development of useful life-prediction models for infrastructure systems would require some of the following advances (NRC 1999):
- a better fundamental understanding of infrastructure materials and systems, including interfaces and degradation modes and spanning all size scales
- a better understanding of the relevant characteristics of the operating environment
- development of standardized test methods and databases
- development of sensors for monitoring systems during construction and use
- incorporation of economic models in life-cycle cost analyses.

The workshop also suggested that major obstacles to adaptation of life-cycle prediction models and accelerated test procedures for infrastructure applications are two interrelated factors:

- poor integration of the relevant engineering community into materials-based infrastructure research, and
- concerns about risk and liability
- It expressed the opinion that "practicing engineers have little opportunity to develop the same level of trust in simulation models and accelerated laboratory tests as they have in their many years of empirical field observations."

The workshop concluded that..."life-prediction models and accelerated-testing procedures have the potential to increase the deployment of new materials in infrastructure applications and to improve traditional materials..."

Predictably, it also suggested that NSF should do more to "...support materials research directed toward understanding the combined effects of degradation mechanisms and applying that understanding to quantitative predictions of system life..." and, because of the large variations between sectors, NSF should "...evaluate each infrastructure sector and attempt to organize its research..." for easy formulation of research needs and ready technology transfer to practice.

THE NSF DURABILITY INITIATIVE

The NSF Initiative, NSF 98-42 (NSF 1998) actually predated the Academy workshop by a few months. It was developed principally in the Civil and Mechanical Systems Division (CMS) of NSF's Engineering Directorate but with significant input from colleagues at the Federal Highway Administration and several State Departments of Transportation (DOTs), especially CALTRANS in California, as well as the Air Force Office of Scientific Research (AFOSR).

Its stated aim was: "...developing innovative short-term laboratory or in situ tests which allow accurate, reliable prediction of long-term performance of materials, machines and structures..." based on better understanding of the fundamental nature of deterioration processes and innovative ways to model these processes as they affect life and long-term performance and as they, in turn, are affected by time and size effects.

The Initiative aimed to give preference to high risk/high-payoff research by individual investigators or small groups. While most of the deterioration processes of interest clearly were those associated with environmental effects and exposure to overloads, over speeds and other unanticipated operating conditions, there was no clear or implied limit on the phenomena or materials groups to be studied. It was intended, in line with NSF's general approach, to encourage innovative thinking in the community. Similarly, application areas were not specified, although it was suggested that some relevant ones were units of the constructed infrastructure, transportation systems and units, and manufacturing machinery. Some possible research topics were suggested (NSF 1988):

- multiple interactive effects and deterioration mechanisms
- accelerated techniques, related instrumentation and model validation to long-term field data
- determination of service life from wear tests and modeling
- deterioration of structural materials and protective coatings (e.g. polymeric coatings on bridges) as a function of environment
- failure mechanisms of composite materials (e.g. reinforced-concrete failure and corrosion protection systems)
- size effects in testing, instrumentation and modeling
- relevant statistical methods and reliability
- comparison of models with long term field data.

It was further suggested that "...topics could include evaluation of existing data on long-term performance in light of short-term tests and using relevant models, and development of completely new testing, instrumentation and modeling techniques". Finally, the Initiative description reminded of NSF's commitment to integration of education and training with the research, as well as encouragement of formation of interdisciplinary teams, where relevant, in this case possibly by inclusion of researchers from mathematics or statistics.

Response to the initiative was substantial; approximately 140 proposals were received, requesting about $ 70 M of funds. They were divided according to topic area and reviewed by a number of peer review panels. Panelists were active researchers from universities, government laboratories, and industry. Because of funding restrictions NSF was unable to support many worthy proposals but in the end some 25 awards were made for a total of about $ 7 M. Co-funding was obtained from other divisions within NSF, including the EPSCoR program, and from AFOSR and FHWA and several state highway departments (California – CALTRANS, Illinois, Oregon and Kansas). Most awards were for a three-year period but a few small 1-2 year awards were also made, specifically for exploratory research. This workshop was scheduled for a time when most projects would have been underway for some time and would have significant results and insights but would still be incomplete and thus able to profit from the discussions and networking.

FUTURE ACTIVITIES

NSF has long been supporting research, which could be classified as durability-related studies, and will continue to do so within the general relevant program areas. The annual investment in durability via this avenue is estimated at roughly $ 2 M. At the present stage there are no plans to repeat a special initiative to focus attention on the durability area. This is not to downgrade the topic but simply a response to demands from many other topic areas for an opportunity for special initiative funding. NSF's programs will continue to encourage high quality, innovative research in all areas, most certainly including the area of durability of materials and structures.

CONCLUSIONS

We have briefly traced the background for the Durability Initiative and its subsequent development. It was issued in response to a perceived need and the response from the community showed that it was the right thing to do at the right time. One may think of the "materials tetrahedron" promulgated by an Academy report in 1989 (NRC 1989) where performance, as the ultimate "materials characteristic", is shown linked to microstructure/composition, properties, and processing in a tetrahedron diagram. Microstructure/composition, processing and property are interlinked to form the base triangle and performance forms the apex of the tetrahedron. Since that time, we have developed very good understanding of most of the interrelationships in the base triangle. Perhaps the response to the Durability Initiative and the results presented in this workshop demonstrate that we are now well on our way to also develop reasonable understanding of the connections of the apex of the tetrahedron, performance, to the topics in the base triangle.

ACKNOWLEDGEMENTS

We wish to acknowledge extensive discussions with Oscar Dillon when he was at NSF and also with a number of our colleagues who participated in formulation of the Initiative, in evaluation of the proposals and in monitoring of some of the funded projects, in particular Delcie Durham and Bruce MacDonald from NSF; Jim Chang, Ozden Ochoa, Charles Lee and Tom Hahn from AFOSR; as well as Wes Lum and Lee Barton from CALTRANS; Paul Vermani and Dick Livingston from FHWA and others.

REFERENCES

Boresi, A.P. and Chong, K. P., (2000), *Elasticity in Engineering Mechanics*, John Wiley, NY.

NRC. (1989). *Materials Science and Engineering for the 1990s*: Maintaining Competitiveness in the age of Materials, National Materials Advisory Board, National Research Council, Washington, D.C.: National Academy Press.

NRC (1999). *Research Agenda for Test Methods and Models to Simulate the Accelerated Aging of Infrastructure Materials.* Report of a Workshop, Board on Infrastructure and the Constructed Environment, National Materials Advisory Board, Commission on Engineering and Technical Systems, National Research Council, Washington, D.C.: National Academy Press.

NSF (1998). *Long Term Durability of Materials and Structures: Modeling and Accelerated Techniques*; Initiative Announcement for FY 1998, Directorate for Engineering, National Science Foundation, Arlington, VA. NSF 98-42.

Structures

Long Term Durability of Structural Materials
P.J.M. Monteiro, K.P. Chong, J. Larsen-Basse, K. Komvopoulos (Eds)

APPROACHES TO ENHANCING CONCRETE BRIDGE DECK DURABILITY

V. C. Li and J. Zhang

The Advanced Civil Engineering Materials Research Laboratory
Department of Civil and Environmental Engineering,
University of Michigan, Ann Arbor, MI48109-2125, USA

ABSTRACT

This paper reviews recent results on the mechanisms of durability enhancement in concrete bridge decks via the use of fiber reinforced cementitious composite (FRCC). The deterioration of concrete bridge decks due to shrinkage crack formation and fatigue crack propagation is briefly reviewed. As an approach to enhancing fatigue resistance, fiber addition, and the mechanism of fatigue crack propagation in FRCC is studied. Crack bridging degradation phenomenon is discussed and a fatigue life prediction model based on crack bridging and bridging degradation in FRCC under cyclic loading is presented. Second, a ductile strip concept is investigated for reducing and/or eliminating shrinkage cracks in concrete bridge decks. This approach is introduced and some preliminary experimental results are presented.

KEYWORDS: Durability, Concrete bridge decks, Fatigue, Fiber reinforced cementitious composite, Ductile strip, Shrinkage crack

INTRODUCTION

According to information provided by FHWA, out of 583,349 bridges in the 1996 National Bridge Inventory (NBI), 333,641 have cast-in-place concrete decks and another 38,844 have precast concrete deck panels. The average life of a concrete deck is determined by many factors including initial design, material properties, traffic, environment, salt application, presence and effectiveness of protective systems and maintenance practices among others. Deterioration of the deck is the most common cause requiring repair, rehabilitation or replacement of bridge superstructures. Extensive cracking and large potholes directly affect traffic safety. Therefore, determining the mechanisms of deterioration and developing efficient technologies for resisting and/or eliminating such mechanisms are very important research needs for durability enhancement of concrete bridge decks. Concrete slabs are subjected to considerable fatigue loads. Average daily truck traffic (ADTT) can vary from site-to-site, from less than 500 to over 5,000 trucks per lane, or 200,000 to 2 million trucks per year. Passage of each axle or closely spaced group of axles can be considered as a load cycle. Over the years, a bridge deck slab can be subjected to multi-millions of load cycles. Durability can be considered as the ability to retain an original property, or resistance against long-term deterioration. Often this terminology is used in connection with different kinds of deterioration in materials and structures, under a complex combination of environmental and mechanical loads. For example, concrete durability is considered against chloride ions, carbonation, alkali-aggregate reaction, freeze-thaw cycles, and fatigue, and durability of steel rebars is considered against corrosion and fatigue. Reinforced concrete structures are subjected to these multiple deterioration factors, and structural durability is dependent on each of these factors, as well as their combined effects.

Recent studies show that the service life of reinforced concrete (RC) bridge decks is controlled not only by the corrosion of steel reinforcements, but also by fatigue cracking of concrete slabs [1-4]. The failure mechanism of RC bridge decks is revealed by fatigue loading tests with a moving wheel. The failure progresses through the following five stages [1]. First, cracks are developed on the bottom face of a deck in a transverse direction to traffic. These cracks are mainly due to concrete shrinkage and temperature changes which develops tensile stress (due to restrain) in the longest dimension of the deck. Shrinkage and temperature induced stress together with bending stress due to traffic loading can combine to form these cracks, but, in some occasions, shrinkage stress by itself is high enough to form cracks. Second, longitudinal cracks are developed on the bottom while transverse cracks are developed on the top. On the bottom face of the deck, due to the transverse cracks formed in the previous stage, the deck slab loses load transfer in the longitudinal direction so that flexural cracks are formed in the longitudinal direction. Together with the first set of shrinkage induced transverse cracks, this new set of longitudinal cracks forms a network of cracks. On the top face of the deck, the repeated traffic loading leads to the initiation and growth of transverse cracks starting from the location of girders to the middle of the deck. Since these top transverse cracks are formed in weak sections, they are certain to join the bottom transverse cracks, forming through cracks. Third, water penetrates into the through cracks. The asphaltic topping does not drain easily but tends to retain the water for a long period (e.g. as long as one week after one hour of raining). The water migrates downward through the cracks, creating efflorescence on the bottom surface of the deck. Fourth, the through cracks are gradually worn out under repeated traffic load. The loss of aggregate interlocking leads to the loss of load transfer in the longitudinal direction. As a result, the deck slab does not behave as a plate any longer, but acts as transverse 'beams'. The presence of water accelerates the wearing out of the cracks. Finally, the transverse 'beams' fail in shear fatigue due to the insufficient amount of transverse steel reinforcement of the deck. This leads to the spalling of concrete, and the depression of the deck slab takes place, leading to service termination. Shear punching failure was also observed in the fatigue test conducted on FRP reinforced concrete deck slabs [4].

The above studies indicate that, in RC bridge decks, the importance of fatigue durability is as important as that of corrosion durability. Furthermore, previous investigations [1, 5] show that the integrity of RC bridge decks is actually the key for improving the durability of bridges. Therefore, the prevention of fatigue failure in RC bridge decks is crucial, and the damage sequence described above has to be interrupted before final deck failure. Specifically, the failure mechanism of RC bridge decks shows that the formation of through cracks under repeated traffic loading plays a major role in the sequence of five stages. The formation of through cracks is completed relatively early in the service life, and the rest of the life is spent for erosion and wearing-out of concrete cracks and fatigue of steel rebars [6]. This implies that increased fatigue crack resistance of concrete leads to improvement of the service life of bridge decks, since the progressive crack growth is caused by low fatigue crack resistance of concrete. Therefore, research efforts are needed to investigate and improve the fatigue durability of RC bridge decks and, in turn, the fatigue crack resistance of concrete materials.

Fiber reinforced cementitious composites (FRCCs), typically fiber reinforced concrete (FRC) and fiber reinforced mortar (FRM), are promising materials for fatigue resistant structural elements. With fiber addition, improvements on various mechanical properties, including toughness, impact resistance and fatigue strength have been experimentally demonstrated, e.g. [7-13]. These studies suggest that the use of fiber produces significant improvement not achievables with adjustments of the concrete mix design itself. In the present paper, first recent theoretical and experimental studies on FRC fatigue resistance are reviewed. Second, a newly developed technique for eliminating shrinkage and temperature cracks in concrete bridge decks by inserting ductile strips which are made of FRCC is presented.

FATIGUE RESISTANCE OF FRCs

Experimental Findings

Fatigue of FRCs has been investigated experimentally using the Stress-Life Approaches, and FRCs are shown to have improved fatigue performances. Stress level and fatigue life diagrams (S-N curves) have been obtained for many kinds of FRCs: steel [7-10], polypropylene [11], carbon [12], polyethelene [13], and hybrid (hooked end steel + polypropylene) [10]. The effect of fiber addition to concrete on fatigue strength is positive. An FRC showed 4 times increase in fatigue strength with 3% of straight steel fibers [7]. Hooked end steel FRCs showed 2-3 times increase with less than 1% fiber content [8, 9]. Polypropylene FRC improved fatigue strength by 1.2 times with 0.32% of fiber content [11]. A typical comparison on the fatigue resistance of plain concrete and FRC under bending load is shown in Fig. 1 in terms of maximum flexural stress and fatigue life diagrams [10].

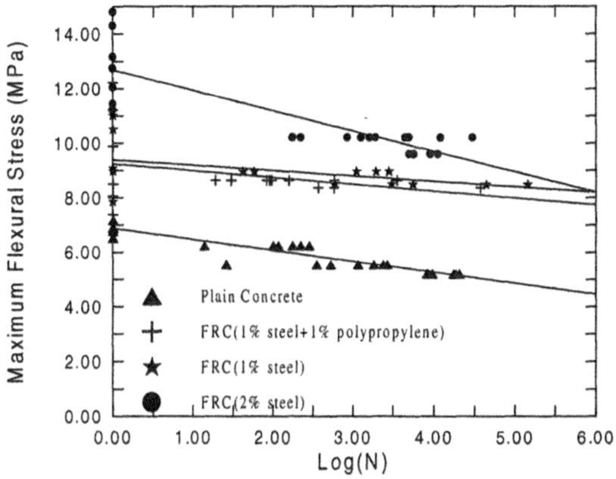

Fig. 1. Experimental maximum flexural stress and fatigue life diagrams of plain concrete and FRCs

Mechanism of Fatigue Crack Propagation

Normally, it can be said that fatigue is a process of progressive, permanent internal structural changes occurring in a material subjected to repetitive stress. The progressive fatigue damage on material constituents is responsible for fatigue life of a material. For FRC, the material phases can be broadly classified as matrix (cement paste and aggregates), fibers, as well as the interfaces of fiber/matrix and aggregate/hydrated cement paste. The fatigue loading causes these physical phases to undergo microscopic changes, such as opening and growth of bond cracks, which exist at the interface between coarse aggregate and hydrated cement paste prior to the application of load, reversed movement of fiber along the interface, fiber surface abrasion and damage of interface in repeated sliding processes. These microscopic changes in turn cause some detrimental changes in macroscopic material properties. Typically, the aggregate bridging force as well as fiber bridging force decreases with number of cycles due to the interfacial damage [14, 15] or fiber breakage due to the surface abrasion [13]. The damages on interfaces of fiber/matrix and aggregate/matrix, which are generally the weakest phase in concrete and FRCs, as well as on soft polymer fibers are likely responsible for fatigue crack initiation and growth in concrete and FRCs. The rate of fatigue crack growth in concrete and FRCs are highly dependent on the crack bridging law governing the zone behind the cement matrix crack and on the law governing the degradation of the crack bridging with the number of load cycles. Fatigue crack growth behavior, in turn, governs the fatigue life of concrete and FRC structures.

Crack Bridging Degradation

As stated above, crack bridging behavior of FRCs under cyclic loading has a significant importance for understanding and predicting fatigue crack propagation. Zhang et al. carried out an experimental study on the crack bridging behavior of FRCs under uniaxial tensile fatigue load [15]. In this study, a series of deformation controlled fatigue tensile tests with constant amplitude between maximum and minimum crack openings were carried out on two side pre-notched specimens. Two types of FRCs, reinforced with commercially available smooth and hooked steel fibers, respectively, are investigated. In this paper, only the results on straight steel fiber reinforced concrete (SSFRC) are presented. The test procedures and results are summarized as follow.

A testing method for measuring the stress-crack width relationship developed by Stang et al. [16] is used in the current tests. The test set-up and the geometry of the test specimen are shown in Fig. 2(a). The test takes place in specially designed grips, one fixed to the load cell and the other fixed to the actuator piston with standard Instron fixtures. The grips consist of a permanent part and an interchangeable steel block, which is fixed to the permanent part through 4 bolts. The specimen is glued to the blocks. The glued surfaces of the interchangeable steel blocks and the specimen are sandblasted before gluing to enhance the bond between steel and specimen. A fast curing polymer which attains 90% of its maximum strength in about 4 minutes was used. The deformation was measured using two standard Instron extensometers (type 2620-602) with 12.5 mm gauge length mounted across each of the two 9 mm deep and 3 mm width notches. The tests were performed in a 250 kN load capacity, 8500 Instron dynamic testing machine equipped for closed-loop testing. The uniaxial fatigue tensile test was conducted under

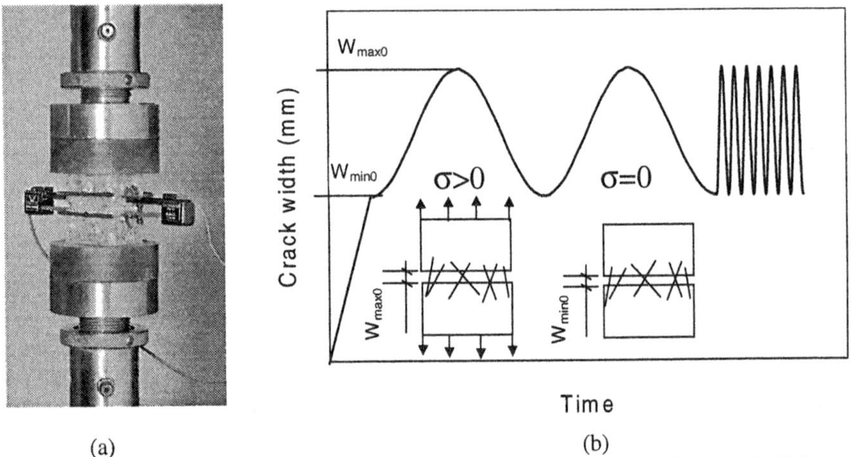

Fig. 2. (a) View of the test set-up for fatigue tension and (b) Deformation-time diagram in fatigue test

displacement control with constant amplitude between maximum and minimum crack widths. The minimum crack width value was obtained by a single loading-unloading tensile test and measured at zero loads on the unloading branch. The fatigue test commenced with a ramp to the minimum crack value at a rate of 0.01 mm/second followed by a sine waveform fatigue loading in deformation control. In order to control the accuracy of the maximum crack width value, different load frequencies of 0.25 Hz in the first two cycles and 3.5 Hz for all the rest of cycles were adopted. This fatigue loading procedure is shown in Fig. 2(b). The fatigue tensile test results on the SSFRC material is shown in Fig. 3 and Fig. 4. Fig. 3 demonstrates a typical bridging stress-crack width curve (load-unload loops) during fatigue loading under deformation control. From this figure, it can be found that the secant stiffness ($\Delta\sigma/\Delta W$) of reloading branches reduces gradually with the number of load cycles, therefore the bridging stress at the maximum crack width decreases gradually. The diagrams of bridging stress at maximum crack width versus number of load cycles for a typical maximum and minimum crack widths (W_{max0} and W_{min0}) are shown in Fig. 4, where the results in the range of 1 to 10^5 cycles and 1 to 10^2 cycles are displayed respectively in the figure. Here, the average result and all the individual test results are displayed together.

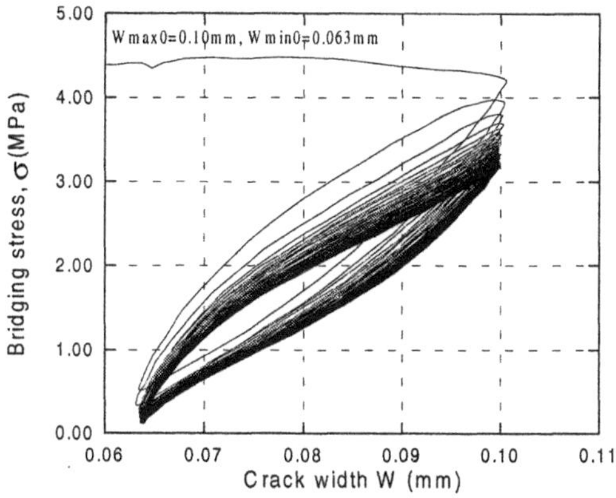

Fig. 3. Typical stress-crack width curve of a fatigue tensile test

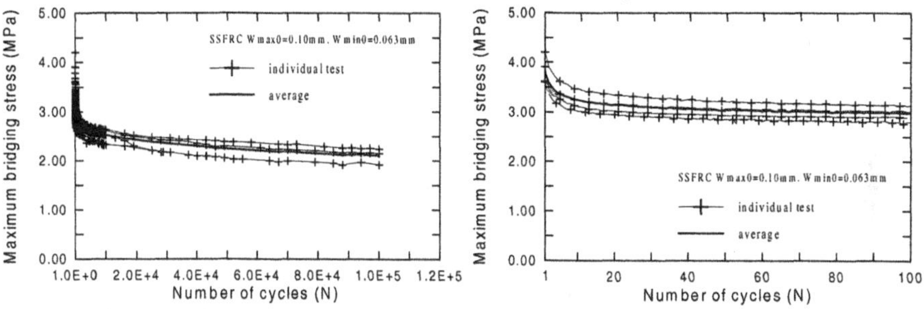

Fig. 4. Typical relation of maximum bridging stress and number of cycles

From these results, it is evident that the maximum bridging stress decreases with number of fatigue cycles for the SSFRC under deformation-controlled fatigue load. The behavior of the stress degradation in the material can be generalized as a fast dropping stage (within first ten to fifteen cycles) with a decelerated rate of stress degradation followed by a stable decreasing stage with an almost constant degradation rate within the experimental period. The bridging stress reduces 7, 15, 23, 17, 16 and 13 percent of their

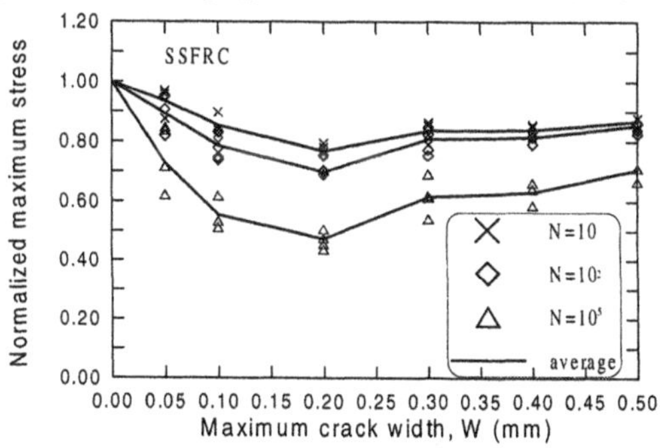

Fig. 5. Relations of normalized maximum bridging stress and maximum crack width of SSFRC, showing the results after 10, 10^2 and 10^5 cycles respectively.

original values when the maximum pre-cracked values are 0.05 mm, 0.10 mm to 0.50 mm respectively after 10 cycles (see [15] for details). This indicates that the rate of bridging decay is affected by the maximum crack width, as shown in Fig. 5 here showing the relations between maximum stress, which is normalized with the stress at first cycle, and maximum crack width after 10, 10^2 and 10^5 cycles respectively. According to Fig. 5, the largest stress degradation in SSFRC occurs at a certain point of the maximum crack width between 0.1 to 0.3 mm. Beyond this maximum crack width, the stress decay diminishes. The largest reduction on crack bridging stress can be more than 50 percent of its original value after 10^5 cycles. From the experimental results described above, we can conclude that the bridging fibers and aggregates in cement-based composites suffer from fatigue damage, exhibiting bridging stress degradation with number of fatigue cycles.

FATIGUE LIFE PREDICTION OF FRC BEAMS UNDER FLEXURAL LOAD

Fatigue strength approach based on experiments requires time-consuming test data collection and processing for a broad range of design cases which, in principle, is not applicable to other design cases. Therefore, a mechanism based fatigue model that is capable of both predicting the fatigue life for a given FRC structure and designing an FRC material for a given fatigue life is needed. Recently, fatigue models based on crack bridging degradation have been developed [17]. The model is able predicting fatigue crack propagation and further predict fatigue life of FRC structures. The model can be summarized as follow.

The fatigue crack growth process in concrete or FRC materials can be broadly divided into two stages: the crack initiation period and the development period. Now considering a simply supported rectangular beam loaded in bending fatigue load with a constant amplitude between maximum and minimum moment M_{max} and M_{min}. When M_{max}<M_{fc}, where M_{fc} is the first crack moment, the fatigue life of beam can be given by:

$$N_t = N_{ci} + N_{cg} \tag{1}$$

When M_{max}≥M_{fc} , the fatigue life is:

$$N_t = N_{cg} \tag{2}$$

where N_t is the total fatigue life, N_{ci} and N_{cg} are the fatigue life component for the crack initiation and growth respectively. The first term, N_{ci}, is dependent on the microcracking in material which is highly influenced by the microstructure of concrete matrix, such as water/cement ratio, aggregate properties as well as pore structure, size distribution and content. The second term, N_{cg}, is strongly dependent on the bridging performance within the fracture zone under fatigue loading.

The present model focuses on the fatigue life prediction on N_{cg}, i.e. the case of maximum load M_{max} is larger than the first crack load M_{fc}. Based on the above discussions, some basic assumptions for fatigue modelling on N_{cg} can be stated: (1) after a dominant fatigue crack is created, the bridging behavior within the fracture zone governs the rate of fatigue crack advancement; (2) the stress at the crack tip remains constant and is equal to the material tensile strength; (3) material properties outside the fracture zone are unchanged during fatigue loading. It is further assumed that concrete and FRC materials essentially show a linear response in tension up to peak load. After peak one discrete crack is formed. And the discrete crack formation is described by the crack bridging law (or stress-crack width relationship) under both monotonic and cyclic loading. Thus the following material parameters are fundamental in the constitutive relations of concrete and FRC in fatigue tension: the Young's modulus E, the tensile strength σ_t and the cyclic stress-crack width (σ-W,N) relationship of both aggregate bridging and fiber bridging. In compression the behavior of concrete and FRC materials is assumed to be linear elastic and the Young's modulus in compression is the same as in tension. With the above assumptions, a semi-analytical method for predicting fatigue behavior of unreinforced concrete and FRC beams under bending load had been developed. In the model, the cyclic bridging law (or cyclic stress-crack width relationship) was incorporated in integration form which can easily be replaced by other bridging models for different kinds of FRC materials with different fiber types, volume concentration and matrix properties. The complete theoretical curves, in terms of fatigue crack length or crack mouth opening displacement (CMOD) with number of cycles diagrams, as well as the classical S-N curves are obtained and compared with experimental results. The details of the model derivation can be found in the paper by Zhang et al [17].

In the numerical simulation, a specific fatigue loading procedure with M_{min} equal to zero corresponding to that the condition of fatigue tension and bending tests is assumed. The geometry of the specimen in the numerical model is the same as that used in the fatigue bending tests. A fit based cyclic bridging laws for different types of FRC (including concrete) based on the experimental results introduced in the previous section are used in the simulation. The detailed expressions of the monotonic and cyclic crack bridging models as well as related material parameters used in the model can be found elsewhere [17]. In order to compare the results between monotonic loading and fatigue loading, the monotonic bending behavior is simulated first in terms of the load-CMOD relation. Fig. 6(a) shows the predicted monotonic flexural stress-CMOD curves of plain concrete (PC) and SSFRC respectively, together with experimental results for SSFRC. On inspecting the numerical results for the load-CMOD diagrams of the two types of concrete beams under three point bending, several features can be distinguished: (1) load level I: The flexure stress increases linearly with deformation up to tensile strength of the materials, 5.2, 5.4 MPa for PC and SSFRC respectively. In this stage, material behavior obeys elastic constitutive relations and no fictitious crack is formed, therefore CMOD is equal to zero. (2) load level II: The flexural stress increases up to 7.1 and 9.1 MPa for PC and SSFRC. In this period, the deformation increases a little more than proportionally with respect to the stress. A fictitious crack develops in the middle of beam and grows with the load increasing; (3) load level III: The flexural stress increases up to the maximum values, the flexural modulus of beam, about 10 MPa for SSFRC. At this stage, the deformation increases much more than proportionally with respect to the stress. Fatigue behavior is commonly represented by stress-fatigue life curves, normally referred to as S-N curves. In the case of fatigue in bending, S refers to the maximum flexural stress according to classical elastic theory. Fig. 6(b) shows the predicted S-N curves for these two types of concrete, where the fatigue life is presented in the form of logarithm. Some test results are shown together with the theoretical results. It can be seen that model predictions agree well with the test results. First, the S-Log(N) curve of plain concrete is almost linear which agrees with a number of experiments [8-10]. For steel fiber reinforced concrete the S-Log(N) curve becomes curved. Second, the present

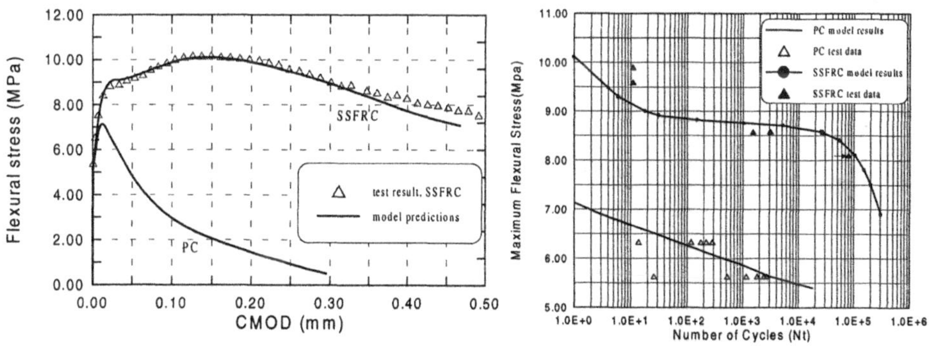

(a) (b)

Fig. 6. (a) Monotonic flexure stress versus CMOD curves for plain concrete and SSFRC beams and (b) relation of maximum flexure stress with fatigue life, shown together with experimental data

model predicts that the steel fibers can significantly improve the bending fatigue performance of concrete structures, which has been demonstrated by many researchers [7-10]. For steel fiber concrete beams, with maximum flexural stress between 9.00 to 10.00 MPa (Level III), the fatigue life is very short, within 1 to 30 cycles. The reason for this short fatigue life is a combination of a large initial crack length and significant bridging degradation due to large crack openings. With the maximum flexural stress between 5.4 to 9.00 MPa (Level II), the fatigue life increases notably with decreasing maximum flexural stress. The longer fatigue life is a product of both the shorter initial crack length and the smaller crack openings. When the maximum flexural stress is lower than 5.4 MPa (Level I), no dominant macro fatigue crack occurs after first cycle. However, fatigue crack initiation will not be treated in the present study.

INTRODUCING DUCTILE STRIP FOR DURABILITY ENHANCEMENT OF CONCRETE BRIDGE DECKS

General Introduction

As pointed out in the Introduction section, fatigue cracking in concrete bridge decks appears to be preceded by the formation of cracking due to concrete shrinkage in the transverse direction. Thus deterioration due to fatigue can be curtailed if shrinkage cracks in concrete is minimized or even eliminated. In the present study, attempts at localizing shrinkage induced deformation into designated strips, where an engineered fiber reinforced cementitious composite (ECC) material with strain-hardening and high strain capacity (up to 5%) is used, were carried out. As a result, while microcrack damage exists in the ECC strip, the concrete remains intact. This concept has recently been demonstrated by simulating the shrinkage in concrete under restrain condition as tensile load acting on a specially designed specimen. Experimental results show that it is possible to achieve the targeted deformation mode with certain design on the ECC/concrete interfaces. Due to the special material properties of ECC, the strain energy produced by shrinkage (under restrain condition) of hardened cement and temperature changes can be released by the high strain ability of ECC material so that cracking in plain concrete can be avoided. Thus the fatigue durability of concrete slabs can be improved, resulting in a longer service life. The proposed design concept may be implemented by placing ECC as periodic special joints or "ductile strip" between stretches of concrete slabs. By replacing standard joints with ductile strips, common deterioration problems associated with joints may be also eliminated. The current concept of introducing ductile strip in concrete bridge decks will be summarized as follows. The details on this work can be found in the paper by Zhang, et al [18].

Design of Concrete Slab with Ductile Strips

Assume that a concrete bar is composed of two kinds of materials, ductile ECC material and plain concrete with length l_I and l_{II} respectively. The bar has the same cross section along the length. Further assume that the two materials are perfectly joined together without failure at the interface under tensile load. The general dimension of the bar and the corresponding stress-strain relationship under tensile load of individual materials are shown in Fig. 7, where the concrete tensile strength is higher than that of the ECC material. Under uniaxial tension, the overall strain capacity of the bar, ε_c, is reached when the load reaches the tensile strength of the ECC material. Hence ε_c is

$$\varepsilon_c = \varepsilon_I\left(\frac{l_I}{l}\right) + \varepsilon_{II}\left(\frac{l_{II}}{l}\right) \tag{3}$$

where ε_I is the strain capacity of ductile material and ε_{II} is the strain value of plain concrete corresponding to the tensile strength of ECC material. l is the total length of the bar. Therefore, the composite strain capacity, ε_c, (strain at peak stress in curve labeled I-II in Fig. 7(b)), is a function of ε_I, ε_{II} and l_I or l_{II}. For given material properties, ε_c is influenced only by the individual element length l_I or l_{II}. Fig. 8 demonstrates the overall strain capacity, ε_c as a function of l_I with different given strain capacity of ECC material, ε_I. It clearly shows that for a given ductile strip width, the higher the strain capacity of ductile ECC material, the higher the overall strain capacity of the composite bar. In addition, a high composite strain capacity can also be obtained through adjusting the length of the ductile strip. With a reasonable combination of plain concrete and ductile ECC strips, it is possible to achieve a prescribed strain capacity requirement, which may be twenty or thirty times the strain

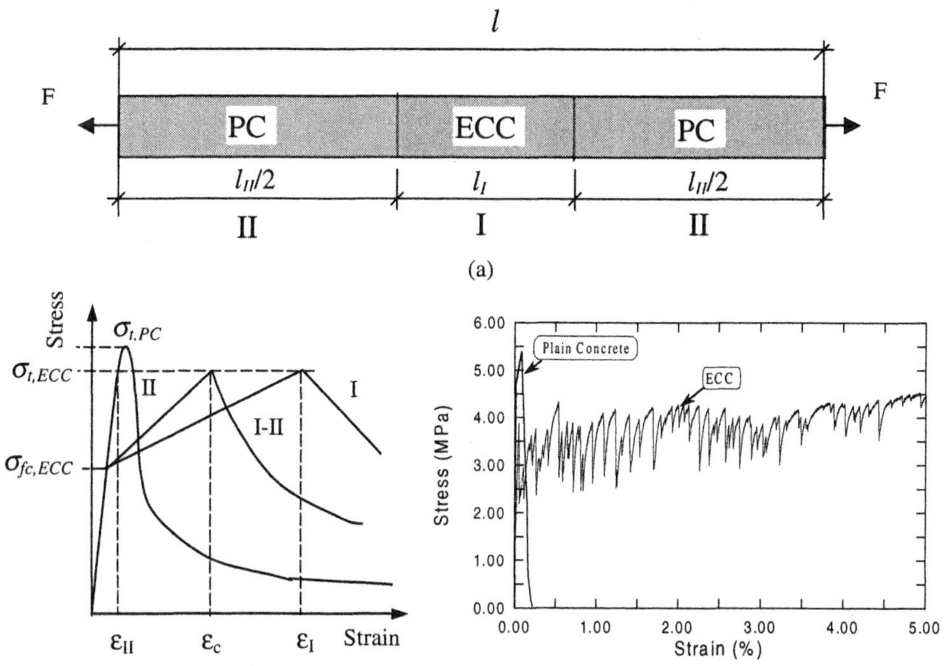

Fig. 7. (a) A PC/ECC/PC tensile bar, (b) schematic stress-strain behavior of ECC and concrete under uniaxial tensile load and (c) experimental obtained stress-strain curves of plain concrete and ECC

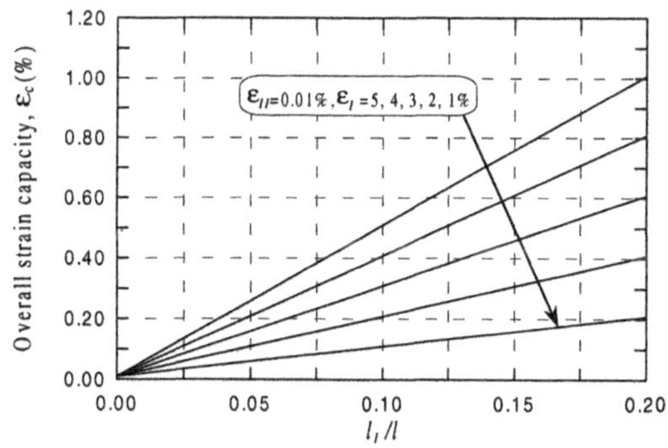

Fig. 8. Overall strain capacity of ECC-concrete composite bar as a function of ECC strip length, l_I

capacity of plain concrete without lost of load carrying capacity. In this case, cracking can be avoided within the plain concrete section when the structure is subject to tensile stress, such as shrinkage stress. For localization of deformation into the ECC strip, sufficient strength difference between ECC and concrete must be guaranteed, i.e. $\sigma_{fc,ECC}<\sigma_{t,ECC}<\sigma_{t,PC}$, as shown schematically in Fig. 7(b). The experimentally determined tensile stress-strain curves of the designed ECC and concrete, which satisfy the above requirements, are shown in Fig. 7(c). In addition, ECC material and plain concrete can be considered as two kinds of cementitious materials with different properties. It is necessary to design the ECC/concrete interface to ensure that damage under tensile stresses occurs inside the ductile strip instead of at the interface area. If first cracking occurs at or close to the ECC/concrete interface, the multiple cracking phenomenon cannot be developed due to the "fiber-end" effect at the interface area, i.e. the fiber bridging is weak at the interface area. Therefore, the present interface design principally overcomes the fiber-end effect along the ECC/concrete interface. As an example, a geometrical method to enhance the ECC/concrete interface will be given below. Fig. 9(a) demonstrates the idea of the ECC/concrete interface geometric design. From stress element analysis at interface, the normal and shear stresses at interface, σ_t and τ_s are given by

$$\sigma_t = \sigma \sin^2(\phi)$$
$$\tau_s = \frac{1}{2}\sigma \sin(2\phi) \qquad (4)$$

where σ is the overall tensile stress acting on the slab. ϕ is the angle between ductile material/concrete interface and horizontal line. From Eqn (4), we can see that the normal stress σ_t is a function of ϕ for a given stress level σ. Fig. 9(b) shows the relationship between the interfacial normal stress and the angle ϕ under some typical stress level. σ_t is reduced significantly by lowering the angle ϕ. This indicates that it is possible to prevent interfacial failure with a reasonable interfacial angle for a given interfacial tensile strength. For example, for hot joining, the general interfacial tensile strength should be equal to the minimum value of the tensile strength of concrete and the first crack strength of ECC material. Therefore, this interfacial tensile strength at least can achieve 2-3 MPa after 28 days curing. In this case, as ϕ is selected to be less than 38 degrees interfacial failure can still be prevented even as 5 MPa overall stress is acting on the structure since the interfacial tensile stress is less than 2 MPa. On the other hand, the tensile strength of ECC is normally less than 5 MPa. Therefore, the above selection of ϕ, i.e. 38 degrees, is still conservative [18].

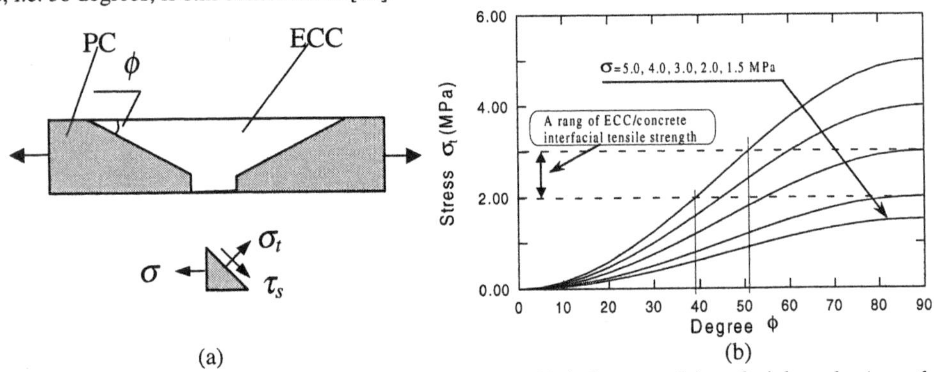

Fig. 9. (a) ECC/concrete interface geometric design and (b) influence of interfacial angle ϕ on the normal stress at ECC/concrete interface

Experimental Simulation and Results

In this section, the validity of the concept will be experimentally verified through uniaxial tensile test on ECC-concrete composite specimen. In the present work, ϕ=30 degree was adopted. The size of specimen used in the tensile tests is shown in Fig. 10(a). In order to cast the concrete sections and the ECC strip at the same time and to ensure an inclined angle of 30 degrees for the ECC/concrete interface, a special casting device was developed. The details on the specimen casting device and procedures can be found elsewhere [18]. The overview of the specimen after casting is shown in Fig. 10(b). The above tensile specimens were cured in water at 23^{0}C and tested at 28 days after casting. The tensile test results on the specimens with ECC strip are shown in Fig. 11(a) in terms of tensile stress versus strain diagrams. As the specimen was loaded in tension, first cracking occurred in the ECC element at around 2.5 MPa followed by multiple cracking due to sufficient difference between the first crack strength of ECC, ultimate tensile strength of ECC and tensile strength of concrete. Deformation was successfully localized into the ECC strip instead of in the concrete section. The width of the microcracks in the ECC strip is less than 0.1-0.2 mm. Furthermore, in the present dimension of the specimen, the strain attained 1.4% at peak load (3.5 MPa). A view of the specimen after cracking in ECC section is shown in Fig. 11(b).

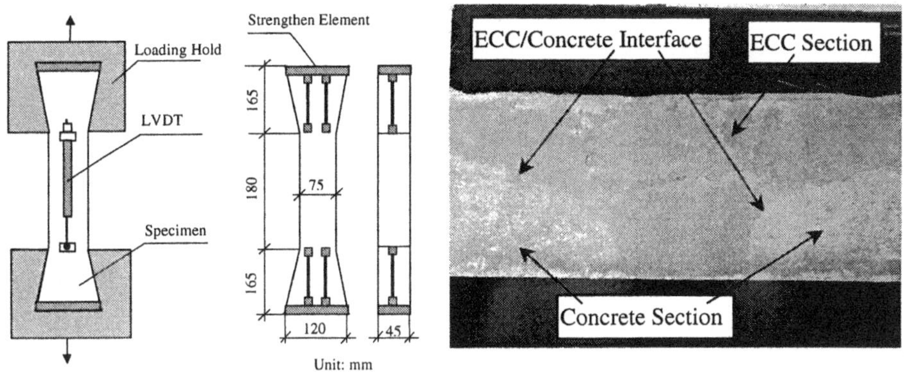

(a) (b)

Fig. 10. (a) Test set-up and geometry of specimen and (b) view of the specimen after casting, the ECC has a slightly darker color than the concrete

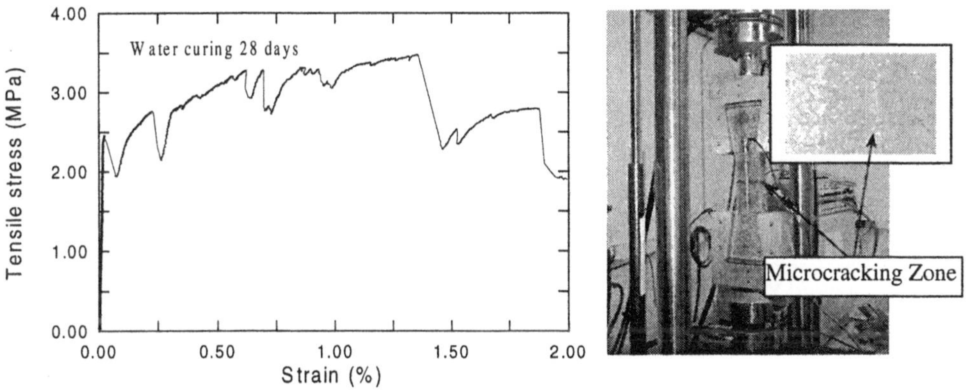

(a) (b)

Fig. 11. (a) Tensile stress strain curve of the ECC-concrete composed specimen and (b) a view of cracking in ECC strip, the close-up view of the microcracking zone is shown in insert

CONCLUSIONS

This paper reviews the recent results on the studies of durability enhancements of concrete bridge deck. Fatigue performance of cementitious material can be significantly improved by adding fibers. With the same maximum load level, the fatigue life can be prolonged by several orders of magnitude depending on the fiber types and content. The fatigue crack growth rate in fiber reinforced cementitious composite is found to be governed by the crack bridging degradation behavior of the composite in the fracture zone. A composite slab containing both plain concrete and ECC strips, with proper design at the ECC/concrete interfaces, and careful selection of material properties (i.e. to assure that the tensile strength of concrete higher than that of ECC material), it is possible to localize the tensile deformation into the ECC strip instead of cracking in the concrete section. Due to the strain-hardening performance of the ECC material with high strain capacity (up to 5%), the overall strain capacity and the integrity as well as the fatigue durability of the composite slab can be significantly improved. This concept has been validated with a simple laboratory experimental simulation. Further experiments with larger scale specimens are needed in order to apply the design concept explored in the present work in more realistic field situations.

ACKNOWLEDGMENTS

This work has been supported by a grant from the National Science Foundation (CMS-9872357) to the University of Michigan. Helpful discussions with A. Nowak are acknowledged. Furthermore, knowledge on the fatigue behavior between ECC and concrete interface is not available at present, and should be investigated.

REFERENCES

1. Matsui, S. "Technology Developments for Bridge Decks–Innovations on Durability and Construction." Kyouryou To Kiso(8), 1997, 84-92.
2. Perdikaris, P. C., and Beim, S. "RC Bridge Decks under Pulsating and Moving Load." Journal of Structural Engineering, 114(3), 1988, 591-607.
3. Perdikaris, P. C., Beim, S. R., and Bousias, S. N. "Slab Continuity Effect on Ultimate and Fatigue Strength of Reinforced Concrete Bridge Deck Models." ACI Structural Journal, 86(4), 1989, 483-491.
4. Kumar, S.V. and H.V.S. GangaRao. "Fatigue Response of Concrete Deck Reinforced with FRP Rebars." ASCE J. Structural Engineering, 124, 1, 1998, 11-16.
5. Nishikawa, K. "A Concept of Minimized Maintenance Bridges." Kyouryou To Kiso(8), 1997, 64-72.
6. Okada, K., Okamura, H., and Sonoda, K. "Fatigue Failure Mechanism of Reinforced Concrete Bridge Deck Slabs." Transportation Research Record, 664, 1978, 136-144.
7. Butler, J. E. "The Performance of Concrete Containing High Proportions of Steel Fibres with Particular Reference to Rapid Flexural and Fatigue Loadings." Fiber Reinforced Cements and Concretes-Recent Developments, R. N. Swamy and B. Barr, eds., Elsevier, New York, 1989, 544-552.
8. Ramakrishnan, V., and Josifek, C. "Performance Characteristics and Flexural Fatigue Strength of Concrete Steel Fiber Composites." Int. Sym. on Fiber Reinforced Concrete, Madras, India, 1987, 2.73-2.84.
9. Ramakrishnan, V., and Lokvik, B. J. "Flexural Fatigue Strength of Fiber Reinforced Concretes." High Performance Fiber Reinforced Cement Composites, H.W. Reinhardt & A.E. Naaman, eds., 1992, 271-287.
10. Zhang, J., and Stang, H. "Fatigue Performance in Flexure of Fiber Reinforced Concrete." ACI Materials Journal, Vol. 95, No. 1, 1998, 58-67.
11. Ramakrishnan, V., Gollapudi, S., and Zellers, R. "Performance Characteristics and Fatigue Strength of Polypropylene Fiber Reinforced Concrete." Fiber Reinforced Concrete Properties and Applications SP-105, American Concrete Institute, Detroit, 1987, 159-177.
12. Banthia, N. "Carbon Fiber Cements: Structure, Performance, Applications and Research Needs." Fiber Reinforced Concrete-Developments and Innovations ACI-SP142, J. I. Daniel and S. P. Shah, eds., American Concrete Institute, Detroit, 1994, 91-119.
13. Matsumoto,T. Fracture Mechanics Approach to Fatigue Life of Discontinuous Fiber Reinforced Composites. Doctoral Thesis, Dept. of Civ. & Envir. Engrg, University of Michigan, 1998.
14. Zhang, J. and Stang, H. "Interfacial Degradation in Cement-Based Fiber Reinforced Composites." Journal of Material Science Letter, Vol.16, No.11, 1997, 886-888.
15. Zhang, J., Stang, H. and Li, V. C. "Experimental Study on Crack Bridging in FRC under Uniaxial Fatigue Tension." ASCE Journal of Materials in Civil Engineering, Vol.12, No. 1, 2000, 66-73.
16. Stang, H. and Aarre, T. "Evaluation of Crack Width in FRC with Conventional Reinforcement." Cement and Concrete Composites, 14(2), 1992, 143-154.
17. Zhang, J., Stang, H. and Li, V.C. "Fatigue Life Prediction of Fibre Reinforced Concrete under Flexure Load." International Journal of Fatigue, Vol. 21, No. 10, 1999, 1033-1049.
18. Zhang, J., Li, V.C., Nowak, A. and Wang, S., "Introducing Ductile Strip for Durability Enhancement of Concrete Slabs." Submitted to ASCE Journal of Materials in Civil Engineering, 2000.

LONG-TERM RELIABILITY OF STRUCTURAL SYSTEMS

S. Mahadevan, R. Zhang, P. Shi, H. Mao, A. Dey, and P. Raghothamachar

Department of Civil and Environmental Engineering, Vanderbilt University
Nashville, TN 37235, USA

ABSTRACT

This paper develops the use of limit-state based reliability analysis techniques for long-term durability evaluation of structural systems. Failure modes related to strength deterioration, fatigue, corrosion, creep and wear are considered. Time-dependent reliability analysis is implemented for both component-level and system-level assessment. Combinations of analytical and sampling-based methods are pursued. For system-level assessment, an adaptive importance sampling technique is developed, and implemented in combination with commercial finite element analysis. In addition to computational reliability predictions, the updating of reliability estimates using non-destructive inspection data and Bayesian analysis is investigated. The proposed methodology is demonstrated with numerical examples related to aging civil structures and aircraft structures.

KEYWORDS

Durability, fatigue, corrosion, creep, fretting, multiple site damage, importance sampling, Bayesian method, non-destructive inspection, reliability updating, system reliability, time-variant reliability.

INTRODUCTION

The application of reliability technology to the design and evaluation of civil structures has grown considerably in the past decades. Much of this work, however, considers the reliability estimation problem as time-invariant. The variation of the load over the lifetime of the structure is considered indirectly, through extreme value statistics. This strategy is conservative and satisfactory for the design of new structural systems. However, in the case of existing structures, durability assessment, repair and rehabilitation studies need to consider load histories, resistance degradation, extent of damage etc. The current nation-wide emphasis on infrastructure renewal requires that practical methods be developed to assess the reliability of existing structures, and this requires explicit consideration of the variation of reliability over time.

This paper develops several practical methods to estimate the time-variant reliability of mechanical and structural systems. Currently used analytical methods have several simplifying assumptions regarding the loads and the resistances, which render them inapplicable to practical structures. This paper develops a simulation-based methodology, which has several advantages such as robustness, simplicity of understanding and implementation, and the capability to model realistic situations.

The proposed methodology is general enough to include ductile, brittle, and semi-brittle structural behavior. The methodology is explored for a variety of failure modes related to strength deterioration, fatigue, corrosion, creep and wear. Time-dependent reliability analysis is implemented for both component-level and system-level assessment. Combinations of analytical and sampling-based methods are pursued. The techniques are implemented in combination with commercial finite element analysis. In addition to computational reliability predictions, the updating of reliability estimates using non-destructive inspection data and Bayesian analysis is investigated. The proposed methodology is demonstrated with numerical examples related to aging civil structures and aircraft structures.

TIME-VARIANT RELIABILITY

Consider a single structural component with initial resistance r. Let the resistance of the component deteriorate with time as follows:

$$r(t) = r - g(a, t) \tag{1}$$

where, $r(t)$ is the resistance at time t, r is the initial component resistance, and $g(a,t)$ is the degradation function dependent on the elapsed time and the degradation rate a. For the moment, both r and a are considered as deterministic quantities. Subsequently in this section these assumptions are relaxed, and both r and a are considered random in the final formulation. If the component is subjected to a load process (Poisson pulse process) which produces a sequence of n discrete load events within the time period $(0, t_L)$, the lifetime reliability of the component can be represented mathematically as

$$\begin{aligned} L(t_L) &= P[r(t_1) > s_1 \cap \ldots\ldots \cap r(t_n) > s_n] \\ &= P[r \cdot g(a,t_1) > s_1 \cap \cdots \cap r \cdot g(a, t_n) > s_n] \end{aligned} \tag{2}$$

Extending the above concept of time-dependent reliability to a parallel system of m components, it can be said that for the entire system to survive, the strongest component must survive all the n occurrences of the load. Mathematically this can be represented as:

$$\begin{aligned} L_S &= P[\max_{i=1}^{m} r_i g_i(a,t_1) > c_i s_1 \cap \cdots \cap \max_{i=1}^{m} r_i g_i(a,t_n) > c_i s_n] \\ &= \prod_{j=1}^{n} F_S\left(\max_i^m \frac{r_i g_i(a,t)}{c_i} \right) \end{aligned} \tag{3}$$

where $c_i s_j$ is the structural action induced in the ith component of the system by the jth occurrence of the load, and $F_s(\,)$ is the cumulative distribution function (CDF) of the load intensity. Removing the dependence on the deterministic inter-arrival time of load occurrences, Eqn. (3) is rewritten as

$$L_s = \exp\left[-\nu \cdot \left\{ t_L - \int_0^{t_L} F_S\left(\max_i^m \frac{r_i g_i(a,t)}{c_i} \right) dt \right\} \right] \tag{4}$$

The failure probability of the system can be represented as

$$P_f\,(t_L | \underline{R} = \underline{r}) = 1 - L_S = 1 - \exp - \nu \cdot t_L - \int_0^{t_L} F_S\left(\max_i^m \frac{r_i g_i(a,t)}{c_i} \right) dt \tag{5}$$

As a final step, if $fR(r)$ is the joint pdf of the initial strength of the components and $fA(a)$ is the pdf of the degradation parameter, the system failure probability can be written as

$$p_f(t_L) = \int_0^\infty \{1 - \exp[-\nu \cdot \{t_L - \int_0^{t_L} F_S(\cdot)dt\}]\} f_{\underline{R}}(\underline{r}) \cdot f_A(a) dr da \tag{6}$$

Eqn. (6) represents the failure probability of a parallel system subjected to one time dependent load process. The uppercase letters R and A represent the state variables and the lowercase letters r and a represent actual realizations of the corresponding random variables. Even though Eqn. (6) represents failure probability due to a single load process, it can be adapted to estimate the reliability of a parallel system subjected to two or more time variant loads.

An important concept that can be incorporated in Eqn. (6) is that of periodic repair. It is assumed that after every repair the structure is restored to its original capacity. If within the time period (0, t_L), the structure is repaired at every interval of t_r, the conditional failure probability of the system in Eqn. (5) can be rewritten as:

$$p_f(t_L \mid R = r) = 1 - \exp[-\nu \cdot \{t_L - \int_0^{t_r} F_s(\cdot)dt \cdots - \int_{t_{rn}}^{t_L} F_s(\cdot)dt\}] \tag{7}$$

where $n = t_L / t_r$. The steps for estimating the unconditional failure probability are same as in Eqn. (6).

It is clear that the outer integral in Eqn. (6) is a multidimensional integral (over the whole domain of the component resistance variables). Even for a simple 2-bar parallel system, computation of the integral can be cumbersome. Therefore, an adaptive importance sampling technique has been developed as described below to compute the outer multi-dimensional integral. The inner single dimensional integral over the time domain is computed using a Gauss quadrature scheme.

CORROSION FATIGUE

The damage process considered here consists mainly of pitting nucleation and growth, leading to crack growth under the combination of corrosion and fatigue in aluminum alloys. The total fatigue life may be represented by the sum of the following four phases (Figure 1):

$$t_f = t_{pn} + t_{pg} + t_{sc} + t_{lc} \tag{8}$$

where t_{pn} is the time for pit nucleation, t_{pg} is the time for pit growth, t_{sc} is the time for short crack growth, and t_{lc} is the time for long crack growth. The four phases are modeled as follows:

1. Pit Nucleation: This first stage is related to the electrochemical processes during corrosion which result in the nucleation of a corrosion pit. The proposed method assumes the time to pitting nucleation t_{pn} as a random variable.
2. Pit Growth: The second stage relates to pit growth, which initiates at the constituent particles and involves electrochemical processes affected by clusters of particles. In this model, the pit is assumed to grow at a constant volumetric rate by Faraday's Law.
3. Short Crack Growth: An empirically based probabilistic relationship is used to model the short crack growth, similar to Paris Law. However, the parameters involved are random variables dependent on experimental data.
4. Long Crack Growth: The widely used Paris Law may be used in this stage to estimate the time for long crack growth. A random variable with a mean value of 1mm is assumed to be the transition size from short to long crack. The effect of the coefficient of variation is studied in the numerical example.

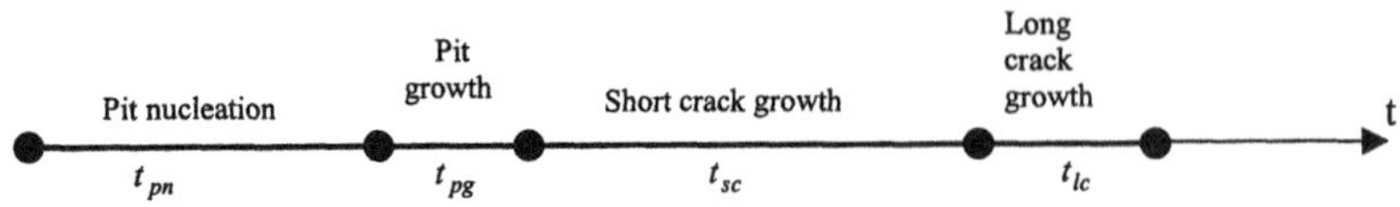

Figure 1: Four phases of pitting corrosion fatigue life

The failure probability at some specified time t may be expressed as:

$$P_f(t) = P(t_f \le t) = P(t_f - t \le 0) \tag{9}$$

A numerical example is considered below. The structure is idealized as an infinite plate with a circular rivet hole. The material considered is an aluminum alloy. An aggressive environment is assumed and pit corrosion occurs on the surface of the hole. The random variables are assumed to be statistically independent. Their coefficients of variation are not available. Therefore, three different values, 0.01, 0.5, and 0.95 are used, and their effects are studied.

Figure 2 shows the CDF curve calculated by the Monte Carlo simulation method and FORM. Notice that both of them are close to each other. It shows that the limit state in this problem is not very nonlinear and FORM can be used well in this problem, providing good computational efficiency. Probabilistic sensitivity analysis is performed at a specified time instant and the sensitivity factors are shown in the Figure 3. The short crack growth parameter C_{sc} and the time to pit nucleation t_{pn} have the most important effect on the failure probability calculation. Parametric analysis has also been conducted with different COV values and distribution types for the random variables.

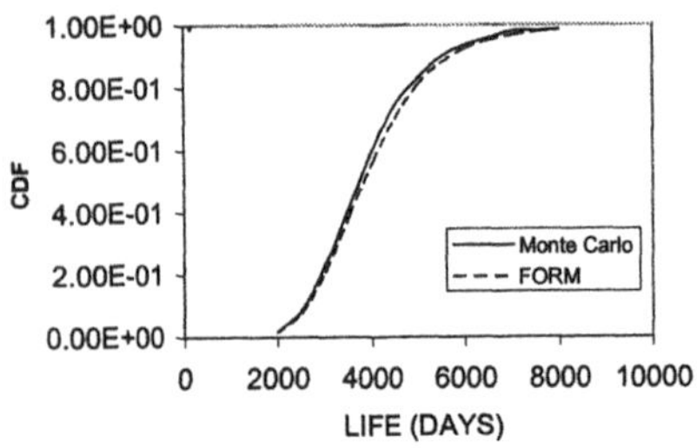

Figure 2: CDF of corrosion-fatigue life

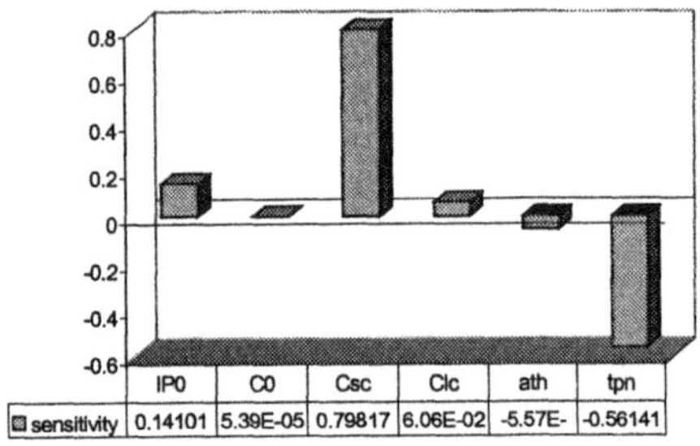

Figure 3: Sensitivity analysis

CREEP FATIGUE

Creep is one of the principal damage mechanisms for materials operating at elevated temperatures. It can produce larger strain deformation, stress relaxation, and crack initiation and growth. For materials under fatigue and creep loading, creep has serious influence on the properties and fatigue life of the material. Several models have been developed to carry out the creep-fatigue life prediction analysis. A popular creep-fatigue damage evaluation method is to use the linear accumulation rule (e.g. Zamrik, 1993). In the ASME Boiler and Pressure Vessels Code (1998), the corresponding failure criterion is expressed in term of load cycle and time fractions employed for damage. The allowable cumulative damage limit is the sum of the cycle or time fraction for fatigue and creep respectively. The criterion consists of a bilinear equation.

Some experimental results (Chen, 1998) show that the experimentally measured creep-fatigue lifetime is slightly lower than that predicted by the linear damage accumulation rule. The fact appears to indicate the compounding effect that occurs when both creep and fatigue are present. The experimental results by Yaguchi (1996) not only show this compounding effect, but also show that the loading sequence affects the creep-fatigue life. Different loading sequences can increase or decrease the lifetime of the material.

A new creep-fatigue failure model is proposed with a continuous function for the creep-fatigue failure criterion. The criterion is a function of two random parameters to model the uncertainty of the failure criterion. As the parameters are given different values, the criterion takes different forms so that it can fit the experimental results better. The creep-fatigue failure criterion is given in terms of creep damage D_c and fatigue damage D_f with two experiment-obtained parameters θ_1 and θ_2 as

$$g(N_c, N_f, n_c, n_f, \theta_1, \theta_2) = D_{cr} - (D_c + D_f) = 2 - e^{\theta_1 D_c} + \frac{e^{\theta_1} - 2}{e^{-\theta_2} - 1}\left(e^{-\theta_2 D_c} - 1\right) - D_f \quad (10)$$

The probabilistic model for reliability analysis of creep-fatigue life based on the linear damage accumulation rule and the proposed creep-fatigue failure function is illustrated with a numerical example. The failure probabilities based on the proposed failure function and bilinear function for different cases are calculated and shown in Fig. 4. These cases have different amounts of creep and fatigue damage, but with the same cumulative damage (0.6). The two failure functions underestimate or overestimate the failure probability for different ranges of creep damage.

Fig. 5 shows that different amounts of creep damage have different effects on the failure probability, though the overall accumulated creep and fatigue damage is the same (D = 0.6). Creep damage between 0.1 and 0.4 increases the failure probability significantly. The reliability has a large reduction due to the creep-fatigue interaction with creep damage over this range. It is also seen that FORM and Monte Carlo simulation method agree very well for this problem.

The proposed failure criterion curve relaxes the symmetry assumption with respect to the line $D_f = D_c$, which is used in the ASME Code. With different values of θ_1 and θ_2, the criterion can be made to fit the experimental results to account for creep-fatigue interaction and also the effect of different amounts of creep damage on the residual fatigue life. Another advantage of the proposed failure function is that unlike the bilinear model whose derivatives are discontinuous at the junction of the two linear segments, its first and second derivatives are all continuous over the range of D_c (0,1). Therefore SORM analysis can be carried out without difficulty, and the uncertainty in the failure criterion can be modeled explicitly with two parameters.

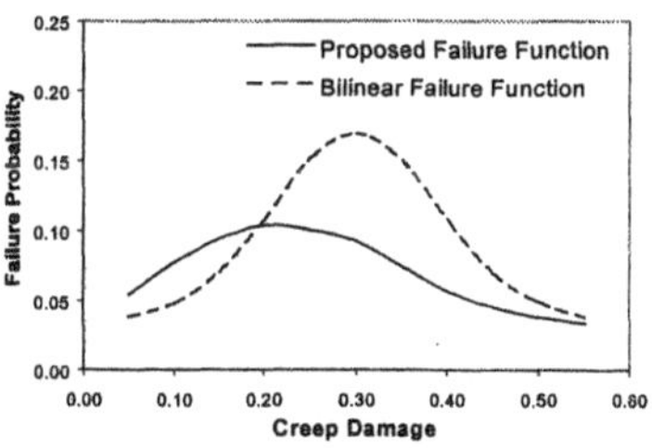

Figure 4: Failure probability based on two failure functions

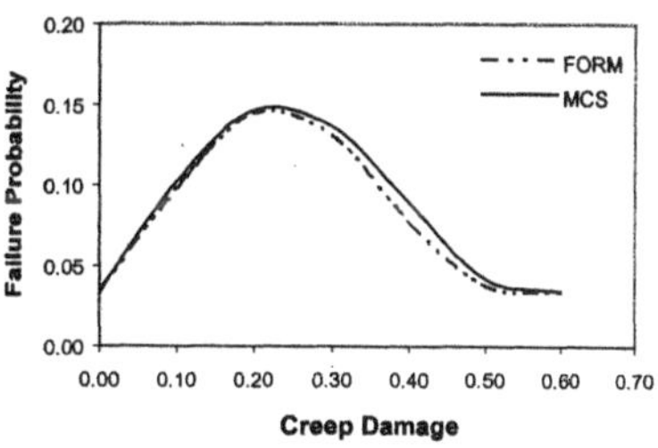

Figure 5: Effects of creep-fatigue interaction.

FRETTING FATIGUE

Fretting fatigue is one of the main mechanisms of the formation of cracks in riveted lap joint assemblies in aging aircraft. Like plain fatigue, fretting fatigue damage has two stages: the initiation of a crack from a surface suffering severe distress, and the propagation of the crack. Hills (1994) and Szolwinski (1995) concluded from experiments that for aircraft components, the "majority" of the fatigue life involves the nucleation of a crack. Therefore this study is focused on the probabilistic prediction of the crack initiation life of fretting fatigue.

A deterministic fretting fatigue parameter k proposed by Ruiz (1984) provides a possible means to predict the site and likelihood of crack initiation during a given number of loading cycles. However, since this parameter does not incorporate material properties except using them for stress and displacement calculation and is not able to predict fretting fatigue life, it is difficult to use as a reliable design criterion. Similarities between multi-axial fatigue crack nucleation concepts and observations of the formation of fretting fatigue cracks lead to the application of multi-axial fatigue concepts to the fretting fatigue problem. The empirical relationship between total strain amplitude and load cycles to failure is described by Smith-Matson-Topper (SWT) equation:

$$\sigma_{\max}\left(\frac{\Delta\varepsilon}{2}\right) = \frac{(\sigma_{f'})^2}{E}(2N_i)^{2b} + \sigma_{f'}\varepsilon_{f'}(2N_i)^{b+c} \tag{11}$$

where $\sigma_{f'}$ is the fatigue strength coefficient, b is the fatigue strength exponent, $\varepsilon_{f'}$ is the fatigue ductility coefficient and c is the fatigue ductility exponent. N_i is the number of load cycles to a 1 mm surface crack nucleation and $\Delta\varepsilon$ is the applied strain range equal to ($\varepsilon_{\max}$ - $\varepsilon_{\min}$). With an experiment of the curved fretting pads clamped into contact with the flat surface of the specimen, Szolwinski (1995) proved the validity of the application of the SWT model in the Hertz-contact fretting fatigue problem.

Define the limit state function: $g(\underline{x}) = N_i - n$, ANSYS analysis can be combined with FORM and is run repeatedly in each iteration step in locating the MPP. To illustrate the proposed approach, fretting conditions in a pinned connection under cyclic loading having dimensions typical of riveted panels is considered. Figure 6 shows a strip of a wide panel with many pins with a repeat distance. The fretting condition in the neighborhood of contacts between pin and panels is analyzed using ANSYS finite element software. Figure 6 shows the refined element mesh of the pin and the area of the panel adjoining the interface. The Ruiz parameter and the probabilistic SWT life model are calculated respectively using the integration of FORM and FEM. It is seen from the result that the nucleation sites of fretting fatigue crack predicted using these two methods are approximately at the same position. In addition, the probabilistic SWT model gives the quantitative CDF of the fretting fatigue life, as shown in Fig. 7. This is a quantitative demonstration of the qualitative "likelihood of crack initiation" predicted by Ruiz parameter. In principle, the proposed approach provides an outline for the combination of mechanics-based fretting fatigue nucleation modeling, finite element analysis and reliability method to estimate fretting fatigue reliability of structural components.

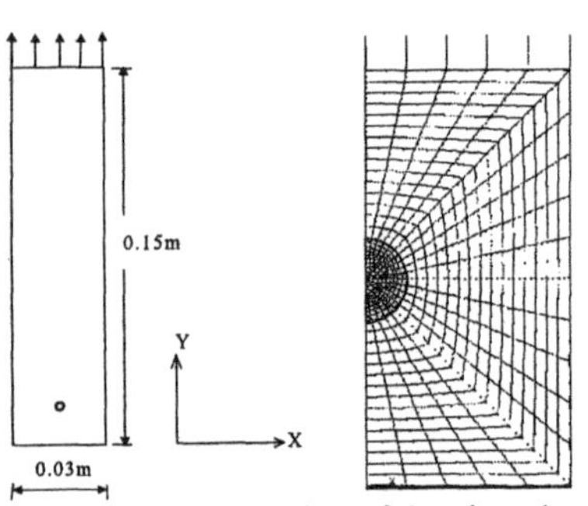

Figure 6: Configuration of the pinned connection and FEM model

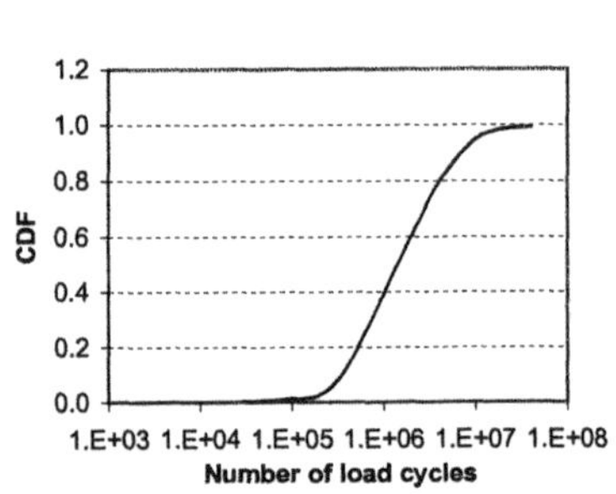

Figure 7: CDF of number of load cycles

ADAPTIVE IMPORTANCE SAMPLING

The above sections considered reliability computation for individual limit states. Realistic structures consist of multiple components, multiple limit states, and multiple failure paths. An adaptive importance sampling method has been developed by the authors to compute time-dependent system reliability. The importance sampling method attempts to generate most of the samples in the failure domain and computes the failure probability. In theory, this is a very efficient approach compared to basic Monte Carlo simulation where most of the samples are likely to be in the safe domain for high reliability problems. However, in practice, for any importance sampling technique to be effective one must have some prior knowledge of the system failure domain. Several methods have been proposed for the selection the sampling domain. These include multi-modal (Melchers 1989, Karamchandani et al 1989) and curvature-based methods.

The importance sampling method that is used here is based on the adaptive multi-modal sampling technique proposed by Karamchandani et al (1989). The method has been previously demonstrated for single and multiple limit state problems. However, the system reliability estimation of practical structures involves multiple sequences with both ductile and brittle component failures. Therefore this paper extends and implements Karamchandani's method to problems with multiple failure sequences. In this method, the initial sampling density function is chosen to have the same form and variance as the original density function but centered at an initial starting point in the failure domain. Once several samples have been obtained in the failure domain, a multi-modal sampling density function is constructed which emphasizes multiple points in the failure domain, each in proportion to the true probability density at the point. However, not all the sample points are emphasized; only one representative point from a cluster of points is chosen.

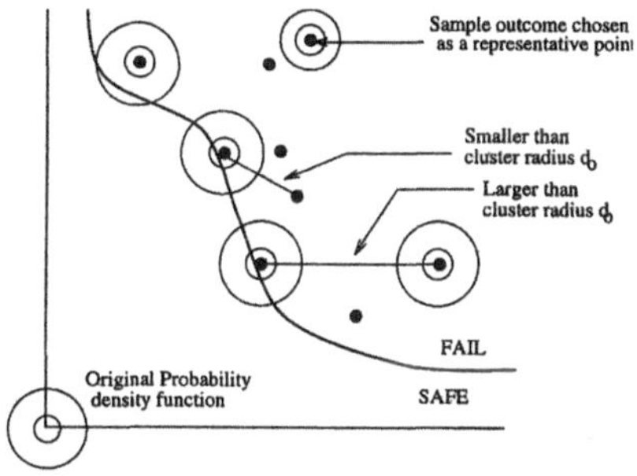

Figure 8: Adaptive importance sampling

The representative points are separated by a distance greater than the "cluster radius" d_0 (see Fig. 8). Usually the value of d_0 is taken to be the average distance between the mean and the sampling points. The multi-modal sampling density to generate the *i*th sampling point is:

$$h^i_{\tilde{X}}(\tilde{x}) = \sum_{j=1}^{k} \hat{\omega}^j_i f^{(j)}_{\tilde{X}}(\tilde{x}) \tag{12}$$

where $\hat{\omega}^j_i$ = importance attached to the *j*th sampling point

$$= p_{f|\tilde{X}=\hat{x}^{(j)}} f_{\tilde{X}}(\hat{x}^{(j)}) / (\textstyle\sum_{r=1}^{k} p_{f|\tilde{X}=\hat{x}^{(j)}} f_{\tilde{X}}(\hat{x}^{(r)})) \tag{13}$$

where $f_X(x)$ is the original density function, $f_X^{(j)}(x)$ is the original density function with the mean shifted to $\hat{x}^{(j)}$, and $\hat{x}^{(1)}...\hat{x}^{(k)}$ are the representative points. Each representative point has a larger probability density than the other points in its cluster.

After i sample points, the estimated failure probability is given by:

$$p_f = \frac{1}{i}\sum_{j=1}^{i}\frac{p_{f|\tilde{X}=\tilde{x}^{(j)}} f_{\tilde{X}}(x^{(j)})}{h_X^j(x^{(j)})} \tag{14}$$

In the adaptive simulation technique used in this study for system reliability analysis of redundant structures, the branch and bound technique is used once to determine the first complete failure sequence. The first sequence is used to define the initial failure domain for starting the adaptive importance sampling. Once the first failure sequence is identified, an initial sampling is done to generate a few samples (about 10 samples) that follow the first failure sequence to system failure. From these initial samples, representative points are selected as mentioned earlier, and a multi-modal sampling density function is constructed as in Eqn. (12).

Adaptive importance sampling is done next. Samples are obtained with the multi-modal density function, and all samples that lead to structural failure by any sequence are accepted. After each sampling, the set of representative points, and therefore, the multi-modal sampling density, are modified to include the new system failure samples, thus refining the failure domain approximation. Thus more system failure sequences in addition to the first one are included, and the sequences are weighted according to their probability.

Sampling is continued till the failure probability converges to an accepted level of accuracy. Two convergence tests are used in this paper. The relative change in failure probability estimate with each additional sample is computed as

$$\delta = \frac{|p_{f_k} - p_{f_{k-1}}|}{p_{f_{k-1}}} \tag{15}$$

This is compared with a preset tolerance limit ε. The simulation converges if the criterion $\delta \le \varepsilon$ is satisfied for 10 successive simulations, and if the COV of the failure probability at the end of the 10th such simulation is less than ε_c (= 0.05, for example; the value depends on the degree of accuracy required).

The tasks in the proposed adaptive importance sampling method can be divided into three distinct groups: (1) First failure sequence search; (2) Adaptive sampling; (3) Structural analysis. Different tasks are used in each of these groups. Hence these three groups can be kept separate from each other. The first two groups make use of the third group for all the structural analyses. Figure 9 shows the organization of the various groups of tasks in the computer implementation of this technique.

Figure 10 shows a fourteen-bar truss. The load processes for the structure are assumed to be Poisson pulse processes. The horizontal load process $S_1(t)$ has a mean occurrence rate of 0.5/yr with a mean duration of 0.3 yr. The vertical load process $S_2(t)$ has a mean occurrence rate of 1.0/yr with a mean duration of 0.2 yr. According to the proposed method, the first failure sequence identified for the truss was 4-5. Therefore, as discussed earlier, the initial sampling is centered around this failure sequence. During the course of further simulations, it was observed that the other dominant failure sequences were included in the failure domain. The structure was analyzed for failure under three different degradation schemes, namely linear (n = 1, representative of corrosion of reinforcement in reinforced concrete structures), square-root (n = 0.5, diffusion related degradations such as leaching of sodium

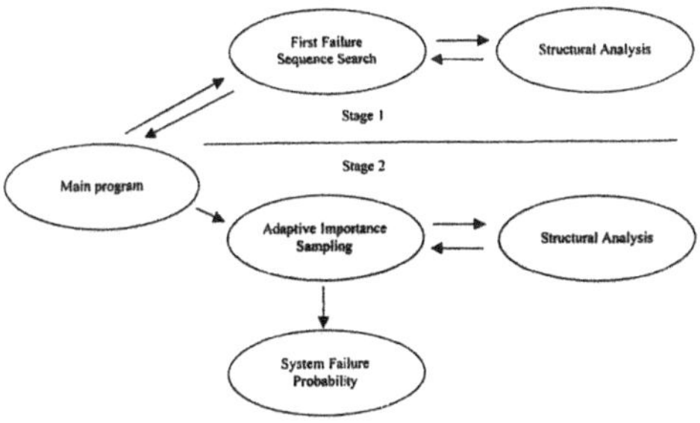

Figure 9: System reliability estimation using adaptive importance sampling

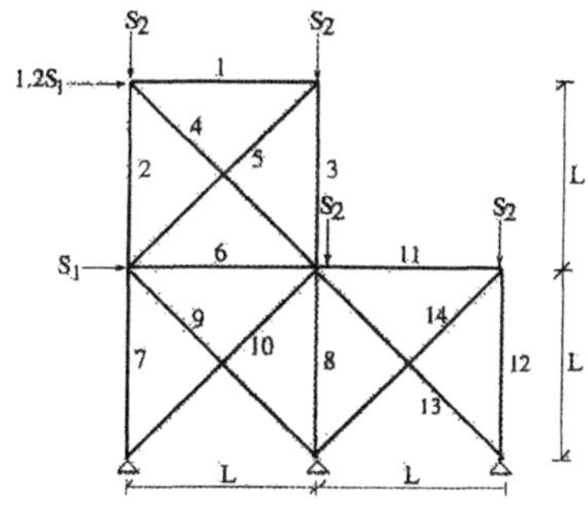

Figure 10: Fourteen-bar Truss

hydroxide) and parabolic ($n = 2$, sulfate attack) (Mori and Ellingwood, 1992) expressed as

$$R(t) = R_0(1 - at^n) \tag{16}$$

The degradation parameter, a, is assumed to be a normal random variable for each degradation scheme, and its mean value is determined by assuming that the structure will carry 80% of its original load carrying capacity at 50 years. A COV of 0.15 is assumed for a. Even though the final residual strengths for all three schemes have been assumed to be the same, the initial degradation for the square root mechanism is faster than the other mechanisms. This implies that resistances will initially decrease at a much faster rate and the failure probability will initially increase at a higher rate for the square-root degradation scheme. The results of the simulation as shown in Figure11 are consistent with this.

Additional similar analyses were also performed assuming that there is complete repair of the structure every 5 years. Even though this is an unrealistic assumption, it was used to demonstrate the proposed method. In reality, only partial repair is done, i.e., if damage is detected in a member, only that member is repaired. The proposed method can be extended to include this condition by restoring the resistances of the repaired members only. It is seen in Figure 11 that for the truss square root degradation (diffusion related processes) is the most critical. Also, for several example problems (Dey and Mahadevan, 2000, Mahadevan and Raghothamachar, 2000), it was observed that estimates of the proposed method (< 400 samples) agreed very well with those of basic Monte Carlo simulation (100,000 samples).

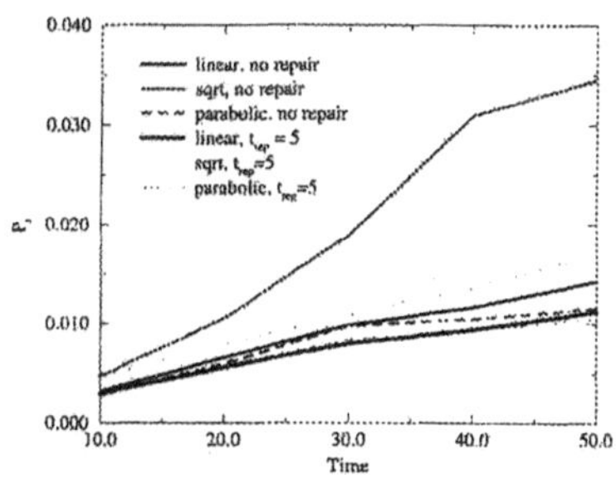

Figure 11: Reliability estimation with repair / no repair

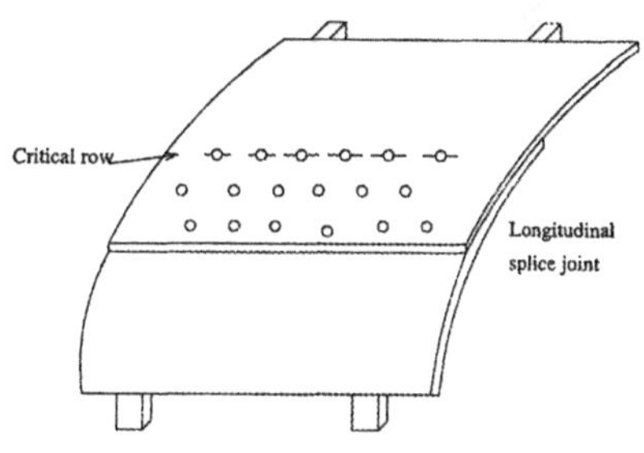

Figure 12: Multiple site damage of lap joints along fuselage

MULTIPLE SITE DAMAGE

The reliability analysis of corrosion fatigue under multiple site damage is a system reliability problem. The proposed adaptive importance sampling approach in the previous section is used to evaluate the system failure probability of the structure under MSD.

In the presence of MSD cracks, the corresponding fatigue parameters need to be computed from the finite element method. It is assumed that interactions among the MSD cracks are not significant during the short crack growth stage until the short crack reaches a critical size. For computational implementation, the element size in the finite element analysis may be made to equal to this interaction critical size. Upon this assumption, the total corrosion fatigue life is divided into two parts from the perspective of structural analysis: non-interaction stage and interaction stage. Before the short crack propagates to the interaction critical size, the fatigue life can be evaluated from the same analytical expression as for the case of single site damage. After that, the load cycles are applied incrementally to investigate the corresponding crack growth at multiple sites. Using the above assumption, the time for propagation to the interaction critical size may be obtained for all the MSD cracks. It is obvious that when the slowest crack site reaches the interaction size, all the other cracks would have propagated further into the interaction stage. The corresponding propagating length of these cracks are considered to be the initial crack sizes for finite element analysis. The time for MSD structure to get to this state is the time spent on the no-interaction stage. After that, the interaction between MSD cracks is considered through incremental analysis. The stress intensity factor from all the crack sites is obtained from the finite element analysis. Applying a fixed number of load cycles on the structure, the updated crack sizes at all the crack sites are calculated. The load cycles are incrementally added until the linkage of two neighboring cracks. The total time to failure for this coalescence is then obtained.

The structure of interest is a lap joint along the fuselage in an aircraft structure with multiple circular rivet holes. The fuselage pressurization is the main fatigue load on the longitudinal fuselage splice, causing the hoop stress on the lap joints. Three assumptions are used: (1) The MSD cracks are assumed to occur at the outer row in the outer sheet of the lap joint. (2) The corrosion pits are assumed to initiate along the edge of the rivet holes due to lack of coating. (3) The crack initiated from the pits will propagate towards the nearest crack initiated from the neighboring rivet holes. The component failure is assumed to be the crack propagation towards the adjacent rivet hole leading to the linkage of two sites. The system failure is defined to be the linkage of any two neighboring cracks.

A two-dimensional finite element model is adopted for the analysis of lap joints. The loading is assumed to be the cyclic hoop stress existing in the fuselage. Twelve cracks of different initial size, caused by pitting corrosion, are assumed to emanate from six equally spaced rivet holes on the upper row, as shown in Figure 12. It takes about 380 simulations for the adaptive simulation to converge to the failure probability estimate, compared to 10,000 samples needed in the basic Monte Carlo method to converge to the same level of accuracy. The variation of system failure probability and the variation of COV are shown in Figures 13 and 14.

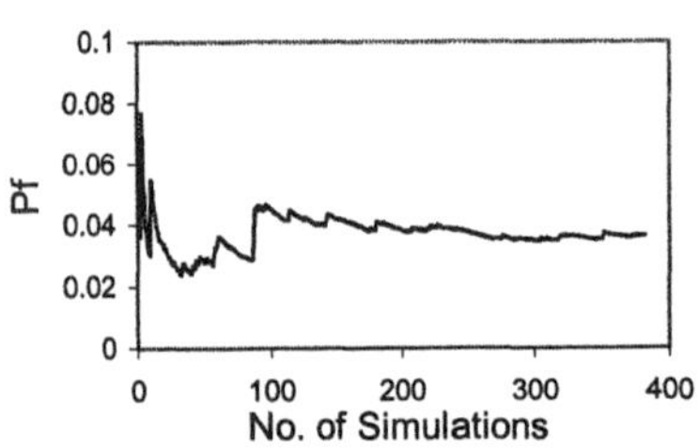

Fig. 13: Variation of system failure probability Pf

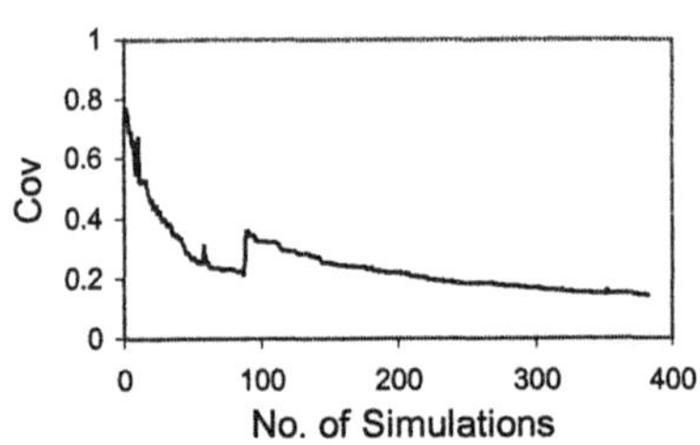

Figure 14: Variation of COV

RELIABILITY UPDATING WITH INSPECTION

Reliability prediction based on mechanical degradation model and regular non-destructive inspection (NDI) are two complementary ways for ensuring the safe performance of aging structures. Based on the information provided by the outcome of NDI, the original life prediction model associated with many uncertainties can be updated to give a more accurate prediction for the remaining life. While periodic nondestructive inspection gives additional information on the in-service condition of the structure, it also adds uncertainty to the damage evaluation process. Two criteria for the assessment of uncertainty of the NDI technique---detectability and accuracy---are considered in the framework of probabilistic methods for fatigue and corrosion fatigue reassessment. Detectability is defined as the damage size below which the inspection can not detect. It is a random variable and its distribution can be obtained from the function of probability of detection (POD). Due to the large uncertainty and error in the interpretation of the measured signal, the measured damage size may not indicate the actual damage size. In many cases, it is assumed that the measured damage size is a normal distribution with a mean value equal to the actual damage size. Fatigue damage is quantified through crack size, and the corrosion damage is quantified in terms of percent material loss (PML), which measures the percentage of component thickness that is absent over a corroded area and is a function of the pit radius and the thickness of the component for a hemispherical pit.

Defining the limit state function of the degradation problem as g, and the inspection outcome as I, using the Bayesian theorem, the updated reliability is defined as $P_{f,up} = P(g \leq 0|I)$. For the corrosion fatigue problem, three possible outcomes for an inspection at time t are considered:

1. No corrosion, no crack detected: $r_t - r_d \leq 0$
2. Corrosion damage measured with radius R, but no crack detected: $r_t - R = 0 \cap a_t - a_d \leq 0$
3. Crack detected with size A: $a_t - A = 0$

where r_t is the pit radius, r_d is the detectability of the corrosion inspection technique, R is the measured pit radius, a_t is the crack size, a_d is the detectability of the crack inspection technique, A is the measured crack size.

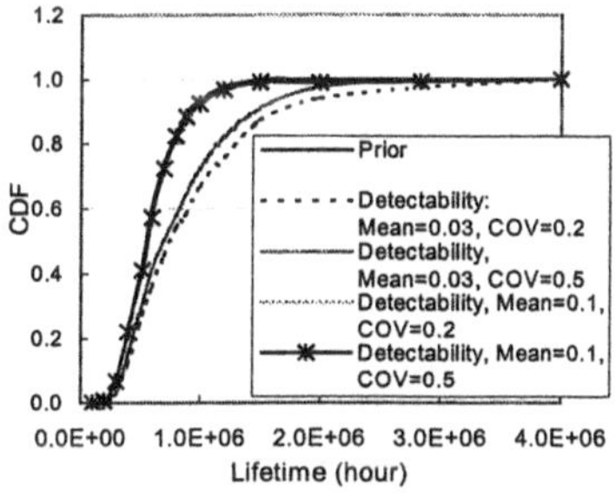

Figure 15: CDF updating using NDI

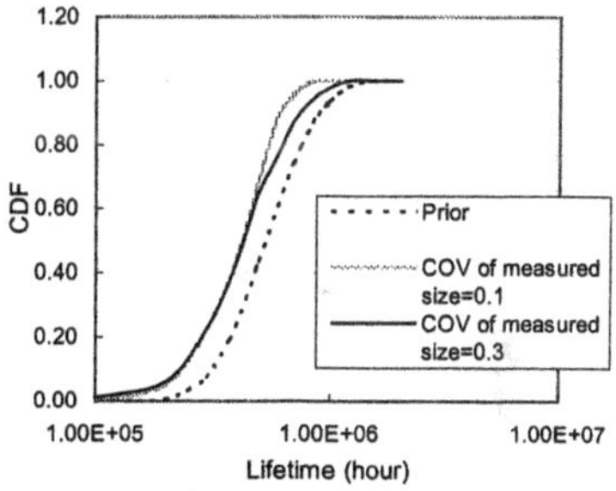

Figure 16: Use of measurement size

A 1.6mm thick flat sheet of Al 2024-T3 alloy subjected to a cyclic load with stress range $\Delta\sigma$ =100 MPa and frequency ν = 0.5 cycles/hour is considered. It is assumed that an eddy current inspection is performed after the aircraft has experienced 50,000 hours of operation. For illustration purposes, two cases of inspection outcome are considered herein: no damage detected and a crack of 0.6 mm detected. NDI techniques with different detectability and accuracy are considered. The predicted and updated CDF of corrosion fatigue life are shown in Figures 15 and 16. As expected, the inspection technique with high detectability and accuracy is seen to offer useful and critical information for the evaluation of the state of the structure and the prediction of the remaining life. A similar updating procedure, with the inclusion of model uncertainty, has also been demonstrated for fatigue reliability updating of butt welds in the tension flange of a steel bridge structure (Zhang and Mahadevan, 2000).

CONCLUSION

This paper presented several mechanics-based models for the reliability prediction of aging structures, under strength deterioration, corrosion, creep, fretting wear and fatigue. An adaptive simulation technique has been developed for the assessment of structures with multiple components, failure modes, and failure paths. Interactions between the failure modes have been considered. The methods have been implemented in combination with a commercial finite element code, and have been demonstrated for application to civil and aerospace structures. A Bayesian approach has been developed for the updating of the reliability model with field inspection information.

ACKNOWLEDGEMENT

The study was supported by the National Science Foundation under grant no. 9872342 (Program Director: Dr. S.C. Liu). The support is gratefully acknowledged.

REFERENCES

Chen, L.J., Gao, G., Tian, J.F., Wang, Z.G. and Zhao, H.Y. (1998). Fatigue and Creep-Fatigue Behavior of a Nickel-Based Superalloy at 850C, *International Journal of Fatigue*, **20:7**, 543-548.

Dey, A., and S. Mahadevan, (2000). Reliability Estimation of Brittle Structures with Time-Varying Loads and Resistances, *J. Structural Engineering,* ASCE, **126:5**.

Hills, D.A. (1994), Mechanics of Fretting Fatigue, *Wear*, **175**, 107-113.

Karamchandani, A., Bjerager, P. and Cornell, C. A. (1989). Adaptive Importance Sampling, *Proceedings ICOSSAR 1989*, (Eds. Ang, A. H.-S., Shinozuka, M. and Schuëller, G.I.), ASCE, 855-862.

Mahadevan, S., and Raghothamachar, P. (2000). Adaptive Simulation for System Reliability Analysis of Large Structures, *Computers and Structures*, **77:6**, 725-734.

Melchers, R. E. (1989). Importance Sampling in Structural Systems, *Structural Safety*, **6**, 3-10.

Mori, Y., and Ellingwood, B.R., (1992). Reliability-Based Service Life Asessment of Aging Concrete Structures, *Journal of Structural Engineering*, ASCE, **119:5**, 1600-1621.

Oswald, G.F., and Schueller, G. I., (1984). Reliability of Deteriorating Structures, *Engineering Fracture Mechanics,* **20:3**, 479-488.

Ruiz, C., Boddington, P.H.B., and Chen, K.C. (1984). An Investigation of Fatigue and Fretting in a Dovetail Joint. *Experimental Mechanics*, **24:3**, 208-217.

Smith, K.N., Watson, P. and Topper, T.H. (1979). A Stress-strain Function for The Fatigue of Metals. *Journal of Materials,* **5:4**, 767-778.

Szolwinski M.P. and Farris, T.N. (1995). Mechanics of Fretting Fatigue Crack Formation. *Structural Integrity in Aging Aircraft*, ASME.

Yaguchi, M., Nakamura, T., Ishikawa, A. and Ashada, Y. (1996), Creep-Fatigue Damage Assessment by Sequent Fatigue Straining, *Nuclear Engineering and Design*, **162**, 97-106.

Zamrik, S.Y. (1993). Damage Models for Creep-Fatigue Interaction. *Technology for '90s . A decade of progress*. Au-Yang (ed.), The ASME Pressure and Vessels and Piping Division, New York.

Zhang, R., and Mahadevan, S. (2000), Model Uncertainty and Bayesian Updating in Reliability-Based Inspection,"*Structural Safety*, **22**, 145-160.

Long Term Durability of Structural Materials
P.J.M. Monteiro, K.P. Chong, J. Larsen-Basse, K. Komvopoulos (Eds)

DEVELOPMENT OF AN INTELLIGENT STRUCTURAL DAMAGE ASSESSMENT SYSTEM: PRELIMINARY RESULTS

R.M.V. Pidaparti[1] and M.J. Palakal[2]

[1]Department of Mechanical Engineering
[2]Department of Computer Science
Indiana University Purdue University Indianapolis
723 W. Michigan Street
Indianapolis, Indiana 46202-5132

ABSTRACT

The overall goal of this project is to develop a structural damage assessment system to quantify the damage due to different sources in aging structures, estimate the severity of the quantified damage, and integrate the developments into an intelligent system so that it can be used to empirically predict fatigue failure and fatigue life of aging materials and structures. The proposed system will provide a fatigue "safety index" to assess the long-term durability and size effects on aging structures. A multi-disciplinary approach consisting of materials, damage/fracture mechanics, artificial intelligence, computer vision, pattern recognition techniques, and engineering optimization is being pursued to quantification and prediction of damage in aging structures. The intelligent system and the associated developments are validated through a series of carefully selected problems from aging aircraft structures. This paper discusses some of the developments up to date and the progress of the proposed intelligent structural damage assessment system.

KEYWORDS

Structural damage assessment, Corrosion, Artificial neural networks, Image processing, Signal analysis, Wavelet analysis.

INTRODUCTION

Structural damage quantification and estimating its severity is needed in many aging structures in aerospace engineering (aircraft wings, fuselages, rotating and manufacturing machinery) and civil engineering structures (bridges, building, pressure vessels). The damage may be due to fatigue, corrosion and/or wear of materials resulting from operating conditions and the environment. Some of the major problems of aging military and commercial aircraft include, for example, in-service cracking of the aircraft wing upper surface, widespread fatigue damage of the various structural components, uncertainty in variable amplitude loading and overload effects of aircrafts, discrete source damage induced by foreign objects, and repairs of metallic components with composite counterparts to extend the service life.

Given the modern day requirements for extending fatigue life, maintenance personnel are required to inspect and ensure the safety of the structures. Periodic inspections of critical areas using appropriate non-destructive evaluation (NDE) techniques are carried out for ensuring safety. The inspection intervals are calculated based on damage tolerance predictions of crack-growth for aircraft and rotorcraft structural components (Bates, 1995). Structural integrity prediction tools are needed to estimate the severity of the damage in many aging structures in aerospace engineering (aircraft wings and fuselage) as well as in civil engineering (bridges, buildings and pressure vessels). The current study deals with the fatigue damage predictions in aging aircraft structures. Recently, Pidaparti et. al. (2000) developed a structural integrity simulation program for aging aircraft panels in Matlab environment.

The long-term durability assessment of structures should involve NDI/NDE techniques integrated with prediction methods for in-situ tests and validation. However, current approaches do not attempt to integrate both these methodologies. Our focus therefore, is to develop such an integrated system which will provide capabilities for reliable damage assessment and prediction using existing NDE techniques. Such a system will result in reduced maintenance and lower cost.

OVERVIEW OF THE INTELLIGENT STRUCTURAL DAMAGE ASSESSMENT SYSTEM

The approach proposed in this research attempts to quantify the damage due to different sources in aging structures, estimate the severity of the quantified damage, and integrate the developments into an intelligent system so that it can be used to empirically predict fatigue failure and fatigue life of aging materials and structures. The objective is to develop a fatigue "safety index" using the intelligent system to assess long-term durability and size effects on aging structures. The intelligent system and the associated developments will be validated through a series of carefully selected problems for which other alternate or experimental solutions are available in the literature.

Figure 1 shows the organization of the Intelligent Structural Damage Assessment System (ISDAS). The development of the system involves interfacing an NDI system with a database, quantification and classification of damage, estimation of the severity of the quantified damage and prediction of the safety index in terms of fatigue life and residual strength. As shown in Fig. 1, the ISDAS system consists of five major components: (i) a database that manages information obtained from various NDI systems; (ii) modules for damage quantification and classification using computer vision and pattern recognition techniques; (iii) an intelligent learning system based on artificial neural networks and fuzzy logic for severity estimation; (iv) an integrated decision maker using expert system methodologies to report the safety index; and (v) a graphical user interface which allows the users to interact with the system. The outcome from this intelligent system will be a safety index which reflects the long-term deterioration of the structure. The intelligent system and the associated developments are being tested and validated through a series of carefully selected sample problems in aging aircraft structures. Details of the two specific modules in ISDAS, damage classification/quantification and severity estimation, are described below.

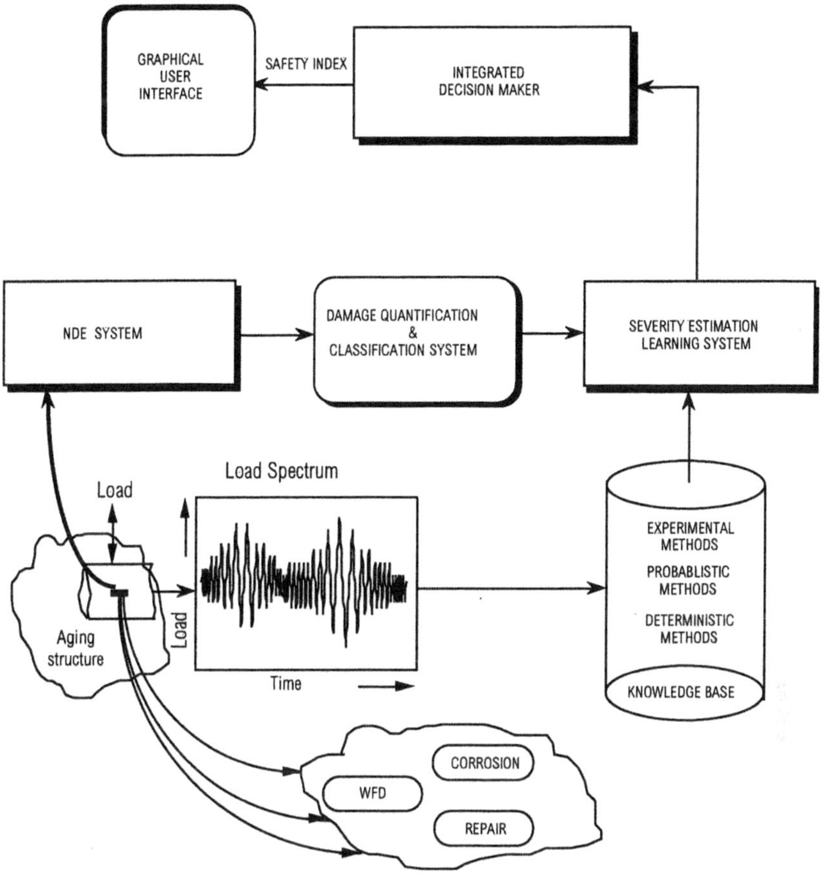

Figure 1: Overview of the Intelligent Structural Damage Assessment System (ISDAS)

Damage Quantification and Classification Module

Imaging has become an increasingly important tool to enhance detection and characterization of damage from the existing NDI techniques (acoustic imaging, infrared imaging, eddy current imaging, impedance imaging and X-ray radiography). Images obtained using NDI techniques can be effectively used to assess the damage more accurately than conventional methods. Image analysis-based techniques are developed for the identification and quantification of corrosion damages.

The overall process of identification and quantification of corroded regions from NDI images is shown in Figure 2. The process essentially involves two stages: first, classification of various regions in the image as corroded or uncorroded, and second, prediction of the material loss of the corroded regions.

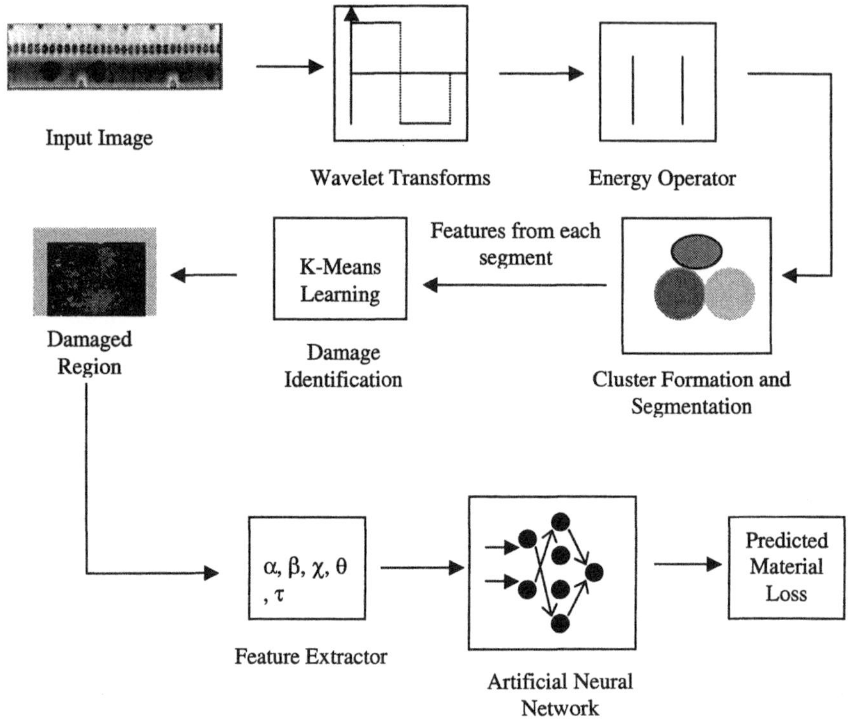

Figure 2: The Damage Analysis and Quantification Process

The classification process involves segmenting the image into various regions. Multi-resolution wavelet analysis is performed on the NDI images to obtain a set of wavelet coefficients as feature vectors. These features were used for the identification of the damaged regions on the NDI images using clustering techniques. Each of the segments on the segmented image would correspond to a damaged region or an undamaged region as shown in Figure 3. Some of the recent results on segmentation algorithms are reported in Rebbapragda et. al. (1999).

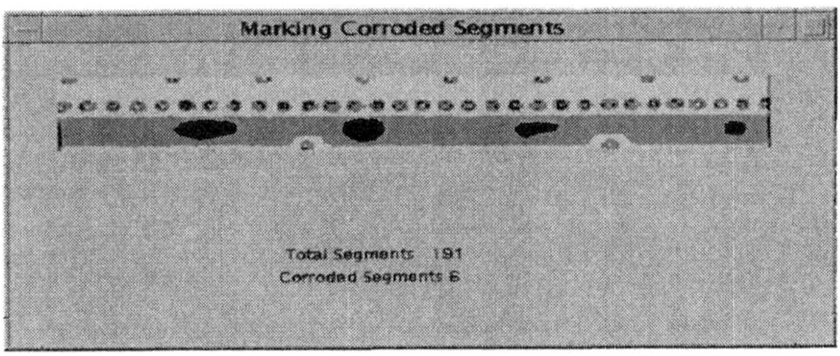

Figure 3: Segmented and classified regions of a damaged panel

Once the damaged segments are identified, first-order and second-order features are extracted from each identified segment. First order statistical features are computed using the histogram of the NDI images. These include *mean, standard deviation, skew, energy,* and *entropy.* The second order features such as *angular second moment, inverse second moment, entropy,* and *contrast* are calculated using a co-occurance matrix. The co-occurance matrix is an estimate of the second order joint probability density. A back-propagation neural network is then used to quantify the damage. Neural networks are capable of realizing a variety of non-linear relationships of considerable complexity and are effectively used in this research. Figure 4 shows results of using different number features for predicting the material loss for the same specimen and Figure 5 shows the material loss predicted by the neural network compared with experimental data. It can be seen from Figure 4 that 15 features were sufficient for the neural network to generalize and predict material loss fairly accurate. The quantification of damage is based on the extent of material loss. For further results on material loss prediction, see Palakal et. al. (2000).

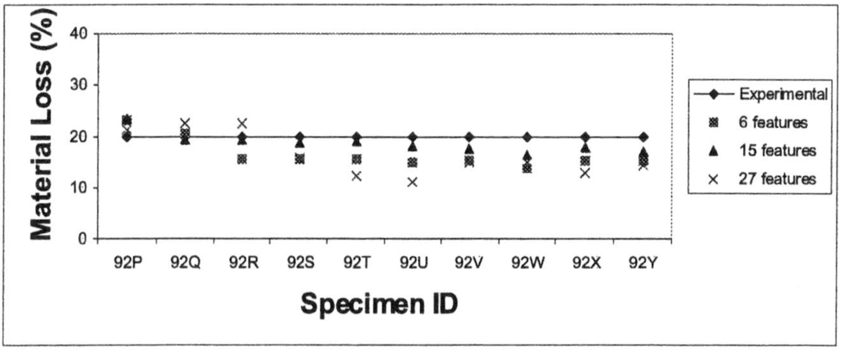

Figure 4: Comparison of different numbers features used for predicting material loss

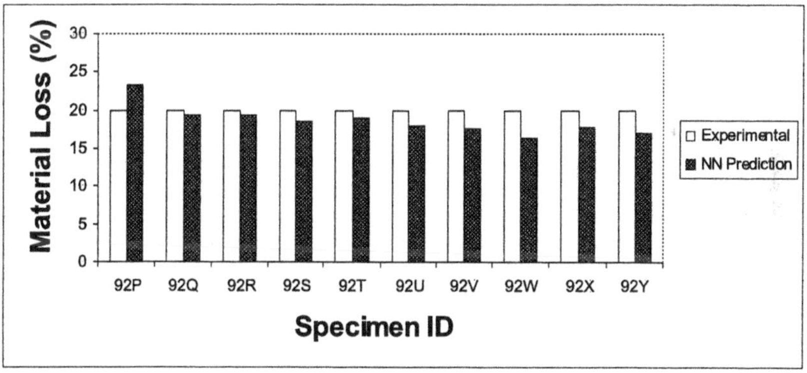

Figure 5: Neural network prediction of material loss compared with experimental data

Severity Analysis and Estimation Module

Severity of the damage assessment is based on various factors such as the quantitative value of the damage, the area where it occurred and other peripheral information. The severity of the damage will be estimated through a learning and prediction model that is based on artificial neural networks and fuzzy logic. During the learning phase, the models learn to predict various properties such as fatigue life, material property, residual strength, and crack growth. As an example, the residual strength and corrosion rate predictions of aging aircraft panels is presented.

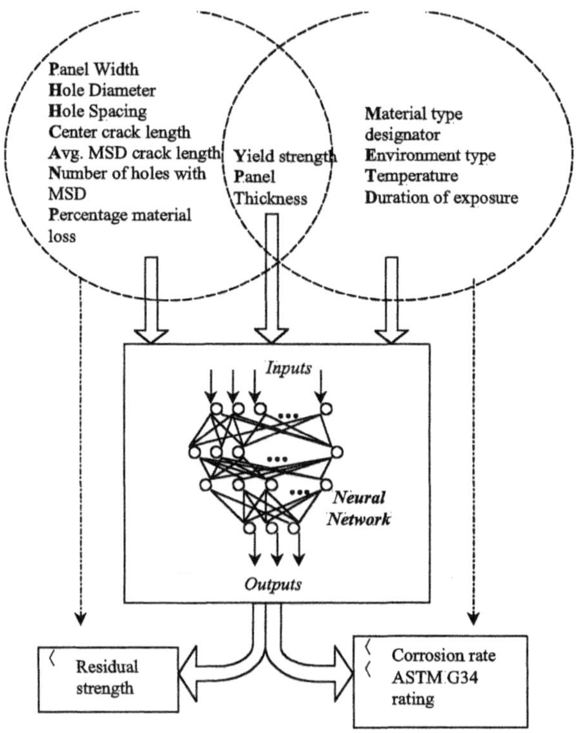

Figure 6: Neural network model to predict corrosion and residual strength behavior

A neural network model is developed for predicting the residual strength and corrosion parameters of MSD panels of aging aircraft. A multi-layer, feed-forward neural network with back-propagation learning algorithm was used in this study. Figure 6 shows the parameters affecting the corrosion behavior and residual strength of MSD panels. A total of 13 parameters were used to model both the phenomena. All the parameters except material type designator and corrosion environment are continuous variables. Material type designator can take integer values from 1 to 4 depending on whether the material belongs to the 2xxx, 3xxx, 6xxx or 7xxx series of Aircraft Aluminum, respectively. Similarly, the corrosion environment can take integer values from 1 to 5, depending on the type of environment.

Table 1 presents the predictions from the different analytical models and the neural network models along with the experimental data (Sivam & Ochoa, 1999; Sheuring & Grant, 1995; Moukawsher, et. al., 1996; Smith et. al., 2000; Luzar, 1998) for the testing set. Although, the analytical methods predict better than the neural network model for a few panels, overall the predictions from neural network are consistently close to the experimental data. The neural network is able to predict the corrosion rate and the ASTM rating for the panels, fairly well. Figure 7 compare the corrosion rate and the ASTM rating predicted for the panels in the training set, with the experimental data. As observed from these figures, the network model captures the corrosion phenomena fairly accurately.

TABLE 1

PREDICTION OF RESIDUAL STRENGTH FOR TESTING SET FROM VARIOUS METHODS

Reference	*Specimen ID*	*Experimental strength (kN)*	*Neural Network model (kN)*	*Net Section yield method (kN)*	*Swift's Linkup load (kN)*
Luzar	2024-T3	112.23	109.11	123.52	69.78
Luzar	7075-T6	92.73	86.22	325.74	143.96
Moukawsher	RS-01a	156.72	173.29	115.99	116.05
Moukawsher	RS-04c	160.94	159.58	93.76	93.12
Smith	WSU 12	112.18	107.65	158.62	114.04
Smith	WSU 19	94.94	92.54	141.80	87.15
Smith	WSU 26	83.70	76.37	124.99	85.42
Sivam	10	27.59	30.00	28.83	-
Sivam	17	20.06	16.77	21.80	-

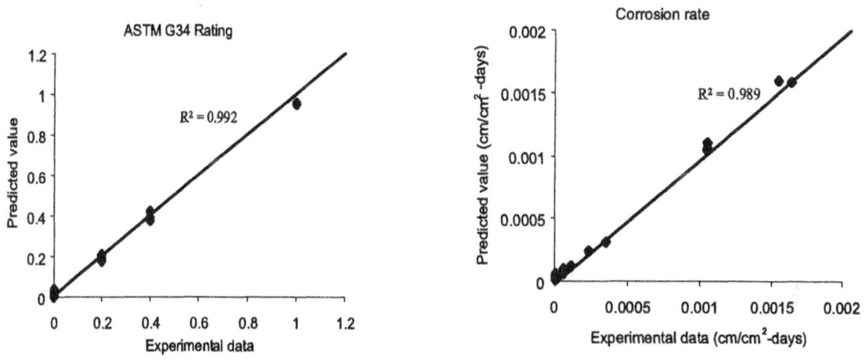

Figure 7: Comparison of neural network results of corrosion rate and rating against experimental data for the training set

The Damage Assessment Model

The damage assessment model is based on an optimization process in which different networks, analytical model and experimental data will interact in a dynamic process to obtain the key parameters for developing the safety index. The optimization model tries to minimize the total energy in the system with physical constraints based on mechanical behavior of the material and physics, similar to the approach by Pidaparti & Palakal (1998). The outcome from the damage assessment model will be corrosion rate, fatigue crack growth behavior and residual strength. These two parameters along with other uncertainties will be combined to obtain the safety index. Currently, this aspect of the research is being carried out and the results will be reported in the future.

SUMMARY

An intelligent structural damage assessment system (ISDAS) is being developed for the purpose of estimating the structural integrity of aging aircraft panels with damage. The ISDAS program uses analytical/neural network solutions to predict the residual strength, fatigue crack-initiation, fatigue crack-growth, and fatigue life based on several user defined failure criteria. The framework of the ISDAS program is designed such that it is user friendly and has limited graphics capabilities. The developed system is tested against the experimental and analytical data and preliminary results were found to be in good agreement. Currently, this system is being extended to include an optimization method to determine the safety index of an aged structure. The overall software system is written in JAVA environment and can be easily portable.

Acknowledgements

The authors thank the National Science Foundation for supporting this work through a grant CMS-9812723 with Dr. Ken Chong as the Program Manager. The authors thank Dr. Jones of FAA/NDI Validation Center, Dr. Peeler of AFRL, Dayton, Ohio, and Dr. Sivam of Raytheon Systems, Texas. Thanks also due to Mr. Rebbapragada, Mr. Jayanti, and Dr. Q. Wang for their contributions.

References

Bates P.R. (1995). Technical Considerations for Managing Aging Rotorcraft. *ASME Structural Integrity in Aging Aircraft* **47:1**, 21-34.

Koch G.H. (1995). On the mechanisms of interaction between corrosion and fatigue cracking in Aircraft Aluminum alloys. *Structural Integrity of Aging Aircraft*, Chang C.I. & Sun C.T. (eds), *American Society of Mechanical Engineers* **47**, 159-169.

Luzar J. (1998). Pre-corroded fastener hole multiple site damage testing. *Boeing Technical Report* **EA 96-135**, 1-46.

Moukawsher E.J., Heinimann M.B., and Grandt Jr. A.F. (1996). Residual Strength of Panels with Multiple Site Damage. *Journal of Aircraft* **33: 5**, 1014-1021.

Palakal M.J., Pidaparti R.M. and Rebbapragada S. (2000). Intelligent Computational Methods for Corrosion Damage Assessment. *AIAA Journal*, (under review).

Pidaparti R.M., Palakal M.J. and Rahman Z.A. (2000). Simulation of Structural Integrity Predictions for Panels with Multiple Site Damage. *Advances in Engineering Software* **31**, 127-135.

Pidaparti R.M. and Palakal M.J. (1998). Fatigue Crack-growth Predictions in Aging Aircraft Panels using Optimization Neural Network. *AIAA Journal* **36:7**, 1300-1304.

Rebbapragada S., Palakal M.J., Pidaparti R.M. and Jones C.R. (1999). Corrosion detection and quantification using image processing for aging aircraft panels. *Third Joint FAA/DoD/NASA Conference on Aging Aircraft*, Albuquerque, New Mexico.

Sheuring J.N. and Grandt A.F. Jr. (1995). An evaluation of Aging Material Properties. *Structural Integrity of Aging Aircraft*, Chang C.I. & Sun C.T. (eds), *American Society of Mechanical Engineers* **47**, 99-110.

Sivam T.P. and Ochoa C.M. (1999). Aircraft Corrosion inspection and evaluation technique using ultrasonic scanning methods. *Second Joint FAA/DoD/NASA Conference on Aging Aircraft*, Williamsburg, Virginia.

Smith B.L., Saville P.A., Mouak A., and Myose R.Y. (2000). Strength of 2024-T3 Aluminum Panels with Multiple Site Damage. *Journal of Aircraft* **37:2**, 325-331.

Long Term Durability of Structural Materials
P.J.M. Monteiro, K.P. Chong, J. Larsen-Basse, K. Komvopoulos (Eds)

ACCELERATED TESTING AND MODELING OF CONCRETE DURABILITY SUBJECTED TO COUPLED ENVIRONMENTAL AND MECHANICAL LOADING

Y. Xi, K. Willam, D.M. Frangopol
Ababneh, A. Nakhi, J. S. Kong, and C.L. Nogueira

Department of Civil, Environmental and Architectural Engineering
University of Colorado, Boulder, CO 80309-0428

ABSTRACT

At present, there exist several standard methods for accelerated testing of concrete durability, such as AASHTO T277 (or ASTM C1202) for rapid chloride permeability; ASTM C666 for resistance of concrete under rapid freezing and thawing; ASTM C1260 for alkali-silica reaction; and ASTM C452 for sulfate attack. Each of these accelerated test procedures is designed for evaluating one specific durability aspect of concrete. In reality, however, concrete is exposed to the combined attack of more than one type of environmental and mechanical loading. Therefore, there is a pressing need to develop a comprehensive and accelerated testing procedure for the realistic assessment of durability of concrete under combined mechanical and environmental loading. On the other hand, innovative models need to be developed in conjunction with the new testing method for predicting long term durability of concrete, taking into account the multiple interaction effects. Moreover, both material parameters and environmental load parameters, which determine the deterioration processes of concrete, exhibit significant random variations. For this reason, a reliable prediction of long term performance of concrete structures needs to be developed and combined with probabilistic models for predicting the uncertainties in materials, environmental, and modeling parameters.

There are three objectives of the present study: (1) Develop a new testing procedure to study the coupling between environmental and mechanical load effects; (2) Develop a theoretical model to predict the long term performance of concrete under environmental and mechanical coupling; and (3) Develop a novel method for reliability-based evaluation of deteriorating concrete structures. The focus of this project is the coupling among temperature, humidity, chloride penetration and fatigue load effects. This paper summaries some of the recently obtained results and ongoing research activities.

KEYWORDS

Accelerated testing, Chloride permeability, Ultrasonic testing, Durability, Damage, Drying shrinkage, Moisture diffusion, Hygrochemomechanical, Probabilistic Analysis

EXPERIMENTAL STUDIES

Hygrochemomechanical Effect on Chloride Permeability of Concrete

An innovative concrete specimen with a hollow square (or circular) cross section was designed to study the response under the simultaneous action of mechanical loading, moisture diffusion and chloride penetration (Nakhi et al. 2000a). There are three major influential parameters when the loading effect is to be included: loading level, loading cycles, and loading frequency. Therefore, there are three objectives in this part of the study. The first one is to identify the critical loading level at which the chloride diffusion process is expedited drastically; the second is to find the lowest number of loading cycles (at the critical loading level) within which the diffusion processes can be accelerated; and the third is to find the highest loading frequency (at the critical loading level and the lowest number of loading cycles) that can effectively accelerate the diffusion processes.

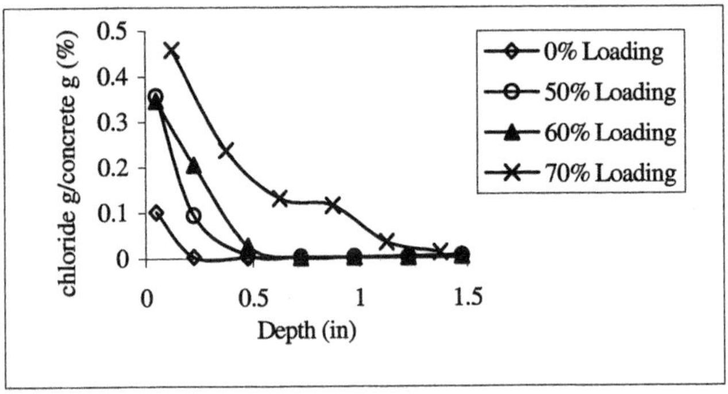

Figure 1. Profiles of chloride concentration at different loading levels.

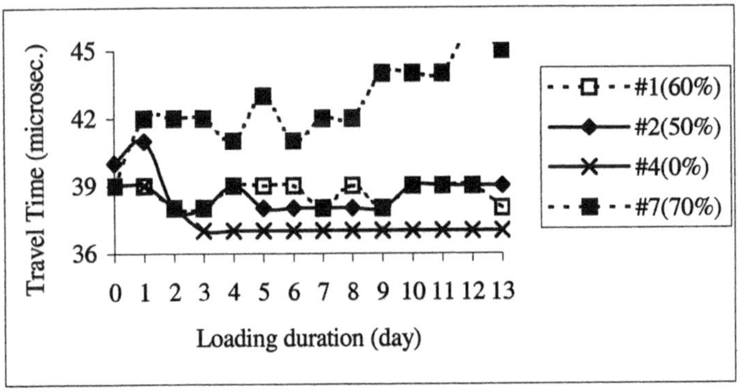

Figure 2. Travel times measured at the samples center after loading.

Concrete specimens were loaded daily to several specified loading levels. During each loading cycle, the load was held for about 20 min. Totally, 13 loading cycles were applied, and thus the total testing period was 13 days. The internal elastic damage in the concrete specimens was measured by an ultrasound technique (V-meter). The surface damage was monitored by image analysis. After the final loading cycle, the chloride solution was removed from the specimen, then profiles of chloride concentration along the thickness of concrete wall were determined by chemical analysis of the dust collected from each specimen.

Fig. 1 shows the chloride concentration profiles under various loading levels. One can see that chloride ions penetrate into deeper part of concrete at higher level of loading. This is mainly due to the microcracking formed in concrete, which increase permeability of concrete. Fig. 2 shows the effect of loading levels as well as the effect of loading repetitions on the ultrasonic measurement, i.e. the travel time of the ultrasonic signal under through-transmission. One can see that the higher the loading level, the longer the travel time, more specifically, above 60% of compressive strength, the number of loading cycles has major impact on the development of internal damage. Up to 70% of loading, no surface cracks were detected by the image analysis.

These test results showed that cyclic loading accelerates chloride penetration through concrete, and that a significant increase in concrete permeability occurs when the concrete is loaded above 60% of its compressive strength. So, 60% can be considered as the critical loading level. The applied mechanical loading on concrete creates internal elastic damage, which can be detected by the measured travel time of ultrasonic signals. At 70% of ultimate strength, significantly longer travel times were measured in the last few days, especially on the last day of testing. Further details of the experimental study were described in Nahki et al. (2000b).

Ultrasonic Testing of Damage in Concrete in Axial Compression

In order to use ultrasonic technique to quantitatively describe internal damage of concrete, a systematic experimental study was performed (Nogueira, 2000). Ultrasound tests were conducted to evaluate microcrack propagation and degradation of elastic properties in cement-based materials under increasing axial compression. A total of 15 prismatic specimens (7.62 x 7.62 x 15.24 cm) were loaded until failure while longitudinal and transverse ultrasonic pulses were transmitted and captured using ultrasonic transducers. Pulse velocity, attenuation, and frequency contents of the digital signals were analyzed and correlated to the level of applied load. The loading was monotonically applied to the specimens at a very low rate and held constant while the signals were recorded. The measurements were taken while the load was sustained, therefore, effects of the applied stress and damage were taken into account in the ultrasonic measurements.

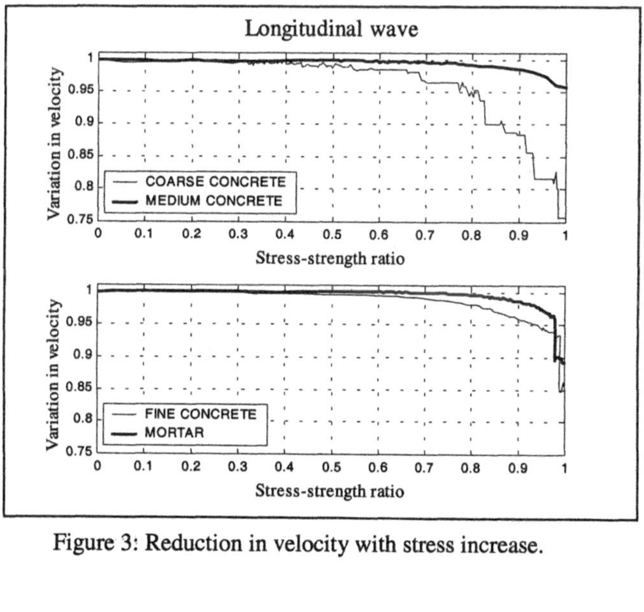

Figure 3: Reduction in velocity with stress increase.

As shown in Figure 3, until up to 75% of the compressive strength the decrease in the longitudinal wave velocity is very small – less than 5%. The same trend can be observed for the transverse wave velocity. This result is consistent with the results obtained previously, where low cycle fatigue loading at 70% of compressive strength was applied.

In order to establish, whether elastic damage in concrete may be described by a single parameter scalar damage model, two damage models were examined. The first model is based upon the degradation of the secant modulus of elasticity only, while other properties, like Poisson's ratio and mass density, are considered to remain constant. The second is a two-parameter scalar damage model, which takes into account the independent variation of the elastic properties in terms of the secant modulus of elasticity and the secant shear modulus. The test data showed very good agreement of the uni-modular damage concept – when the secant values of the modulus of elasticity and the shear modulus are compared with the velocity-based damage measurements of the ultrasonic wave transmission experiments.

Peak-to-peak amplitude, defined as the difference between the first positive and negative peaks of the ultrasonic signal, is a measure of the ultrasonic pulse attenuation while it propagates through the specimen. Attenuation is a frequency-dependent phenomenon primarily due to scattering and absorption of the ultrasonic wave by each grain in the microstructure of the concrete and also by microcracks. In the case of concrete subjected to mechanical damage, attenuation is associated with the formation of cracks around the aggregates. As shown in Figure 4, the increase of the stress-strength ratio is followed by a decrease in the peak-to-peak amplitude.

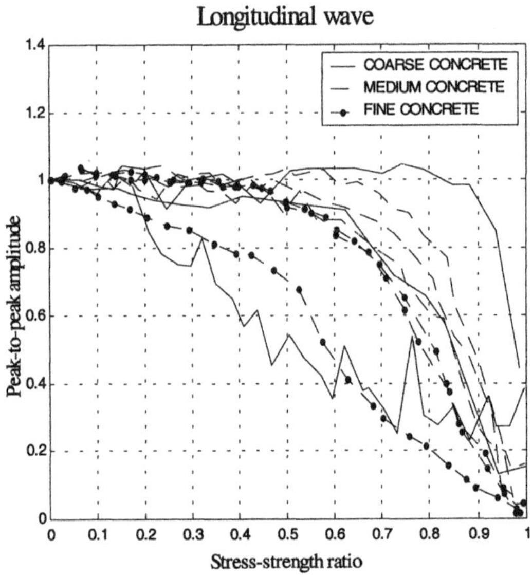

Figure 4: Decrease of the peak-to-peak amplitude with the increase in the stress.

The response characteristics of concrete specimens in the frequency domain were analyzed by using Fourier transform. Figure 5 shows the frequency spectra of the longitudinal and transverse waves transmitted through a concrete specimen (fine concrete) for several stress levels. It can be noted that no frequency shifts take place with increasing axial compression. The relative values of the magnitudes, for both longitudinal and transverse pulses, are approximately constant for a given stress.

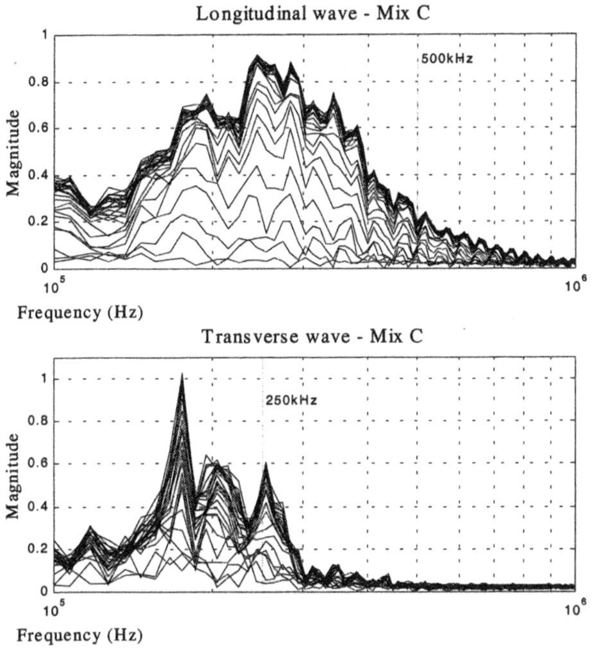

Figure 5: Frequency spectra (fine concrete).

THEORETICAL STUDIES

Deterioration of concrete is induced by the interaction between the different material properties and the environmental factors, which include temperature, humidity, and various types of aggressive chemicals. The interaction and thus the deterioration of material properties is governed by coupled diffusion processes (heat conduction, moisture diffusion and mass transfer). On the other hand, damages induced by mechanical loadings interact with the environmental factors and accelerate the deterioration process. The multiple interactive phenomena have created a completely new research field for mechanics and computational research communities, because hygro-thermo-chemo-mechanical coupling is considerably different from conventional mechanics problems that we have been dealing with so far.

A General Thermodynamic Framework for Environmental and Mechanical Coupling

In order to predict and simulate the interactive process of coupled temperature, moisture, aggressive chemicals, and mechanical loading, there are two major tasks that must be completed. The first one is to establish the governing partial differential equations which characterize these coupling effects. The second is to develop proper material models for the parameters in the differential equations. The conventional approach to formulate coupling among the different diffusion processes is by introducing additional terms observing the principle of equipresence and the reciprocal relations of Onsager. For example, in the case of heat conduction, a new term can be added in the equation to account for the effect of moisture transfer on heat conduction, which is the so-called Dufour effect. In the case of moisture transfer, on the other hand, a new term can be introduced to reflect the effect of temperature change on moisture migration, which is the so-called Soret effect. The focus of this study is to develop an unified framework, based on which the coupled differential equations can be established and the material models involved in the equations can be developed.

In order to include the effects of temperature, moisture, and various aggressive chemicals, the conventional potential energy and complementary energy approaches need to be modified. In this study, the Helmholtz free energy is used to expand the traditional strain energy potential, and the Gibbs free energy for the complementary strain energy. Starting from the first and the second laws of thermodynamics, the final form of the diffusion equation taking into account temperature fluctuation, moisture migration, and strain variation is developed (Xi et al., 2000)

$$A_\theta \dot{\theta} + A_\eta \dot{\eta} + A_\varepsilon \dot{\varepsilon}_{ij} = \rho r - q_{i,i} - \mu_{,i} J_i \tag{1}$$

in which A_θ, A_η, and A_ε are material parameters associated with temperature fluctuation θ, moisture variation η, and strain field ε_{ij}; ρ = density; q_i = heat flux; r = heat source per unit mass, such as the heat of hydration for concrete; μ = chemical potential of the diffusion component; J = mass flux of the diffusion component. This last term in right hand side of Eq. 1 represents internal entropy production, which is zero for reversible processes. Considering a simple case of reversible heat conduction with no moisture transfer, no effect from stress and strains, and no internal heat source, Eq. 1 becomes $A_\theta \dot{\theta} = -q_{i,i}$. With Fourier law for isotropic materials, $q_i = -k\theta_{,i}$, where k is the conductivity; Eq. 1 becomes $A_\theta \dot{\theta} = (k\theta_{,i})_{,i}$; and if the conductivity k is a constant, then we have $A_\theta \dot{\theta} = k\theta_{,ii}$. This is the conventional heat conduction equation.

Although equations similar to Eq. 1 were used in the literature without rigorous derivation (Bazant, 1988; Majorana and Mazars, 1997), the above formulation provides a generalized approach to derive the coupled diffusion equations. A similar approach can be used for other coupling processes, in addition to the coupling among moisture transfer, heat conduction and stress variation. Considering chloride diffusion in fully saturated concrete under a temperature field, the moisture transfer will not be a concern in this case, but the concentration gradient of free chloride will be one of the driving forces and should be considered as a state variable. As a result, the moisture change η in Eq. 1 can be replaced by the concentration variation of chloride, and related material parameters will need to be replaced as well.

Modeling the Coupling Effect of Moisture Diffusion and Drying Shrinkage

In recent years, many researchers have investigated the interactive effect of the mass diffusion and the cracking resulted from moisture gradient and mechanical loading. Drying shrinkage of concrete is caused by the loss of moisture, and thus it is controlled by the moisture diffusion process. On the other hand, shrinkage causes cracking of concrete and affects its moisture diffusion properties. Therefore, the moisture diffusion and drying shrinkage are two coupled processes and their interactive effect is very important for the durability of concrete structures. The moisture distribution in concrete can be characterized by using nonlinear moisture diffusion equation based on Ficks' law. Two material parameters in the moisture diffusion equation (the moisture capacity and humidity diffusivity) and the drying shrinkage of concrete are modeled based on multiscale methods. In the multiscale method, different theoretical models are used at different scale levels. Diffusion mechanisms and shrinkage mechanisms are considered at the nanoscale; a composite mechanics method (generalized self-consistent method, in particular) is used at the microscale and mesoscale; and continuum approach is applied at the macroscale. The effect of drying shrinkage on the moisture diffusion is characterized by the scalar damage model based on continuum damage mechanics. The coupled problem of moisture diffusion and drying shrinkage is solved using a finite difference method (Ababneh et al. 2000).

The isothermal moisture diffusion in concrete can be formulated:

$$\frac{\partial w}{\partial H}\frac{\partial H}{\partial t} = div\ (D_H grad\ H) \tag{2}$$

in which, H is the relative humidity, D_H is the humidity diffusivity, $\partial w/\partial H$ is the moisture capacity and t is the time. Drying shrinkage of the cement paste and concrete may be characterized by a multiscale model (Xi and Jennings 1997). The key issue in this part of the study is how to describe the coupling effect.

The interactive effect of drying shrinkage and moisture diffusion can be studied using one of the two alternatives. The first alternative is to consider that the stresses or strains induced by drying shrinkage are one of the driving forces, and thus, there will be an additional term in the diffusion equation, which corresponds to the effect of stresses and strains (Majorana and Mazars 1997). The second alternative is to consider the effect of the damage on the diffusion parameters, i.e. moisture capacity and moisture diffusivity in Eq. 2. In this case, the stresses or strains are not considered a driving force and appeared explicitly in the diffusion equation, but as internal parameters. The effect of damage on transport properties of concrete can be incorporated in a similar way for the degradation in the secant modulus of elasticity.

With the increase of damage, the concrete diffusivity increases. So, the effect of damage on the moisture diffusivity, D_H, can be expressed in terms of the scalar damage parameter, D (Ababneh et al. 2000):

$$D_H(H,D) = \frac{D_H(H)}{1-D} \tag{3}$$

in which, $D_H(H)$ is the diffusivity of intact concrete, and $D_H(H,D)$ is the diffusivity with damages due to drying shrinkage.

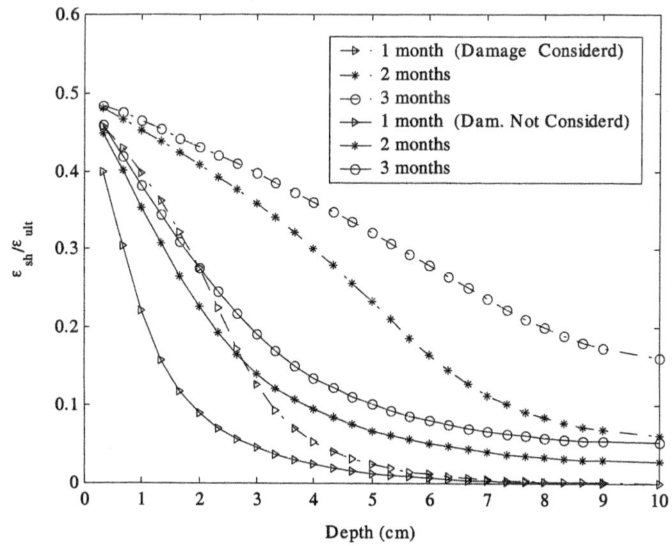

Figure 6: Effect of Damage on shrinkage of concrete.

The effect of drying shrinkage on the damage of concrete is taken into account by the shrinkage strain induced by moisture loss. On the other hand, the effect of damage of concrete on the diffusion of moisture is taken into account by modifying its diffusion properties. Figure 6 shows the variation of the shrinkage strain with the depth from the exposed surface at different exposure times with and without considering the effect of damage of concrete due to shrinkage. The figure shows that, the damage of concrete accelerates the diffusion process and increases the rate of drying. Basically, there is an active interplay between the drying process and the corresponding damage, that is, the moisture loss causes drying shrinkage, which activates damage. In turn, the increase of damage enhances the diffusivity of concrete, which accelerates the moisture transfer.

Modeling the Coupling Effect of Moisture Diffusion and Chloride Penetration

There are mainly three driving forces for the diffusion of chloride ions in non-saturated concrete. The first driving force is the non-uniform distribution of chloride ions, which is important for both saturated and non-saturated concrete. The second driving force is the diffusion of moisture, which is important in the case of non-saturated concrete. The third driving force is the ionic migration driven by an electric potential gradient, which is important only in some special cases. A systematic study was performed recently focusing on the effect of moisture diffusion on the chloride penetration. Preliminary results were published by Ababneh and Xi (1999) in a conference paper, in which the diffusion equation that incorporates the first two driving forces was written as:

$$\frac{\partial C_t}{\partial t} = \frac{\partial C_t}{\partial C_f}\frac{\partial C_f}{\partial t} = div\ [D_{Cl}\, grad(C_f)] + \frac{\partial w}{\partial t} C_f \qquad (4)$$

where, C_t= total chloride concentration (in gram of total chloride per gram of concrete, g/g); w = moisture content of concrete (in gram of moisture per gram of concrete, g/g); D_{cl} = chloride diffusivity (cm^2/day), C_f= free chloride concentration (in gram of free chloride per gram of concrete, g/g). Eq. 4 for chloride penetration must be solved together with Eq. 1 for moisture diffusion.

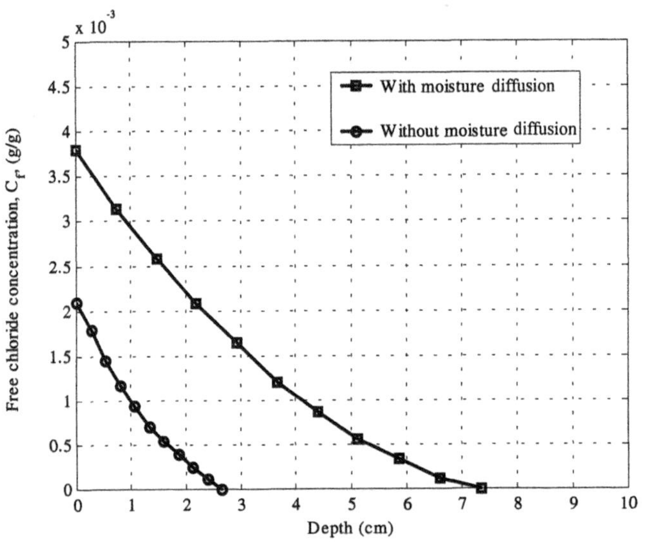

Fig. 7: Effect of the moisture diffusion on free chloride penetration.

The effect of moisture diffusion becomes very important in partially saturated concrete. The ingress of moisture can serve as a carrier of chloride ions. To study the effect of the moisture diffusion on the diffusion of chloride ions, a 10 cm concrete slab is exposed to 3% chloride solution on the top surface and 80% initial pore relative humidity is assigned. The slab is analyzed for two moisture conditions. In one case, the moisture exchange is allowed by assuming H=100% on the top surface. In the other case, the moisture exchange is excluded by keeping the humidity level inside the concrete slab constant at 80%. The plots of depth versus free chloride concentration after one year of exposure for the two cases are shown in Figure 7. From the graph we can see that the moisture diffusion almost doubles the magnitude of the free chloride concentration and contributes significantly to the chloride ion ingress into the partially saturated concrete. A numerical solution based on the Alternating-Direction Implicit (ADI) Finite Difference Method is developed to solve the mathematical model (PDE). The numerical solution was implemented using a computer program. The model predicts the effect of moisture diffusion on the chloride ion penetration in partially saturated concrete. Details of the model can be seen in Ababneh and Xi (2000).

PROBABILISTIC ANALYSIS OF CHLORIDE PENETRATION IN SATURATED CONCRETE

The chloride diffusion process involving many variables and some of these variables show strong randomness. The effects of these uncertainties are of significant importance when the lifetime performance of a structural system is considered. In this study, a one dimensional chloride penetration model for saturated concrete was used. The concrete is exposed at one end to the $CaCl_2$ solution with chloride concentration of 10%. The chloride penetration process was measured by using chloride penetration front, which was defined as the position in the concrete where the chloride concentration reaches a certain level.

Table 1
Distributions and parameters for the selected random variables

Random variable	Distribution type	Parameters
Water-to-cement ratio, *w/c*	Triangular	Mode = 0.5 Min. = 0.35 Max. = 0.65
Curing time, t_0 (days)	Normal	Mean = 28 Standard deviation = 2.8

Among many random variables, water-to-cement ratio of concrete mix and curing time were selected as random variables in order to evaluate the effect of the uncertainties on chloride penetration front. The water-to-cement ratio and the curing time were modeled as triangular and normal distributions, respectively (See Table 1). Sample size of 200,000 is used for the Monte Carlo simulation. The computer code for chloride diffusion developed by Xi and Bazant (1999) was linked to a Monte Carlo simulation program developed by Kong and Frangopol for evaluation of the variation of the chloride penetration front, taking into account the variation of the two random variables (Gharaibeh et al. 2000). Figure 8 shows the results of the distribution of chloride penetration front obtained from the Monte Carlo simulation. In the figure, the chloride concentration level for the penetration front is defined as 0.07%.

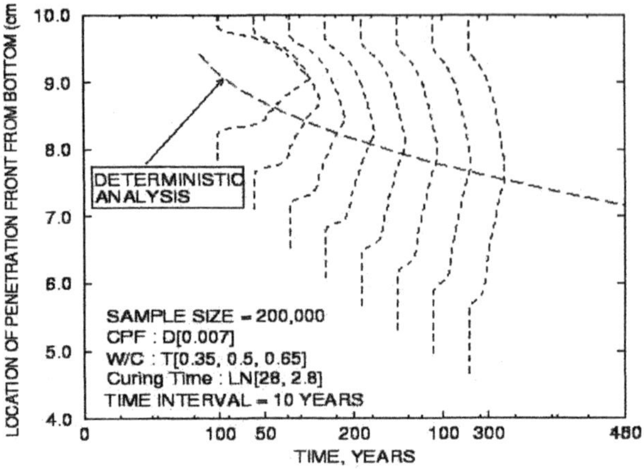

Figure 8 : Distribution of Chloride Penetration Front

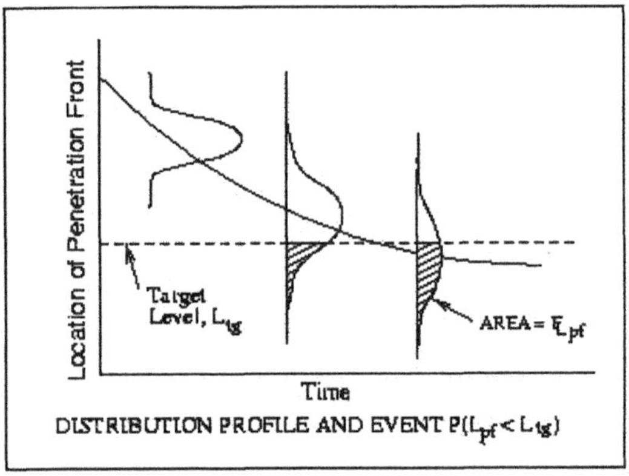

Fig 9 : Evaluating Probability of Event $L_{pf} < L_{tg}$

The main purpose of studying chloride penetration in concrete is for evaluating the corrosion of reinforcing bars in concrete. The corrosion of steel bars starts at a certain level of chloride concentration, which is called the critical chloride concentration. In the literature, the critical chloride concentration varies in a very broad range, due to random nature of related parameters (Xi and Ababneh 2000). One of the important aspects of the present analysis is that the probabilistic feature of the critical chloride concentration at a given depth in concrete can be evaluated. More specifically, the probability of occurrence of the event that the location of the critical chloride concentration is lower than a given target location can be effectively determined. The given target location could be the depth of concrete cover on a steel bar. Similar to reliability analysis for structural failure, we can use the concepts of supply and demand, then the present model calculates the probability for the corrosion to occur considering both the chloride penetration front and the critical value as random variables. To this end, we define the following reliability formulation.

SUPPLY :	Location of Penetration Front, L_{pf} (Distribution)
DEMAND :	Target Location, L_{tg} (Deterministic)
EVENT OF INTEREST:	$L_{pf} < L_{tg}$
PROBABILITY OF EVENT:	$P(L_{pf} < L_{tg}) = F_{Lpf}(L_{tg})$

$$F_{L_{pf}(t)}(L_{tg}) = P(L_{pf}(t) < L_{tg})$$

$$= \frac{\text{Number of Occurences of event } L_{pf}(t) < L_{tg}}{\text{Total Number of Occurences}} \tag{5}$$

$$f_{L_{pf}(t)}(L_{tg}) = \frac{dF_{L_{pf}(t)}(L_{tg})}{dt}$$

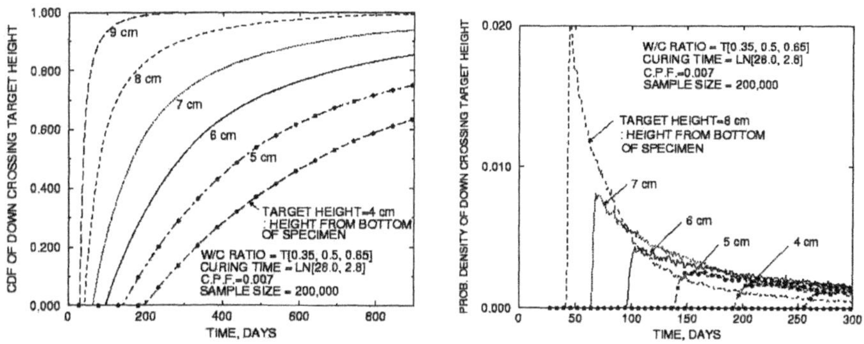

Fig. 10 Cumulative Distribution and Probabilistic Density of Event Lpf < Ltg

Since the distribution of the chloride penetration front location changes over time, the time varying probability $F_{Lpf}(L_{tg},t)$ needs to be determined. A time derivation of $F_{Lpf}(L_{tg},t)$ gives the crossing rate that the chloride penetration front location reaches the given target level with respect to time $f_{Lpf}(L_{tg},t)$ (See Figure 9). The physical meaning of the crossing rate is the probability that chloride concentration is higher than the critical value at the given location.

Figure 10 shows the cumulative distribution and the crossing rate (probabilistic density) of the proposed event with respect to the different target levels (depth). From these graphs, one can see that the chloride penetration profile over time and its crossing rate are determined efficiently. The effects of water-to-cement ratio and curing time on the density distribution of chloride diffusion in concrete at various time steps are also included in the reliability analysis.

ONGOING AND FUTURE RESEARCH

Experimental Studies

After the critical loading level was determined by the innovative testing method, the effect of loading cycles on chloride penetration in non-saturated concrete will be experimentally examined at the fixed loading level 60%. The loading cycles to be tested will be 50, 100, 150, 200, 250. Along the same line, the effect of loading frequency will be examined experimentally.

Although a theoretical model was developed previously, it was based solely on preliminary test data from the literature, a more detailed and systematic study is to be performed for evaluating the effect of shrinkage on moisture diffusion. The effect of moisture diffusion on chloride penetration and the effect of freeze-thaw cycles on chloride penetration will also be investigated. Eventually, the coupling among the four influential parameters, temperature, humidity, chloride, and fatigue loading will be all incorporated.

Theoretical Studies

Coupled diffusion equations will be established for freezing-thawing, moisture diffusion, and chloride penetration. Efficient solution algorithm will be developed. The effect of fatigue loading will be incorporated in the diffusion analysis. The scalar damage model developed mainly for shrinkage induced damage will be further improved by micromechanics-based damage models in order to take into account the effect of mechanical loading on the transport properties. Finally, when the deterministic models are fully developed and validated, reliability analysis of the coupling processes will be performed.

REFERENCES

Ababneh, A., Xi, Y., and Willam, K. (2000) "Multi-Scale Modeling of the Interaction between Drying Shrinkage and Moisture Diffusion in Concrete Materials", Proceedings of 13th Engineering Mechanics Conference: *Engineering Mechanics 2000,* May 21-24, Austin, TX.

Ababneh, A.N., and Xi, Y. (1999) "Diffusion of Chloride in Non-Saturated Concrete", *Proc. of 5th U.S. National Congress on Computational Mechanics*, 528, Aug. 4-6, Boulder, Colorado.

Ababneh, A.N., and Xi, Y. (2000) "Chloride Penetration in Non-Saturated Concrete", submitted to *Journal of Materials in Civil Engineering*, ASCE.

Bazant, Z.P., Editor (1988) "Mathematical Modeling of Creep and Shrinkage of Concrete", John Wiley & Sons, Chichester.

Gharaibeh, E.S., Hanai, T., Xi, Y., and Frangopol, D.M. (2000) "Effects of Uncertainties on Chloride Penetration in Saturated Concrete", *Proc. of Probablistic Mechanics Conference 2000* (PMC2000), July, Notre Dame, 241-244.

Majorana, C., and Mazars, J. (1997) "Thermohygrometric and Mechanical Behavior of Concrete Using Damage Models", *Materials and Structures* (RILEM), 30, 349-354.

Nakhi, A., Xie, Z.H., Asiz, A., Ababneh, A., and Xi. Y. (2000b) "Experimental Study on Chloride Penetration under Coupled Hygromechnical Loadings", *Special Publication of ASCE* "Monitoring of Structures" (in press, will appear in Dec, 2000).

Nakhi, A.E., Xi, Y., Willam, K., and Frangopol, D.M. (2000a) "The Effect of Fatigue Loading on Chloride Penetration in Non-Saturated Concrete", *Proc. of European Congress on Computational Methods in Applied Sciences and Engineering*, ECCOMAS 2000, Barcelona, 11-14 September.

Nogueira, C. L. (2000), Ultrasonic Wave Propagation in Two-Phase Composite Materials and Characterization of Mechanical Damage in Concrete, Ph.D. Thesis, University of Colorado, Boulder.

Nogueira, C.L., and Willam, K. (2000) "Ultrasonic testing of damage in concrete under uniaxial compression" accepted for publication in *ACI Materials Journal.*

Xi, Y., and Ababneh, A. (2000) "Prediction of the Onset of Steel Corrosion in Concrete by Multiscale Chloride Diffusion", *Proc. of International Symposium on High Performance Concrete,* Dec. 10-15, Hong Kong.

Xi, Y., and Bazant, Z.P. (1999) "Modeling Chloride Penetration in Saturated Concrete", *Journal of Materials in Civil Engineering*, ASCE, 11(1), 58-65.

Xi, Y., and Jennings, H.M. (1997) "Shrinkage of Cement Paste and Concrete Modeled by a Multiscale Effective Homogeneous Theory", *Materials and Structures* (RILEM), 30, July, 329-339.

Xi, Y., Willam, K., and Frangopol, D. (2000) "Multiscale Modeling of Interactive Diffusion Processes of Concrete", *Journal of Engineering Mechanics*, ASCE, 126(3), 258-265.

Long Term Durability of Structural Materials
P.J.M. Monteiro, K.P. Chong, J. Larsen-Basse, K. Komvopoulos (Eds)

INTERFACE DURABILITY OF CONSTRUCTION MATERIALS EXTERNALLY REINFORCED WITH FRP COMPOSITES

Julio F. Davalos[1] and Pizhong Qiao[2]

[1]Department of Civil and Environmental Engineering, West Virginia University, Morgantown, WV 26506-6103, USA
[2]Department of Civil Engineering, The University of Akron, Akron, OH 44325-3905, USA

ABSTRACT

Fiber-reinforced plastic (FRP) composites are being used for reinforcement of conventional construction materials (e.g., wood and concrete). Although increased stiffness and strength have been achieved by this reinforcing technique, there is a concern about the reliable performance of the bi-material interface bond, which can be susceptible to delamination. The objective of this study is to develop a qualification program to evaluate the service durability and fracture of FRP-wood bonded interfaces and to initiate a similar study for FRP-concrete interface fracture. Two types of FRP-wood interfaces are studied: Phenolic FRP-wood and Epoxy FRP-wood bonds, and interface fracture of carbon FRP-concrete assemblies is also characterized. The service performance and durability of FRP-wood interface bond is evaluated using a modified ASTM delamination test. An innovative contoured double cantilever beam (CDCB) specimen is used to evaluate Mode-I fracture of both FRP-wood and FRP-concrete interface bonds, and interface fracture toughness data are experimentally obtained. The research presented is useful for evaluation of the in-service performance of bonded hybrid products and qualification of adhesive systems for composite-conventional material interface bonding.

KEYWORDS

Adhesive, Bi-Material Interface, Bonding, Concrete, Durability, FRP Composites, Interface Fracture, Mode-I Fracture, Wood

INTRODUCTION

To improve the performance and durability of structures built of conventional construction materials (e.g., wood and concrete), fiber-reinforced plastic (FRP) composites are increasingly being used as reinforcement for wood and concrete. Current applications of wood reinforcement have focused on the use of FRP strips or fabrics bonded to wood members. Two types of FRP-wood reinforcements are being applied: FRP strips (plates) bonded to wood (used commercially for glulam timber beams) and wood cores wrapped with FRP by filament winding (being investigated for reinforcement of railroad wood crossties or wood utility poles). Similarly, for external FRP reinforcement of concrete structures, FRP laminates or fabrics have been bonded to concrete surfaces for tension and shear strengthening. Although significant increases in stiffness and strength for wood and retrofitting and strengthening for concrete have been achieved by this reinforcing technique, there is a concern about the reliable performance of the bi-material interface bonds (e.g., FRP-wood and FRP-concrete), which can be susceptible to delamination. An inadequate interface bond strength and integrity can lead to delamination and premature failure of a hybrid composite structure and affect the service performance of the product.

In this paper, the performance of FRP-wood bonded interface is characterized by conventional and fracture mechanics tests, and a preliminary study is conducted to evaluate FRP-concrete interface fracture under Mode-I loading. For performance evaluation of FRP-wood bonded interface, modified ASTM standard D2559 test is first used to study the service performance of the bond under moisture and/or mechanical loads, and then a Contoured Double Cantilever Beam (CDCB) specimen (Davalos et al. 1997) is used to evaluate the fracture toughness of bonded interfaces under dry and wet conditions. Based on the results of this study, recommendations and guidelines are given for evaluation and qualification of FRP-wood bonded interfaces. As a part of an ongoing effort, a technique for characterization of the Mode-I fracture of the FRP-concrete interface is discussed, and preliminary results of fracture toughness of FRP-concrete bonded interface are presented.

This paper mainly consists of two parts: the performance evaluation and fracture of FRP-wood bonded interfaces are first presented, followed by the fracture characterization of FRP-concrete bonded interfaces.

FRP-WOOD BONDED INTERFACES

Description of Materials

In this study, two types of FRP-wood interface are evaluated: FRP strips (plates) bonded to wood and wood cores wrapped with FRP by filament winding. The wood material used is Red Maple, and the reinforcing material consists of either E-glass fiber rovings embedded in a Phenolic resin matrix, or E-glass fiber rovings filament wound in an Epoxy matrix. The Phenolic fiber reinforced plastic [Phenolic FRP: 113 yield rovings (15.55/in) and 1 oz CSM surface layers; $V_f = 51\%$] composite material was produced by the pultrusion process; whereas the Epoxy fiber reinforced plastic (Epoxy FRP: PPG High Bond 250 yield rovings at +/-45°; $V_f = 33.6\%$) composite material was produced by the filament winding process. For the Phenolic FRP composite, two types of commercial wood adhesives were used to bond the wood to FRP strips: Resorcinol Formaldehyde (RF, Penacolite® G-1131) and Phenol modified Resorcinol Formaldehyde (PRF, Penacolite® R-400) adhesives. For Epoxy FRP composite, the interface bond between wood and FRP is directly achieved by the Epoxy resin used in the filament winding process. Two distinct coupling agents, HydroxyMethylated Resorcinol (HMR) (Vick 1996) and Resorcinol-Formaldehyde (RF), are used as primers on the wood surface, because of the potential for improving the bond performance significantly.

Durability of Interface by Delamination Tests

The ASTM standard test D2559 was developed for specification of adhesives for structural laminated wood products for use under exterior exposure condition. We used modified test specimens to screen the best combination of parameters to achieve an adequate performance of the FRP-wood interface. Following the ASTM specifications, the specimens were subjected to the following three wet-dry cycles: 1) vacuum/pressure soaking followed by oven drying; 2) steam/pressure soaking followed by oven drying; 3) vacuum/pressure soaking followed by oven drying. The total time required to complete the test was three days, and immediately after the last cycle, the bondline delamination was measured on all end-grain surfaces with the aid of a microscope. The delamination is measured as a ratio of the delaminated (debonded) length to total bondline length for each specimen. For the materials used in this study (*Red Maple* and FRP composites), the 8% delamination limit as specified by the ASTM standard for hardwood species was used to evaluate the bond performance for each manufacturing combination.

Phenolic FRP-wood Interface Performance

For phenolic FRP-wood bonded assembly, the following factors related to the performance of bonded interface are investigated: (1) the influence of coupling agents, (2) the influence of clamping pressure, and (3) the influence of open/closed assembly time. The sizes of the specimens used were smaller than those specified in ASTM D2559. Each laminated FRP-wood assembly for the delamination test consists of four pieces of *Red Maple* wood and two pieces of FRP strips; four wood pieces (each 0.75" x 3" x 24") are placed at the center of the lamination, and FRP strips (each 3/16" x 3" x 24") are located at the top and bottom of the lamination. Wood-wood assemblies for the delamination test were made by bonding six wood pieces (each 0.75" x 3" x 24"). For some of the FRP-wood samples, the wood surfaces adjacent to the FRP strips were primed with the coupling agent (HMR) following the guidelines given by Vick (1996). The HMR primer was spread with a brush at approximately 0.03 lb/ft^2 on the wood surfaces, and the primed surfaces were dried for 24 hours. All wood boards were conditioned to 12% Moisture Content (MC) before bonding. The adhesive, either RF or PRF, was applied with an electronic spreading roller to maintain a constant spread rate of 0.006 to 0.008 lb/ft^2 as recommended by industry. Each of the laminated wood-wood and FRP-wood beam-type members was cut into six 3-inch long specimens, and these specimens were tested following the ASTM D2559 guidelines. To study the influence of key parameters on the bond performance, six wood-wood and either 12 or six FRP-wood samples were tested for each combination of coupling agent, clamping pressure and assembly time.

The effect of HMR coupling agent (Davalos et al. 2000) was first studied. The HMR was applied to the wood surfaces adjacent to Phenolic FRP strips before bonding, and the delamination performance of bonded interfaces was studied. The specimens without HMR primer showed a small percent delamination (< 3.0%) for Phenolic FRP-wood interfaces, and in general, the specimens without HMR exhibited less delamination of wood-wood interfaces, particularly at layers adjacent to the Phenolic FRP. For face-bonding of Phenolic FRP-wood laminates, the manufacturing parameters related to clamping pressure and open/closed assembly time can be easily controlled. The study of these parameters indicated that specimens manufactured with high pressure (p = 210 psi) and intermediate open/closed assembly times (t = 5/30 min) showed the least delamination along both the wood-wood and Phenolic FRP-wood bondlines; therefore, for the RF adhesive used to bond the Red Maple wood and phenolic FRP composite in this study, the combination of 210 psi for clamping pressure and 5/30 min open/closed assembly time is recommended.

Epoxy FRP-Wood Interface Performance

Laminated Red Maple beams were used as the mandrel to apply the Epoxy FRP over the wood by filament winding (Fig. 1). Delamination samples both with HMR and RF coupling agents were cut from the Epoxy FRP-wood beams, and the final dimensions of the samples were 3" x 3" x 4-3/8". The Epoxy FRP-wood interface bond performance under exterior or wet-use exposure conditions (three wet-dry cycles) was evaluated to study the effect of coupling agents (primers) to promote bonding. The Epoxy FRP-wood interface bond was generated during the filament winding process; therefore, the open/closed assembly time and clamping pressure along the Epoxy FRP-wood interface can not be controlled. The influence of two different coupling agents on bond strength (Davalos et al. 2000) was investigated. The Epoxy FRP-wood interface with HMR coupling agent performed well under cyclic wetting and drying delamination tests (no delamination); whereas, the interface with RF coupling agent failed to pass the delamination test (44.2% delamination after the third cycle) (Davalos et al. 2000). It is recommended that the HMR coupling agent be applied to the Red Maple wood surface before wrapping with the Epoxy FRP reinforcement. It is concluded that the HMR significantly improved the bond strength and durability of the Epoxy FRP-wood interface.

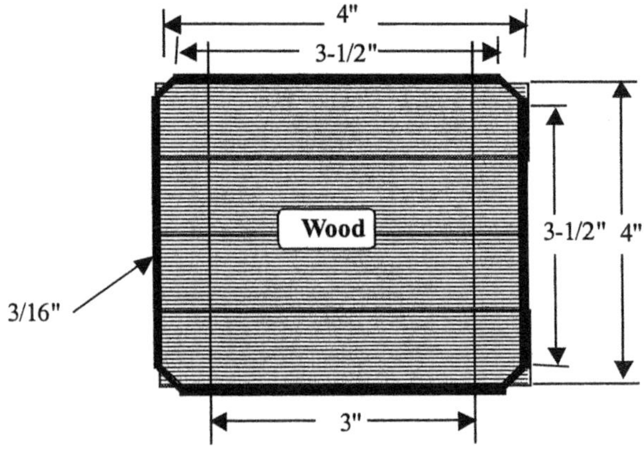

Figure 1: Manufacturing of Epoxy FRP-wood sample by filament winding process

Mode-I Fracture of FRP-Wood Bonded Interface

Once the performance of the interface bond was established by the previous delamination (durability) tests, Contoured Double Cantilever Beam (CDCB) specimens are designed to conduct mode-I fracture tests. In this study, bi-layer CDCB specimens (see Fig. 2) are designed by the Rayleigh Ritz method (Davalos et al. 1997) and used for fracture toughness tests of bonded FRP-wood interfaces under dry and wet conditions. The critical strain energy release rate, G_{Ic}, which is a measure of the fracture toughness, is given as:

$$G_{Ic} = \frac{P_c^2}{2b}\frac{dC}{da} \tag{1}$$

where, P_c = critical load, b = width of the specimen, and dC/da = rate of change of compliance C with respect to crack length a. Using linear-slope CDCB specimens, fracture tests are performed for dry and wet specimens to determine critical loads for crack initiation and crack arrest, from which the critical strain energy release rates (G_{Ic}) are evaluated by making use of experimentally-verified constant compliance rate changes (dC/da) over defined crack lengths.

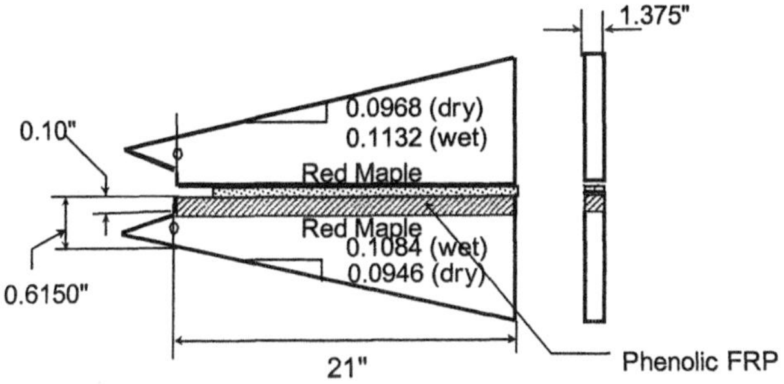

(a) Geometry of Phenolic FRP-wood Specimen

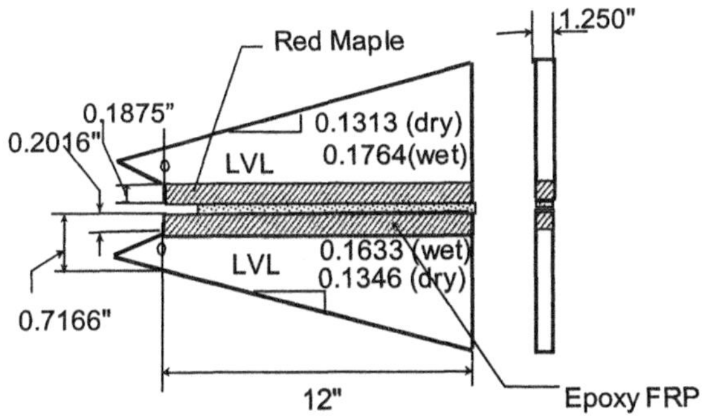

(b) Geometry of Epoxy FRP-wood Specimen

Figure 2: CDCB FRP-wood specimens for Mode-I fracture tests

The geometries of CDCB specimens for dry and wet FRP-wood interfaces are shown in Fig. 2 for Phenolic and Epoxy FRP-wood samples. The Phenolic FRP-wood bonded interface (Fig. 2(a)) consisted of E-glass/Phenolic pultruded FRP, and an integral Red Maple adherend-contour combination; the adhesive used for the bonded interfaces was Resorcinol Formaldehyde (RF, G-1131), and the optimum pressure and assembly times identified above through ASTM D2559 tests (Davalos et al. 2000) were used in the bonding process. The Epoxy FRP-wood bonded interface consisted of E-glass/Epoxy filament wound FRP, and a thin-layer of Red Maple adherend; the material used for the contour was Yellow Poplar LVL (Fig. 2(b)). The adhesive used for bonding the contour LVL and Red Maple adherend interface was Resorcinol Formaldehyde (RF, G-1131), and the contour LVL and Epoxy FRP interface was bonded using Magnobond 56 (a two part epoxy resin system, Magnolia Plastics, Inc., Chamblee, GA).

For dry samples, the specimens were conditioned to 12% wood moisture content (MC) in an environmental chamber. The wet specimens were obtained by submerging the samples in a water bath under 40-minute vacuum and 40-minute pressure soaking cycle, and the specimens were tested immediately after the end of the cycle. The vacuum/pressure soaking cycle was used to saturate the specimens with moisture contents beyond fiber saturation point, and the wet samples obtained by this process exhibited more than 100% moisture contents by weight. Once the geometries of linear-slope test specimens were defined, the specimens were calibrated, experimentally and analytically by the

Finite Element Method, to achieve a linear rate of compliance with respect to crack length (Davalos et al. 1997). In this study, only the Phenolic FRP-wood samples were experimentally calibrated; hence the experimental compliance rate change, *dC/da* (Table 1), is used in Eq. (1) to compute the fracture toughness for Phenolic FRP-wood interface. Based on the accuracy of predicting the compliance vs. crack-length relationship, as verified by experimental and numerical (Rayleigh-Ritz (RR), Modified RR, and FE) studies for Phenolic FRP-wood samples (Qiao et al. 2000), the predicted values of *dC/da* are used in the computation of fracture toughness for Epoxy FRP-wood interfaces (Table 1).

TABLE 1
COMPLIANCE RATE CHANGE OF LINEAR TAPERED SPECIMENS

Specimen types	Slope	dC/da (x 10^{-5} lb^{-1})				% Diff.
		RR	MRR	Exp.	FE	Exp. Vs. FE
Phenolic FRP/Dry	0.0946	24.04	26.79	29.33	27.10	7.60
Phenolic FRP/Wet	0.1084	24.03	26.73	28.43	27.12	4.61
Epoxy FRP-wood/dry	0.1346	12.2	--	--	--	--
Epoxy FRP-wood/Wet	0.1633	12.5	--	--	--	--

Fracture Tests of Bonded Interfaces

The linear-slope CDCB specimens shown in Fig. 2 were used for Mode-I fracture tests of FRP-wood bonded interfaces under dry and wet conditions (Fig. 3). As a summary, the critical initiation and arrest loads and fracture toughness values that were obtained for all of the tested samples discussed are shown in Tables 2 and 3, respectively (Qiao et al. 2000). As indicated in Table 3, an increase in interface fracture toughness due to moisture absorption was obtained for the Phenolic FRP-wood and Epoxy FRP-wood samples. The toughening of the interface under exposure to moisture is mainly due to a much more plastic fracture failure mode. Also, the effect of coupling agent on fracture toughness of Epoxy FRP-wood interfaces under both dry and wet conditions were investigated; the fracture toughness of interfaces with HMR coupling is much higher than of those treated with RF (Table 3). The variability of critical loads for RF-treated wet samples was significant (COV = 80.3%), with values varying from about 10 lbs to 130 lbs. The RF-treated wet samples achieved only about 5% of the fracture toughness of the HMR-treated wet samples. Thus, the primer type used has a major influence on the interface performance.

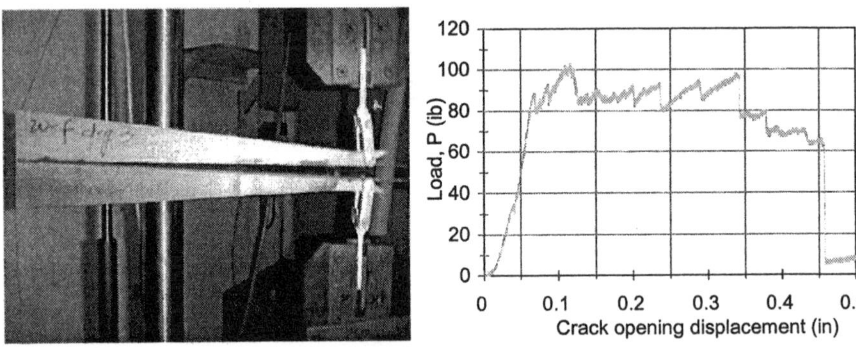

Figure 3: Fracture test of FRP-wood Bonded Interface

TABLE 2
COMPARISON OF CRITICAL INITIATION AND ARREST LOADS

	P_c^i (lb)	P_c^a (lb)
Phenolic FRP-wood/Dry	109.5 (COV = 15.4%)	100.0 (COV = 17.0%)
Phenolic FRP-wood/Wet	157.5 (COV = 13.1%)	146.9 (COV = 15.1%)
Epoxy FRP-wood/HMR/Dry	218.6 (COV = 8.0%)	214.5 (COV = 8.8%)
Epoxy FRP-wood/HMR/Wet	284.0 (COV = 6.4%)	280.9 (COV = 6.5%)
Epoxy FRP-wood/RF/Dry	106.3 (COV = 31.8%)	97.5 (COV = 33.3%)
Epoxy FRP-wood/RF/Wet	62.7 (COV = 78.0%)	60.8 (COV = 80.3%)

TABLE 3
COMPARISON OF FRACTURE TOUGHNESS

	G_{Ic}^i (lb/in)	G_{Ic}^a (lb/in)
Phenolic FRP-wood/Dry	1.28	1.06
Phenolic FRP-wood/Wet	2.57	2.23
Epoxy FRP-wood/HMR/Dry	2.54	2.47
Epoxy FRP-wood/HMR/Wet	4.03	3.94
Epoxy FRP-wood/RF/Dry	0.57	0.48
Epoxy FRP-wood/RF/Wet	0.20	0.18

General Guidelines for Performance Evaluation of FRP-Wood Bonded Interfaces

To evaluate the in-service performance of bonded FRP-wood hybrid products and to qualify adhesive systems for FRP-wood bonding, the following general guidelines for interface bond characterization are suggested:

- The ASTM D 2559 wetting-and-drying cyclic delamination test appears to be sensitive enough to investigate key performance parameters, such as coupling agent (primer) to promote bonding, open/closed assembly time, clamping pressure, bonding surface preparation, etc, and it can be first used as a screening test to evaluate the delamination of bonded interfaces (Davalos et al. 2000).
- Once the best combination of parameters is obtained from the ASTM D 2559 tests, standard block-shear tests (ASTM D 905) can then be used to evaluate average bond "shear" strength (Davalos et al. 2000), which can be applied in design with an appropriate factor of safety.
- Finally, the Contoured Double Cantilever Beam (CDCB) specimen (Davalos et al. 1997) described in this study can be effectively used to obtain interface Mode-I fracture toughness values; these data can be implemented in practical applications to assess the potential growth of a delamination crack at the FRP-wood interface.

FRP-CONCRETE BONDED INTERFACES

FRP-Concrete Interface Bond Behavior and Environmental Effects

Interface bond delamination is a source of major concern requiring special attention when enhancing the structural capacity of a concrete member through the use of external FRP reinforcement. Debonding of the FRP laminate can initiate due to flexural and shear cracking in the concrete, fatigue induced by cyclic loads, environmental degradation, and interface stresses due to the irregular topographical nature of the concrete surface; susceptibility of the latter arises from the presence of holes, surface scratches, and the existence of large shearing stresses at the free edges of the laminate (Russell and Street 1982).

The environment can also play a major role in changing the properties of both the FRP composite materials and the subsequent concrete to which they are adhered. The environment consists of atmospheric related effects, such as varying temperatures and humidity levels, as well as the presence of degrading chemicals, like salts or alkalis - together, this process is known as *aging* (Dutta 1995). The degree by which FRP composites are capable of withstanding such environmental attacks is largely dependent upon the role of the matrix, due to its high susceptibility to aging effects. When exposed to humid air or wet environments, many polymer matrix composites absorb moisture by instantaneous surface absorption and diffusion (Dutta 1995). Though drying helps in reducing the moisture concentrations, the matrix may not be able to completely procure its original state, resulting in drastic long-term changes in material strength. Also, the presence of moisture can lead to debonding stresses across the fiber-resin interface due to osmotic pressure and resin swelling (Karbhari and Engineer 1996a). For this reason, particular attention in the selection of an appropriate resin is imperative.

Karbhari and Engineer (1996a) also indicated the importance of the *type* of fiber to be used in the overall durability of a composite system. They report that degradation in performance levels is greater for glass than for likewise carbon fiber reinforced systems. E-glass fibers, for example, are more susceptible to stress-corrosion and indentation in the presence of water. These negative effects are further exacerbated if *salt* water is involved, as in applications of deicing road salts and near- or in-marine environments. In all cases, composite systems exposed to saltwater have more pronounced stiffness-losses than do their counterparts exposed to freshwater.

The mechanism of water absorption in composites is also greatly influenced by temperature, which can subsequently have irreversible effects. Temperature effects are responsible for the distribution patterns, quantity, and rate of water retention within composites, as shown by Dewimille and Burnsell (1983). Because the coefficient of thermal expansion of the matrix is usually an order of magnitude greater than that of the fibers, residual stresses are induced that can lead to the growth and propagation of microcracks in the matrix. These microcracks can then widen and coalesce to form matrix cracks that may spread in the resin matrix or wander around the matrix-fiber interface (Lord and Dutta, 1988), obviously leading to an overall reduction in stiffness and eventually rendering a matrix as ineffective.

In the application of composite fabrics used as concrete reinforcement, the resin serves concurrently as the matrix for the fibers and the adherent for the composite to bond to concrete. Thus, the integrity of the matrix under environmental exposure is of utmost importance. For example, Davalos et al. (2000) have shown that FRP-wood bonded interfaces are severely affected when subjected to cyclic wetting and drying testing protocols. For concrete externally reinforced with FRP, there is a need for bond-interface characterization methods that can be used to qualify reinforcing systems, establish service performance criteria, and obtain fracture toughness data. With this information then, an effective assessment of the potential for delamination-growth at the interface can be made.

Design of CDCB Specimen for FRP-Concrete Interface Fracture

Similar to the design of FRP-wood CDCB specimens, the contoured shape for a bilayered FRP-concrete specimen is obtained by Rayleigh-Ritz method (Davalos et al. 1997). For a given crack length, the CDCB specimen is modeled as a cantilever beam to obtain its compliance by computing the displacement at the free edge for a unit tip-load. By assuming a constant $K = dC/da$, linear slopes of the contour are determined for discrete crack lengths, a_i, along the entire length of the CDCB specimen. This process is then repeated for different constant K-values. By plotting a graph of the slopes of the contoured shape versus crack lengths (Fig. 4), a range of crack lengths can be determined for which the curves are most nearly linear (in the present case, .8 - 24 inches). A statistical analysis on this range of values, involving calculation of the mean and standard deviation is then performed to determine the K-value corresponding to the smallest coefficient of variation. It is for this $K = dC/da$ value that the contour slopes, k_i, assume most nearly constant values within the prescribed range.

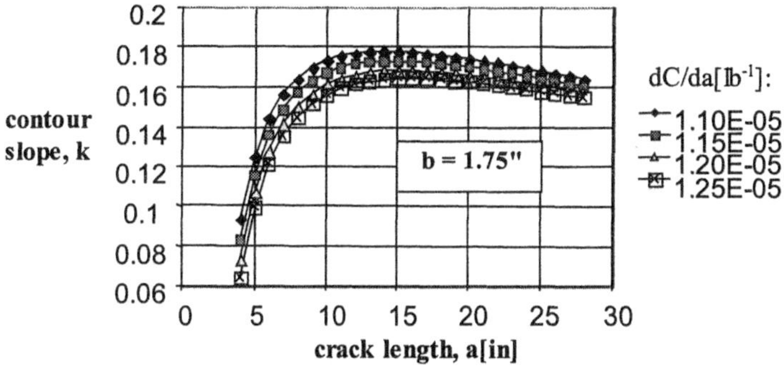

Figure 4: Contour slope vs. crack length for various dC/da values

A major challenge of fracture testing associated with FRP-concrete specimens involves attempting to induce crack propagation at the FRP-concrete interface in order to attain interfacial strain energy release rate measurements. As introduced early, the Mode-I fracture testing of FRP-*wood* interfaces was successfully accomplished using two contours acting as cantilever beams (i.e. the CDCB specimen). Because of the inability of concrete to withstand tensile stresses of any appreciable magnitude, however, we devised an experimental fixture consisting of a steel grip-jacket mechanism (Fig. 5a) that provides confinement of a bulk-concrete substrate (Fig. 5b). Bonded to this substrate is a carbon FRP tow-sheet that is in turn bonded to a *singly* contoured cantilever beam (Fig. 6) made of Yellow Poplar Laminated Veneer Lumber (LVL: $E_{LVL} = 1.958 \times 10^6$ *psi* and $G_{LVL} = 0.063 \times 10^6$ *psi*) produced by Truss Joist MacMillen (Buckhannon, WV). The two edges of the LVL contour corresponding to the bonding-face were chamfered (1/4" x 1/4") slightly in order to induce higher stresses at this surface, thereby promoting interfacial crack propagation. In addition, a special coupling agent known as HMR (Hydroxymethylated Resorcinol) primer (Vick, 1996) was applied to the surface of the wood prior to bonding the carbon composite tow-sheet to the LVL contour, in order to improve adhesion. Master Builders Inc. (Cleveland, Ohio) supplied all the primer and saturant systems used for concrete bonding applications, along with the carbon FRP tow-sheets.

The wooden forms used in casting the concrete substrate were made with a cross-section inverted to that as shown in Fig. 5b. Using this scheme resulted in a very level concrete bonding surface, as it is this surface that was cast flush upon the plywood base. This circumvented the need to trowel and level the surface manually. In addition, the surface of the concrete to be bonded was cast with a 1/8" step (Fig. 5b). This step was recessed 2" from one end of the concrete substrate to allow the steel load-strap to easily slide into place. Moreover, the width of the step was cast to match the width of the LVL section so as to not only aid in guiding proper placement of the contour, but more importantly, to ensure that crack propagation would be maintained at the composite-concrete interface. A piece of tape having a length corresponding to the initial location at which the curves of Fig. 4 begin to assume linearity (i.e. 8" in this case), was applied to this stepped-concrete surface to provide a starter-crack.

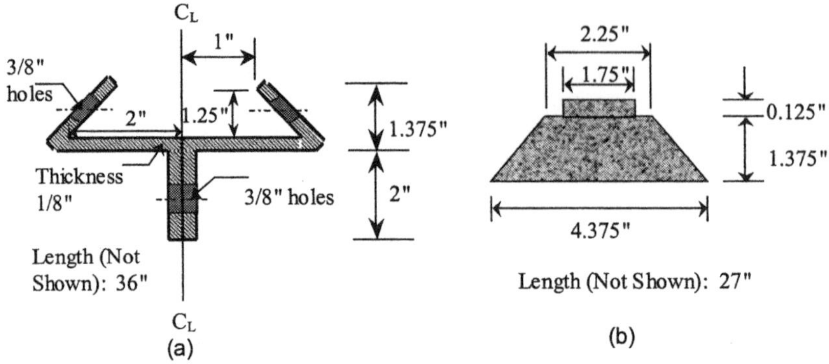

Figure 5: Steel jacket and concrete substrate (Front View)

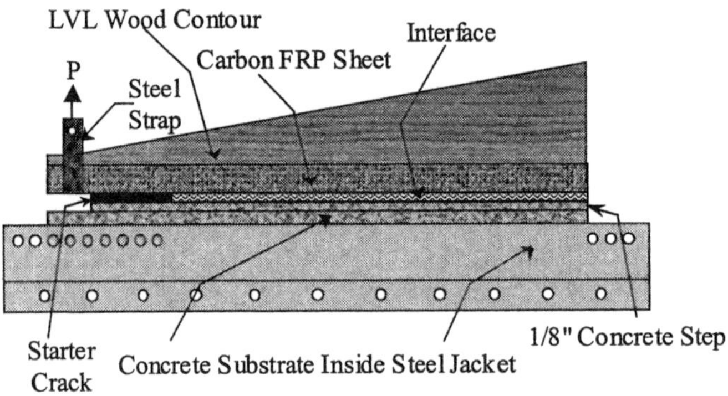

Figure 6: FRP-Concrete test fixture and specimen

The concrete used in casting, as furnished by Hoy Redi-Mix Co. (Morgantown, WV), consisted of a #8 small-stone mix, having a 28-day compressive strength of about 4000 psi. After casting, the curing process was accomplished by laying wet-burlap over the concrete specimens. Next, the specimens were covered with an impermeable sheet of plastic to retain moisture and promote proper hydration of the cement. All this was carried out in an environment maintained at room temperature. Prior to bonding, the concrete substrates were allowed to air-dry and the bonding-surface was cleaned using a brush to remove any loose, or otherwise foreign, particles.

Existing Study on FRP-Concrete Interface Fracture

Giurgiutiu et al. (1999) investigated the fracture toughness of the interface bond between glass-composite overlays and concrete substrates. A DCB approach was used in which the composite layer was assumed to behave as a cantilever beam. After failing the specimen, the effectiveness of adhesion of the composite was determined by observing how much cement-paste residue remained on the composite overlay. They reported that an increase in the strain energy release rate was expected with increased adhesion of the composite to the concrete substrate. They also observed that specimens with high fracture toughness values have a large proportion of the cracks propagating within the concrete substrate, while those with lower values predominantly have cracks propagating at the interface.

Karbhari and Engineer (1996b) used a peel test to determine the fracture toughness of interfaces between either carbon or glass composites bonded to concrete. A number of tests were conducted altering both the peel angle as well as varying the speed of the actuator applying the load. For an optimum actuator speed, cited as 5.08 mm/min, and a peel angle of 45°, the mode-I fracture toughness values they report for the glass and carbon composites are 382.52 J/m^2 and 477.87 J/m^2, respectively.

Preliminary Results of This Study on Fracture Toughness of FRP-concrete Interface

At the time of this publication, three specimens were tested using an MTS servo hydraulic testing machine operating under a displacement-control mode of 0.003 in/sec. Through the use of a computerized system, continuous load and displacement data were recorded for every crack length. While both the behavior and the results of the three specimens were in good agreement relative to each other, the authors deem it necessary that further testing be conducted prior to the disclosure of data.

Most noteworthy regarding the failure behavior of all three specimens, is that crack propagation proceeded strictly at the interface, as designed. This is principally attributed to the confinement of the concrete substrate as imparted by application of the steel grip-jacket mechanism. Moreover, a good distribution of speckled cementitious residue retained on the underside of the carbon composite tow-sheet is indicative of proper adhesion.

Concluding Remarks on FRP-Concrete Interface Fracture

Presently, there exists a need to develop methods to rigorously evaluate the interface bond strength and integrity of FRP-concrete structures. Large-scale tests of structural components used for stiffness and strength evaluations generally do not suitably detect delamination effects. Alternatively, small-scale tests designed to provide average interfacial strength properties do not effectively simulate actual failure mechanisms. Thus, the preliminary investigation reported in this paper is part of an ongoing effort to develop both analytical and experimental fracture mechanics methods that can be efficiently used in evaluating the interfacial bond of FRP-concrete products. In addition, the proposed test specimen provides a basis by which fracture toughness investigations can, furthermore, be extended to include effects resulting from exposure to various environments. Such data, needed for both qualitative and quantitative interfacial evaluations, can then be subsequently used for failure analyses and numerical simulations.

CONCLUSIONS

This paper presented a comprehensive study on service performance and fracture of construction materials externally reinforced with FRP composites. The durability and service performance of FRP-wood interface bonds are evaluated using modified ASTM D2559. Interface fracture toughness data, under both dry and wet conditions, are experimentally obtained by an innovative contoured double cantilever beam (CDCB) specimen. The interface behavior of FRP-concrete assemblies is preliminarily evaluated under Mode-I fracture. The approach developed in this study can be effectively used as guidelines for evaluation of the in-service performance of bonded hybrid products and qualification of adhesive systems for composite-conventional material interface bonding.

ACKNOWLEDGMENT

The study on FRP-wood bonded interfaces was partially sponsored by the USDA under National Research Initiative Competitive Grants Program (NRICGP-CSREES, Grant No. 98-35103-67579). The on-going research on FRP-concrete interface fracture is supported by the National Science Foundation (CSM-0002829).

REFERENCES

Davalos, J.F., Madabhusi-Raman, P. and Qiao, P.Z. (1997). Characterization of Mode-I Fracture of Hybrid Material Interface Bonds by Contoured DCB Specimens. *Engineering Fracture Mechanics*, 58(3), 173-192.

Davalos, J.F., Qiao, P.Z., and Trimble, B.S. (2000). Fiber-reinforced Plastic Composite-Wood Bonded Interfaces, Part I. Durability and Shear Strength. *ASTM Journal of Composites Technology and Research*, 22(4): 224-231.

Dewimille, B. and Burnsell, A.R. (1983). Accelerated aging of a glass fiber reinforced epoxy resin in water. *Composites*, 14, p. 35.

Dutta, P. K. (1995). Durability of FRP Composites. *Proceedings of the Int. Conf. on Fibre Reinforced Structural Plastics in Civil Engineering* at IIT, Madras, pp. 360-370.

Giurgiutiu, V., Lyons, J., Petrou, M., Laub, D. and Whitley, S. (1999). Experimental Fracture Mechanics for the Bond between Composite Overlays and Concrete Substrate. *Proceedings of the International Composites Expo (ICE)*, Cincinnati, OH, pp. 4D(1-6).

Karbhari, V.M. and Engineer, M. (1996a). Effect of Environmental Exposure on the External Strengthening of Concrete with Composites—Short Term Bond Durability. *J. of Reinforced Plastics and Composites*, 15, 1194-1216.

Karbhari, V.M. and Engineer, M. (1996b). Investigation of Bond between Concrete and Composites: Use of a Peel Test. *J. of Reinforced Plastics and Composites*, 15, 208-227.

Lord, H.W. and Dutta, P.K. (1988). On the Design of Polymeric Composite Structures for Cold Regions Applications. *J. of Reinforced Plastics and Composites*, 7, 167-172.

Russell, A.J. and Street, K.N. (1982). Factors Affecting the Interlaminar Fracture Energy of Graphite/Epoxy Laminates. *Progress in Science and Engineering of Composites*, ICCM-IV Tokyo, pp. 279-286.

Qiao, P.Z., Davalos, J.F., and Trimble, B.S. (2000). Effect of Moisture on Fracture Toughness of Composite/Wood Bonded Interfaces. ASTM STP, *Fatigue and Fracture Mechanics*, in press.

Vick, C.B. (1996). Hydroxmethylated Resorcinol Coupling Agent for Enhanced Adhesion of Epoxy and Other Thermosetting Adhesives to Wood. *Wood adhesive 1995*, Forest Products Society, Madison, WI., pp. 47-55.

Corrosion

Experimental and Theoretical Study of Reinforced Concrete Corrosion Using Impedance Measurements

Jieying Zhang, Paulo J. M. Monteiro, and H. Frank Morrison

Department of Civil and Environmental Engineering, University of California at Berkeley, California, CA 94720, USA

ABSTRACT

The ratio of the voltage between two electrodes on the surface of a conducting medium to the current injected between two other electrodes defines the surface impedance of the medium. Its value depends on the distribution of the conductivity in the medium and, for reinforced concrete, on the interface impedance between the concrete and the metal reinforcing rods. The surface impedance for a cylindrcal conductor beneath and parallel to the surface can be modeled analytically. Such a model can simulate the surface impedance measurement for an embedded reinforcing steel bar, which has a complex and frequency-dependent impedance at the interface between the bar and the concrete. A comparison between the interface impedance and the modeled surface impedance shows that the essential characteristics in the interface impedance are preserved in the response measured on concrete surface. This is further verified by measuring on concrete samples using a four-electrode array. The measured surface impedance shows distinct characteristics that are directly related to the corrosion state of the embedded reinforcing steel bar.

KEYWORDS: Concrete; Reinforcing Steel Bars; Corrosion Impedance; Apparent Resistivity.

1. INTRODUCTION

For almost fifty years, the frequency-dependent resistivity mapping of the subsurface, also known as the induced polarization (IP) method, has been applied successfully by geophysicists in the discovery of mineral deposits (Ward, 1990). There is a growing interest in using this technique for engineering and environmental applications (Sharma, 1997). Such non-destructive methods require surface access to study the subsurface structure.

The frequency dependence arises from the complex impedance at the interface between the metallic component (the reinforcing steel bar) and the enclosing medium (concrete). It is well known that the interface impedance between reinforcing steel bar (rebar) and concrete is a function of the state of corrosion of the rebar, therefore a direct extension of the IP method to concrete for directly mapping rebar corrosion is possible. Monteiro et al. (1998). Uses this approach to detect corrosion of reinforced concrete from measurements of impedance at the surface of a concrete structure. The method determines the corrosion state of the reinforcing bar (rebar) through complex resistivity mapping that is conducted by electrodes placed on the concrete surface.

The principle of this corrosion detection method is based on the fact that the embedded rebar and the concrete matrix have distinct electrical resistive behaviors. When current is injected into the concrete matrix, the current distribution is distorted by the presence of the rebar, and the measured voltage response reflects this distortion. The electrical impedance measured between the rebar and electrolyte interface can define its corrosion state; therefore the rebar at different stages of corrosion should produce different voltage responses at the concrete surface as well. In order to study the response measured by this corrosion detection method, an analytical model has been constructed to determine the characteristics of the surface impedance for a single reinforcing rod with an arbitrary interface impedance. For this purpose, the analytical model includes the diameter and depth of the rebar, the electrical resistivity of the material, and the complex, frequency-dependent interface impedance at the rebar and concrete interface that is related to the corrosion kinematics. A few approximations for modeling the complex geometry have been made. The responses are to a point source of current injected into a half space of concrete with an embedded rebar. The concrete, rebar, and the interface properties are assumed to be homogeneous. Rebars at different stage of corrosion are modeled using different interface impedance. The impedance mapping conducted on the concrete surface depends on the desired interface impedance, but it is also strongly influenced by the resistivity of concrete, the depth and diameter of the rebar, and the electrode configuration, all of which must be factored in when measuring the response of rebar and concrete interface.

This analytical study was verified by experimental measurements on concrete samples containing a rebar. In order to simulate the different interface impedances in the experiment, different levels of corrosion rate in the rebar were induced and controlled by an anodic polarization technique each time a measurement was made.

2. INDUCED POLARIZATION TECHNIQUE

The Induced Polarization method (IP method) measures the frequency-dependent resistivity of the subsurface when an ac current is applied directly to the subsurface using a pair of electrodes. The resulting potential differences (voltages) are measured directly by two different electrodes (Figure 1).

When measuring corrosion state, current injected into the concrete causes the rebar surface to become polarized. As shown in Figure 1, charges accumulate on the rebar surface, resulting in a net charge dipole, which adds to any voltage measured at the concrete surface. Because it takes time to return to equilibrium after the rebar surface has been polarized, the measured voltage is also observed as a delayed response relative to the current source and this delay is frequency dependent. This polarization phenomenon is called electrode polarization (Ward, 1990), and it is caused by the current conduction changed from electrolyte (concrete pore solutions) to rebar. The polarization degree varies with corrosion process at the rebar surface, which is the basis for the impedance measurement to assess corrosion state of a metal.

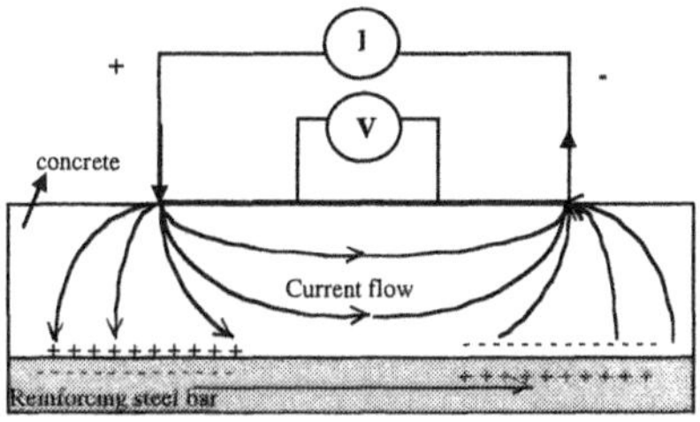

Figure 1: Electrode polarization of a rebar by measurement on the concrete surface

2.1 Apparent resistivity and Measurement Geometry

Electrical resistivity ρ is determined by a 3D form of Ohm's law to normalize for current input and location of a voltage prospecting for the homogeneous half space (Eqn. 1). In reality, two current electrodes are required for the current to flow into the half space via one and to exit via the other. Apparent resistivity ρ_a is introduced taking into account the geometry of electrodes by using a "geometric factor," k (Eqn. 2); ρ_a is the concrete's true resistivity only when measurements are made over a uniform concrete matrix.

$$\rho = \frac{V}{I} \times 2\pi \, | \mathbf{r} - \mathbf{r}_s | \tag{1}$$

$$\rho_a = \frac{V}{I} \times k \tag{2}$$

The value of k , the geometric factor, depends on geometry of the electrodes. The electrodes are arranged in certain pattern called an electrode array. There are various electrode arrays that meet different detection/characterization criteria, such as the commonly used Dipole-Dipole, Wenner, or Shlumberger etc. (Ward, 1990). The four-electrode resistivity meter has been used to measure concrete resistivity on site, of which Wenner array (Figure 2) is preferred (Gowers & Millard, 1999). It includes two current electrodes and two voltage electrodes inline with spacing of *na* between the electrodes. The geometric factor for Wenner array is $k=2\pi na$. This array is often chosen because of its simplicity and high signal to noise level. In IP measurement, various frequencies are employed to study the polarization characteristics, and the measured apparent resistivity is a complex number which can be decomposed into a real part and an imaginary part. As shown in Eqn.3, I_0 and V_0 are the input and output magnitude, ω is angular frequency used, and ϕ is the measured phase lag.

$$\rho_a(\omega) = \frac{V_0(\omega)\times e^{i(wt+\phi)}}{I_0 e^{i\omega t}} k = \left[\frac{V_0(\omega)}{I_0} k\right] e^{i\phi} = |\rho_a| e^{i\phi} = \mathrm{Re}(\rho_a) - i\,\mathrm{Im}(\rho_a) \tag{3}$$

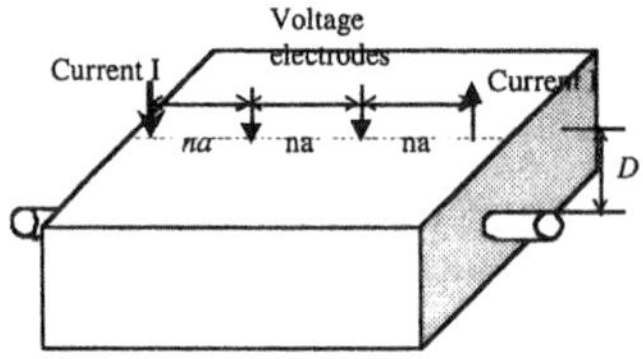

Figure 2: Surface-based electrode array (Wenner)

The measured resistivity is not unique over the same system. It varies with the geometry of the measurement, such as the selected array, the spacing of the array etc. Figure 1 shows that the current distribution is distorted by the embedded, implying that if the spacing of the electrode decreases, the distortion degree will be reduced as more and more current will be carried within the concrete and less amount is carried through the rebar. This is due to the fact that penetration of the current decreases with the electrode spacing. Therefore the response for the embedded rebar, the target of measurement, will be diluted with decreased spacing but with a more focused polarized area. A more focused polarized rebar area is critical for reducing the average effect and for providing precise spatial evaluation of the corrosion.

The concept of dilution effects arises from the influence of the measurement geometry on magnitude of response from an embedded target. Because of the dilution effect, the apparent resistivity measured by the surface array is only indirectly related to the impedance of the reinforcing bar/concrete interface, depending on the polarized area of the reinforcing bar. Analysis of the dilution effect has very important practical applications: 1) it helps optimizing of the electrode spacing so that the measurable response is from the least polarized portion of the rebar; 2) it provides knowledge of response patterns from different measurement geometry; 3) it helps interpreting the measured response in terms of the embedded target response.

2.2 Complex interface impedance

The complex interface impedance between concrete and rebar reflects the degree of corrosion of the rebar, as determined using electrochemical corrosion measurement methods, such as the ac impedance method. During corrosion, there are basically two paths (Ward, 1990) for current to pass from the electrolyte to the metal. First, charges are carried physically through the interface by the chemical reactions, called the faradic path. The corresponding impedance is the change transfer impedance, or the polarization impedance *Rp*. Second, charges do not cross the interface, but current is carried by charging and discharging of the double layer at the interface. This constitutes an electrical capacitance. Thus, the interface impedance can be can be visualized as an equivalent electrical circuit. The simplest and most widely used equivalent circuit is an electrolyte impedance *Re* in series with a parallel double layer capacitance *Cd* and polarization impedance *Rp* (Figure 3).

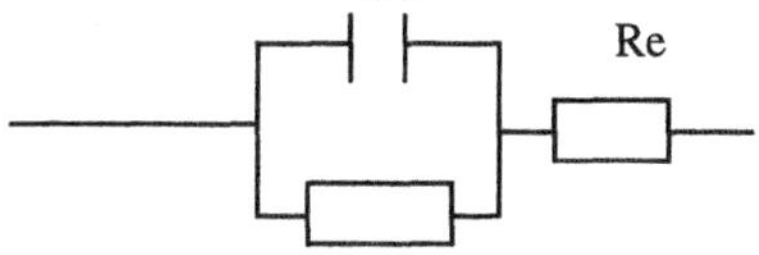

Figure 3: An electrical circuit for concrete and rebar interface impedance

From this electrical circuit, the real and imaginary part of the interface impedance can be displayed in Nyquist plot as a function of frequency (Figure 4). Note that as *Rp* decreases (corrosion rate increasing), the diameter of the curve is reduced. The frequency where the imaginary part is the biggest,.called critical frequency, moves to the right from 0.04 Hz to 0.4 Hz with a decreased *Rp*. From measurement on concrete surface, if the same effect is observed when the corrosion rate changes, the embedded rebar corrosion state can be evaluated.

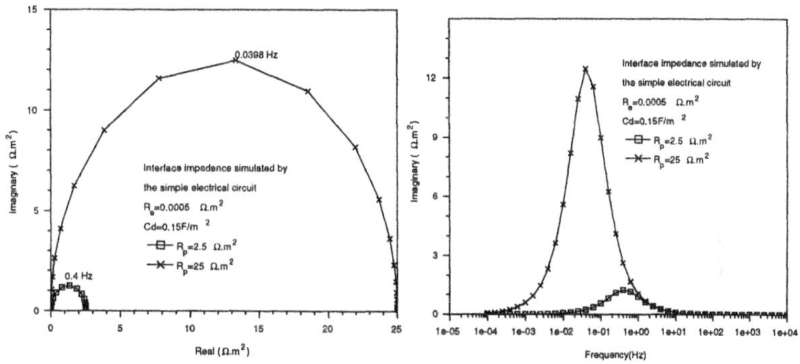

Figure 4: The spectra of two kinds of interface impedance (Figure 3)

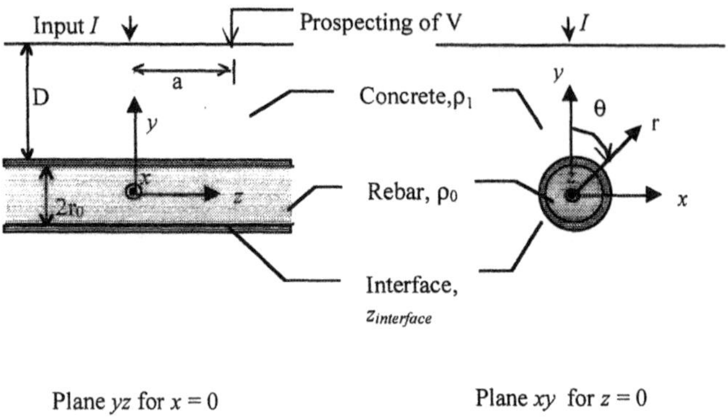

Figure 5: The coordinate system and geometry of the model.

3. MODELING

3.1 Model description

The model calculates the electrical response of an embedded rebar in concrete, where a current is injected at the surface of the concrete, and the voltage is measured also at the concrete surface (Figure 5). In the defined coordinate system, the *z* axis is taken along the axis of the rebar, and the *x-y* plane is along the cross section of the rebar, with origin at location of the injected current. The cylindrical coordinate is defined as (r, θ, z), with the same origin.

The concrete matrix, with electrical resistivity ρ_1, is treated as a half space. The rebar, with electrical resistivity ρ_0, is embedded in concrete at a distance of *D* from the concrete surface. The impedance at the concrete/rebar interface, Z_{inter}, is a quantity controlled by the corrosion processes at the interface which can be produced by an equivalent electrical circuit model. Given a current *I* injected from the concrete surface right above the rebar, the model calculates the voltage response in any other positions in the half space, for example, on the surface where voltage electrodes are located.

3.2 Analytical solution in a whole space

The analytical solution begins with the response of a rebar inside a whole space of concrete. The half space of the above model can be derived by using " image method" whereby an image of the rebar is "drawn" symmetrically in the other half space and the result is derived by principle of superposition. In whole space with a point source *I*, the voltage at any location $\boldsymbol{r}$ is governed by the Poisson equation. It can be written in cylindrical coordinates as

$$\frac{1}{r}\frac{\partial}{\partial r}\left(\frac{\partial V(\mathbf{r})}{\partial r}\right)+\frac{1}{r^2}\frac{\partial^2 V(\mathbf{r})}{\partial \theta^2}+\frac{\partial^2 V(\mathbf{r})}{\partial z^2}=\frac{I}{4\pi}\rho_1\delta(\mathbf{r}-\mathbf{r}_s) \tag{4}$$

where $\delta(\mathbf{r}-\mathbf{r}_s)$ is the Dirac function, and ρ_1 is electrical resistivity of where the point source is (concrete here).

Due to symmetry over the plane *xy* at z = 0, Fourier cosine transform of the Eqn. 4 in the *z* direction yields a modified Bessel equation in the wavenumber domain. It contains the series solutions in the modified functions of the first kind and second kind in regions of concrete and rebar. Based on the expansion of the *1/R* potential in cylindrical coordinates in terms of modified Bessel functions, the solution can be written into two parts: the primary field (without the rebar), and the secondary field due to the rebar (Smyth, 1968). General solutions in the regions of concrete and rebar are given by an infinite series in the modified Bessel functions *Im* and *Km*, weighted by unknown coefficients *Am* and *Bm* as follows: The primary field is expanded as

$$V^P(r,z)=\frac{I}{2\pi^2}\sum_{m=0}^{\infty}(2-\delta_{0m})\left[\int_0^{\infty}K_m(ur_s)I_m(ur)\cos(uz)du\right]\cos(m(\theta-\theta)_0) \tag{5}$$

where $\delta=\begin{cases}1 & if \quad m=0\\ 0 & if \quad m\neq 0\end{cases}$.

If $r>r_s$ interchange r and r_s in Eqn. (5).

Therefore, the solutions in regions of concrete (V_1)and rebar (V_0) are as in the following:

$$\begin{aligned}V_0(\mathbf{r},z)&=\frac{I}{2\pi^2}\sum_{m=0}^{\infty}\left[\int_0^{\infty}A_m(u)I_m(ur)\cos(uz)du\right]\cos(m(\theta-\theta_0))\\ V_1(\mathbf{r},z)&=V^P(r,z)+\frac{I}{2\pi^2}\sum_{m=0}^{\infty}\left[\int_0^{\infty}B_m(u)K_m(ur)\cos(uz)du\right]\cos(m(\theta-\theta_0))\end{aligned} \tag{6}$$

The unknown coefficients can be calculated by matching the solutions at the concrete and rebar boundary where normal current and potential are continuous (Holladay et al.,1984). The interface impedance is considered at $r=r_o$ as in the following:

$$\begin{aligned}&r=r_o;\begin{Bmatrix}V_0(r)+\Delta V=V_1(r)\\ \sigma_0\dfrac{\partial V_0}{\partial r}=\sigma_1\dfrac{\partial V_1}{\partial r}\end{Bmatrix};\\ &\Delta V=Z_{\text{interface}}\times\sigma_1\frac{\partial V_1}{\partial r}=Z_{\text{interface}}\times\sigma_0\frac{\partial V_0}{\partial r}\end{aligned} \tag{7}$$

For each m, the order of Bessel function, the unknowns A_m and B_m are the functions of wavenumbers. The results are substituted into Eqn. 6 and inverse transformed into the space domain.

$$B_m(u) = \frac{\dfrac{\sigma_0}{\sigma_1}\dfrac{[(2-\delta_{0m})K_m(ur_s)I_m(ur_0)]\left[u\times I_{m-1}(ur_0)-\dfrac{m\times I_m(ur_0)}{r}\right]}{\left[I_m(ur_0)+Z_{interface}\sigma_1\left((uI_{m-1}(ur_0)-\dfrac{m\times I_m(ur_0)}{r}\right)\right]} - (2-\delta_{0m})K_m(ur_s)\left[uI_{m-1}(ur_0)-\dfrac{m\times I_m(ur_0)}{r}\right]\left[uK_{m-1}(ur_0)-\dfrac{mK_m(ur_0)}{r_0}\right]}{\left[uK_{m-1}(ur_0)-\dfrac{mK_m(ur_0)}{r_0}\right]-\dfrac{\left(\dfrac{\sigma_0}{\sigma_1}\right)K_m(ur_0)\left[u\times I_{m-1}(ur_0)-\dfrac{m\times I_m(ur_0)}{r}\right]}{I_m(ur_0)+Z_{interface}\sigma_1\left((uI_{m-1}(ur_0)-\dfrac{m\times I_m(ur_0)}{r_0}\right)}} \quad (8)$$

$$A_m(u) = \frac{(2-\delta_{0m})K_m(ur_s)I_m(ur_0)+B_m(u)K_m(ur_0)}{I_m(ur_0)+Z_{interface}\sigma_1\left((uI_{m-1}(ur_0)-\dfrac{m\times I_m(ur_0)}{r_0}\right)}$$

The above solution is for a rebar in a whole space field. The solution of the half space of the model is obtained by superposition of solutions where the embedded rebar and its image in the whole space are considered respectively (Figure 6). We assume that there is no coupling effect between the rebar and its image.

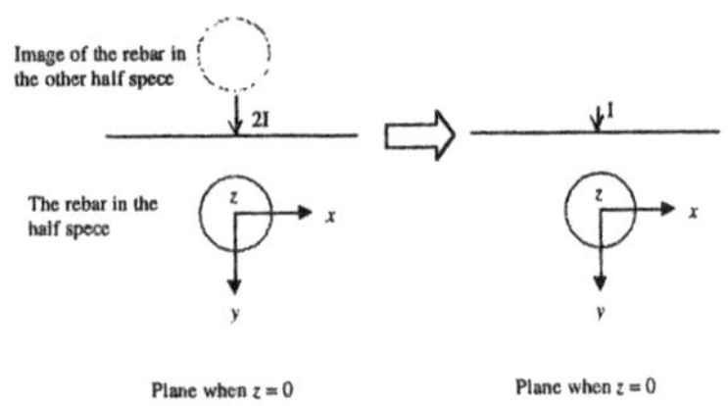

Figure 6: Whole space solution to half space solution

4. MODELING RESULTS

4.1 Geometry effect

Deviation of the measured apparent resistivity from the concrete resistivity indicatess the presence of the rebar. In order to show explicitly that the measured apparent resistivity is a result of both concrete resistivity and rebar resistivity, the complex impedance at their interface is neglected in this case. The apparent resistivity also depends on the measurement geometry: the array type, array geometry, and the electrode position with respect to the rebar. The Wenner response of a rebar (r_0 =0.0127 m, ρ_0 = 0.01Ω.m) embedded in more electrical resistive concrete matrix (ρ_1 = 100 Ω.m) is shown in Figure 7. A conservation high value of resistivity of the rebar is used (usually its value is around 6-9 μΩ.m) because the interface impedance could be high at the measured frequency.

First, it can be seen that the measured apparent resistivity from concrete surface is indeed less than concrete resistivity, displaying the presence of the embedded rebar. Second, the three curves in the plot for different ratios of concrete cover D to rebar radius r_0 show that the apparent resistivity from the shallowly embedded rebar is less and therefore can be more easily detected. In each curve the apparent resistivity decreases with increasing electrode distance *na* in the Wenner array when more current is carried by the less resistive rebar, showing clearly the dependence of the measurement geometry. In the choice of a proper electrode distance in measurement, a balance of a bigger response and a smaller rebar area involved should be considered, with a knowledge of the equipment precision.

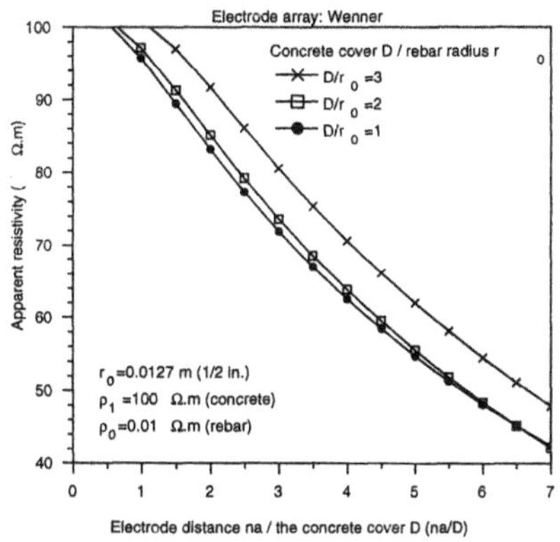

Figure 7: Geometric effect on Wenner response of a rebar embedded in concrete.

4.2 Apparent resistivity response with corrosion rate

The complex resistivity measured on the concrete surface that is due to the concrete/rebar interface impedance (Figure 4) is shown in Nyquist plot (Figure 8). In this simulation, the input of concrete and rebar resistivity is 50 Ω.m 10^{-6} Ω.m, respectively, and the output of the complex resistivity is from Wenner array (na=0.08 m). Here the two curves from different corrosion rates of the rebar intersect with the real axis at different values at the low frequency end, showing different values of *Rp*. The difference, however, is less than that in the pure interface impedance. In the curves of imaginary versus frequency, the critical frequency is around 1 Hz and 1.6 Hz for *Rp* being 2.5 Ω.m^2 and 25 Ω.m^2, respectively, which is bigger than the critical frequency of the interface impedance in Figure 4. That implies that the data can be acquired faster by measuring on concrete surface than using direct method.

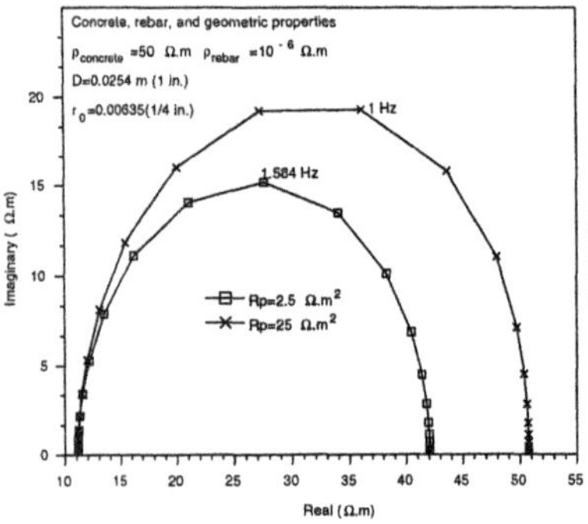

(a) Nuquist plot

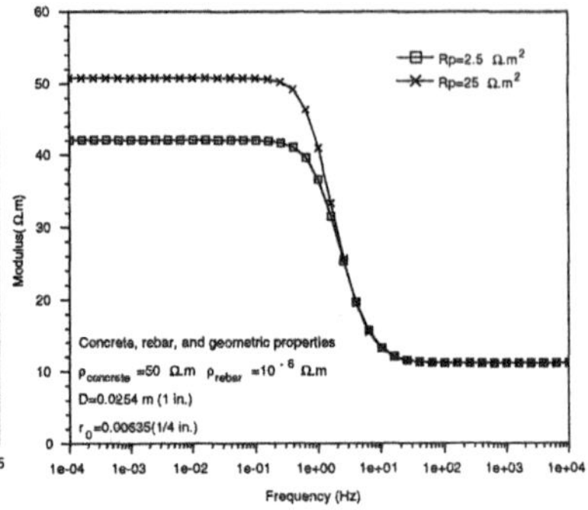

(b) Impedance modulus versus Frequency

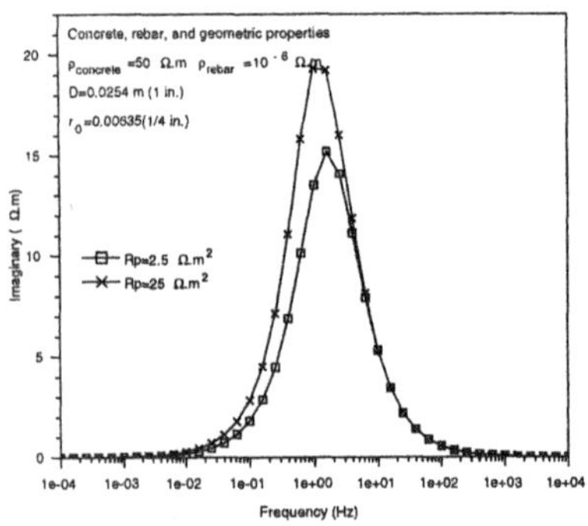

(c) Imaginary versus Frequency

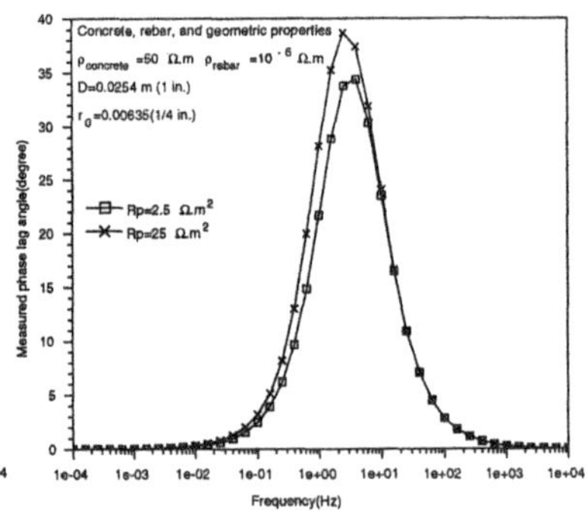

(d) Phase versus Frequency

Figure 8: The measured resistivity at different corrosion rates

4.3 Geometry effect on corrosion measurement

The above simulations by the model show clearly that the embedded interface impedance can be measured on the surface of the concrete. Different interface impedances have different responses, although the measured resistivity does not relate linearly to the interface impedance because of the dilution effect by the measurement geometry. To illustrate this fact, two simulations were done with the same interface impedance (Rp=2.5 $\Omega.m^2$) but different depth of concrete cover over: 0.0257 m and 0.0381 m, respectively. The result is shown in Figure 9. Clearly, the deeper the rebar is embedded, the less frequency dependency is measurable. Although the peak value is more diluted by a thicker concrete cover, the critical frequency is not diluted in the same way. The critical frequency is preserved for the same interface impedance by measuring the concrete surface and independent of the measurement geometry.

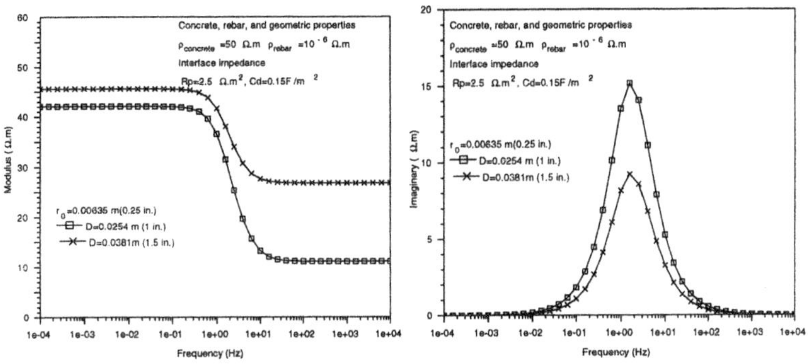

(a) Modulus versus Frequency (b) Imaginary versus Frequency

Figure 9: The surface measured resistivity at different concrete cover

5. EXPERIMENTAL

To confirm the model presented above, two prismatic 0.089 × 0.114 × 0.406 m concrete specimens containing a rebar were tested. The concrete materials included ASTM Type I Portland cement and Kaiser Radum aggregate, with a maximum size of 0.0064 m. The water to cement ratio (w/c) was 0.65, and the cement : sand : aggregate ratio was 1 : 2.26 : 2.26. The configuration of the specimens is shown in Figure 10. The rebar used was ASTM A 615 Grade 60 steel, with a nominal diameter of 0.0127 m.

The specimens were demolded forty-eight hours after casting and cured at 100% RH for ten days before they were placed in the saturated $Ca(OH)_2$ solution. The corrosion states in the rebars were induced and controlled by anodic polarization. The selected corrosion rates in terms of the supplied anodic current density were 0.05, 1, 5, and 10 $\mu A/cm^2$, respectively, as shown in Figure 11. Specimens were measured at these different rebar corrosion rates, respectively.

The apparent resistivity spectra at different corrosion rates are presented in Figure 12. The *na/D* value used was 2.13 in Wenner array. The curve of the resistivity modulus vs. frequency becomes flatened with increasing corrosion rates. The phase angle dropped dramatically with increasing corrosion rates. The real and imaginary parts in Nyquist plot show clearly that the arc diameter decreased with the corrosion rate. These observations are in total agreement with the simulated results in Figure 8.

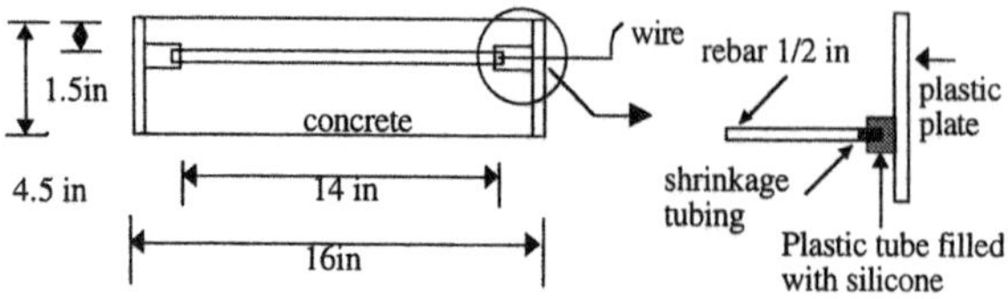

Figure 10: Configuration of the specimens

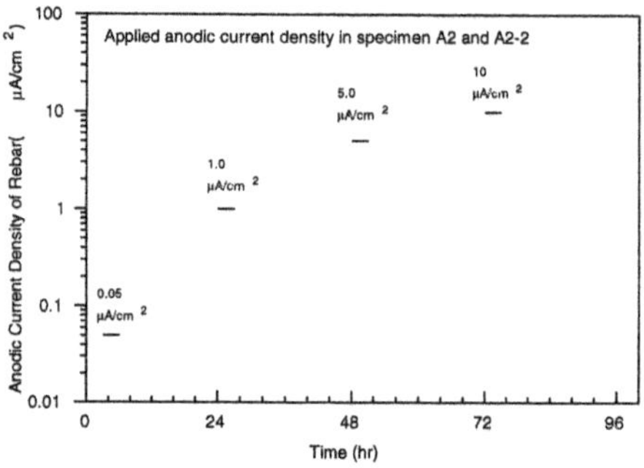

Figure 11: Corrosion rate control history

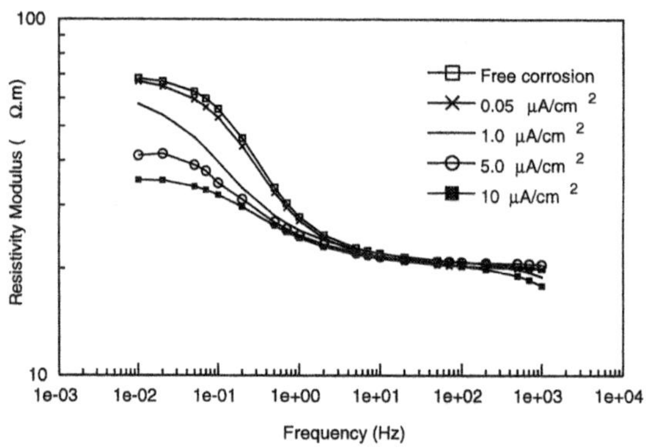

(a) Modulus versus Frequency

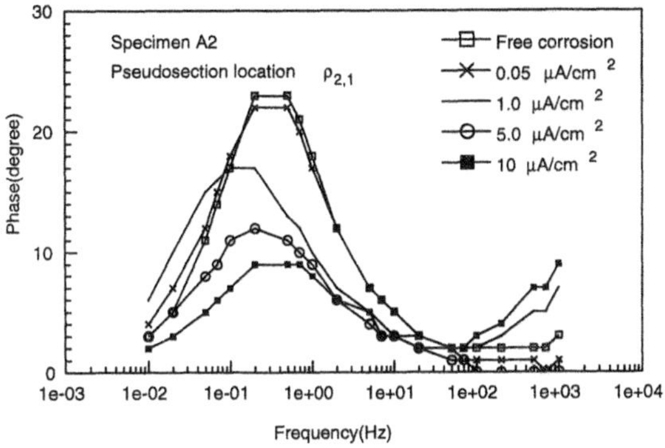

(b) Phase versus frequency

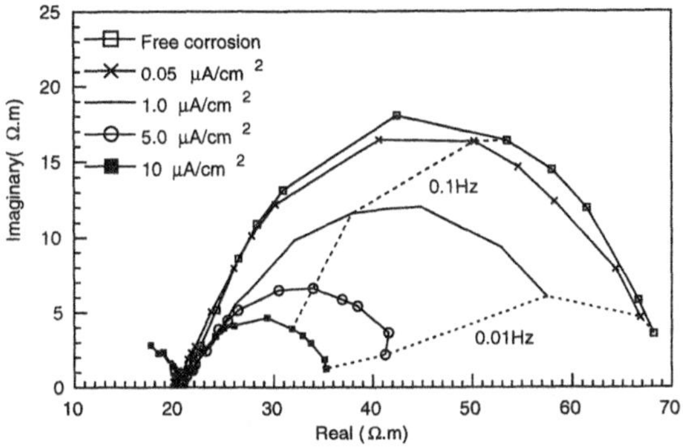

(c) Imaginary versus real

Figure 12: Impedance measured with different corrosion rates

Besides the similarities between the simulation and the experiment, note that the phase response obtained from the experiment is larger than that obtained from the modele. The simulation used a concrete half space with infinite concrete depth, while the experiment used a finite-sized concrete block. Therefore, the experimental measurement amplified the interface impedance because of the limited amount of concrete. Although they both show the essential properties of the interface impedance, the difference must be calibrated before this modeling method can be used as an inverse tool to simulate the experimental data.

6. CONCLUSIONS

1. An analytical solution has been obtained for the surface-based measurement of rebar corrosion in concrete. The complex apparent resistivity has been studied for different concrete and rebar interface impedance and for different measurement geometry.

2. Experimental confirmation has proved that analytical solution can be used to model concrete-surface measurement of the embedded interface impedance for corrosion studies.

3. Both the analytical and experimental study confirmed the capability and feasibility of the new method as a prospective quick, convenient, and quantitative solution to corrosion detection.

REFERENCE

1. Gowers, K. R and Millard, S. G. (1999), Measurement of Concrete Resistivity for Assessment of Corrosion Severity of Steel Using Wenner Technique, *ACI Materials Journal* **96:5**, 536-541.

2. Holladay, J. S. and West, G. F. (1984), Effect of Well Castings on Surface Electrical Surveys," *Geophysics* **49:2**, 177-188.

3. Monteiro, P. J. M., Morrison, F., and Frangos, W. (1998), Nondestructive Measurement of Corrosion State of Reinforcing Steel in Concrete, *ACI Materials Journal* **95: 6**, 704-709.

4. Sharma, P. V. (1997), *Environmental and Engineering Geophysics,* Cambridge University Press.

5. Smythe, W. R. (1968), *Static and Dynamic Electricity*, third edition, McGraw-Hill.

6. Ward, S. H. (1990), Resistivity and Induced Polarization methods, *Geotechnical and Environmental Geophysics*, SEG.

7. Zhang, J., Monteiro, P. J. M., Morrison, F. H. (2000), Non-Invasive Surface Measurement of the Corrosion Impedance of Rebar in Concrete Part I: Experimental Results, submitted and accepted by *ACI Materials Journal.*

Long Term Durability of Structural Materials
P.J.M. Monteiro, K.P. Chong, J. Larsen-Basse, K. Komvopoulos (Eds)

CORROSION AND EMBRITTLEMENT OF HIGH-STRENGTH BRIDGE WIRES

G. Vermaas[1], R. Betti[1], S. C. Barton[2], P. Duby[2] and A.C. West[2]

[1]Department of Civil Engineering and Engineering Mechanics
Columbia University
New York, NY 10027
[2]Department of Chemical Engineering and Applied Chemistry
Columbia University
New York, NY 10027

ABSTRACT

In this paper, the results of an extensive study currently underway at Columbia University on the environmental deterioration of high-strength low carbon wires used in suspension bridge cables are presented. Accelerated corrosion tests were conducted on galvanized and ungalvanized wires, either under sustained load or with no load, using a QFOG cyclic corrosion chamber. After being exposed to such an aggressive environment for a period that varied from a few days to four weeks, the wires were subjected to a tension test and their ultimate strength and strain were measured. Results show a clear reduction of the ultimate strain of the wire as the time of exposure to the corrosive environment increases. A relationship between the ultimate strength and the average corrosion depth is noticed. Different fracture mechanisms were noticed between wires with a slight degree of corrosion and wires with extensive corrosion damage. Progressively brittle behavior was linked to the surface roughness of the wire through nonlinear FEM analyses and through SEM analyses of the fracture surface morphology.

KEYWORDS

Bridge Wires, Hydrogen Embrittlement, Ductility Reduction, Corrosion, Pitting, Accelerated Corrosion Testing,

INTRODUCTION

In the New York metropolitan area, there is probably one of the largest concentrations of long-span cable suspension bridges, some of which are approaching a service life of one hundred years or more. From a complete review (Betti & Bieniek (1998)) of all the cable inspections on these bridges, performed over the last forty years, the following conclusions were reached:

1. There is water penetration into the interior of the main cables, with a water pH as low as 4,

2. On the wires, there are signs of extensive corrosion of zinc coating, evidenced by the presence of zinc hydroxide (white rust).
3. Following the loss of zinc coating, there is evident corrosion of steel, with discernible pitting and loss of steel,
4. There are broken wires inside the main cable. Usually, such broken wires are located on the outer layers of the cables but in some cases, because of the procedures used at the time of the bridge construction, broken wires can also be found in the inner core of the cable.
5. Laboratory tests show significant loss of ductility and of fatigue strength for the deteriorated wires.

However, while the corrosion process has been extensively investigated for structural and other specialty steels, the corrosion process and the associated strength deterioration of high-strength bridge wires has received only little attention. It is believed that the corrosion of high-strength wires is composed of several mechanisms that interact with each other and with the stresses present inside the wire. Such mechanisms are usually indicated as general corrosion, pitting corrosion, intergranular corrosion and stress-corrosion cracking. In addition, another important phenomenon called hydrogen embrittlement could play an important role in the deterioration mechanisms of such wires. In order to effectively prevent corrosion and the associated loss of strength and ductility of bridge cable wires, it is first necessary to understand the specific corrosion mechanisms and their interaction.

TEST SETUP AND PROCEDURES

The purpose of this experimental testing program was to examine and analyze the corrosion process of high-strength steel bridge wires. A cyclic salt spray corrosion chamber was used to simulate in situ conditions. New galvanized and ungalvanized wires were received from the manufacturing plant and samples were made for testing. These samples were corroded in an accelerated corrosion chamber for a certain number of hours and then tested for ultimate strength and ultimate strain. Finally, the sample fractures were observed and the data was analyzed so that the corrosion mechanisms in bridge wires could be better understood.

The typical manufacturing process of a bridge wire involves the wires being sent through a set of dices made of tungsten carbide steel in order to shape the final diameter of the desired cross-section. This process is driven by a set of drums that are each 91.44 cm [36 in] in diameter. The results of cold-working of the steel will alter the wire microstructure, resulting in an increase of the strength of the wire and a reduction of its ductility. Because the grains are squashed and their dislocation density is raised dramatically, residual stresses are introduced. The final wire coil is chemically and thermally processed before being dipped in a hot zinc bath for galvanization. Heating the cold-worked metal (annealing) will reverse the effects produced by the cold-drawing process. The galvanized wire is then wrapped on a reel of 152.4 cm [60 inch] diameter and taken to the construction site.

The bridge wires examined in this experimental program were made of a steel with a nominal ultimate strength of 1.8 GPa [260 ksi]. The composition of these ungalvanized wire samples consisted of 0.770 - 0.780% carbon, 0.680 - 0.690% manganese, 0.190 - 0.200% silicon, 0.008% sulfur, and 0.003% phosphorus with the remainder of the wire being iron. These results fall in the typical ranges given by Stahl (1996) for high-strength low-carbon bridge wires.

A Q-FOG CCT 1100 Corrosion chamber (see figure 1) was used for all cyclic corrosion testing (Barton et al. (1999)). The standard corrosion cycle was a modification of ASTM G85.A2, originally designed for corrosion resistance testing of aluminum coatings (ASTM (1994a)). Because the fog solution specified by this standard was found to be excessively corrosive to bare steel and zinc, the recommended salt concentration was eventually reduced from 5.0% to 0.05%. The following is a summary of the conditions employed:

- Fog solution: 0.05% NaCl or 0.05% NaSu

- pH = 3.0 or pH = 4.6, adjusted with acetic acid
- Cycle Temperature: 45°C
- Cycle Steps:
  - Salt fog (45 minutes)
  - Low humidity (135 minutes)
  - High humidity (180 minutes).

To assess the severity of the cyclic corrosion conditions, general corrosion testing was performed on an ungalvanized set of wire samples. The measurement of mass loss as described in ASTM Standard G1(C.3.1) was employed (ASTM (1994b)). Samples were periodically removed and tested for mass loss, caused by oxidation as described in the test standard.

The samples used in all the experiments conducted in this study were cut to approximately 45.7 cm [18 in] in length. The ends of the galvanized wire samples were dipped in concentrated hydrochloric acid to remove the zinc coating in order to prevent slip during tensile testing. The samples were painted with alkyl-based paint at each end, leaving a 15.2 cm [6 in] length exposed at the center. The corroded length of the wire at the center of the sample would be the weakest section and would force fracture in this location.

For all galvanized wire samples, a 5 mm [0.197 in] strip of zinc was removed from the center of the sample, exposing the steel. This was necessary because, after the first preliminary tests, it was observed that the zinc coating, if intact, will completely protect the inner core of steel and will prevent any corrosion of the wire. Instead, by removal of a small portion of coating, the stripped section will become the site of concentrated cathodic hydrogen evolution, due to the presence of the zinc-iron galvanic couple, and corrosion will occur there, reducing the wire's cross-section and ensuring fracture there. For one set of galvanized wire samples, all of the zinc was removed by immersion in concentrated hydrochloric acid followed by rinsing with water. These wire samples are hereafter referred to as "stripped-galvanized".

During exposure in the corrosion chamber, some wire samples were held under a constant elongation of 0.5 to 1.5 mm [0.02 to 0.06 in] using a properly designed fixture. Such a fixture consisted of circular, notched end-plates of 19 mm [0.75 in] thickness, separated by 13 mm [0.5 in] diameter threaded-rod spacers. Threaded ferrules were attached at each end of the sample wire, which was then placed between the notched end-plates of the fixture. This assembly was inserted into a tensile test machine and the wire sample was loaded through the end-plates to the desired level, which ranged from 8.9 to 22 kN [2000 to 5000 lbs], with the upper limit being about 70% of the ultimate tensile strength of the uncorroded wire. The threaded-rod spacers were then tightened in compression between the two end-plates until the tensile load of the test machine was counteracted. The fixture and sample could then be removed from the test machine and inserted into the corrosion chamber (figure 1). The amount of load on the wire samples was measured before the wire was placed in the corrosion chamber and after it was removed from the chamber, before it was tested for ultimate elongation and load. The loss of load in the wire due to relaxation was considered minimal.

All samples were removed from the chamber at the end of a salt fog step. Using an Instron 4206, wire samples were pulled at a constant extension rate of 2.54 mm/min [0.1 in/min] until fracture occurred, in adherence to ASTM standards (ASTM (1996)). Elongation of each sample was measured using an RDP Type ACT-1000A linear voltage distance transducer which was attached to the wire sample using a 25.4 cm [10 in] gage length Tinius Olsen test bracket. The LVDT output and the applied load signal from the tensile test machine were read using an Omega DAS-800 A/D converter, with Keithly EasyAG software on a microcomputer. Peak load and elongation at peak load were then obtained from the recorded data.

It should be noted that the techniques of the present study differ from those used to analyze hydrogen-stress cracking (HSC). In the present study, the testing of the material properties of the wire samples are measured ex situ as opposed to the in situ conditions of HSC testing. In addition, the strain rate of 0.01/min used in this study is much greater than the 10^{-6}/min rate employed to study HSC (Hartt et al. (1993)). It is well known that embrittlement phenomena are strongly strain-rate dependent above a threshold rate (Seabrook et al., (1950)). However, the chosen strain rate for this testing is well below the threshold suggested by Seabrook et al.

Hydrogen content analysis was conducted on 1 gram samples of galvanized and stripped-galvanized wire, both in uncorroded condition and after 438 hours (73 cycles) of cyclic corrosion. These specimens were packed in dry ice and shipped overnight to Alternative Testing Laboratories (Glassboro, PA.). Analysis was performed by the inert gas fusion-thermal conductivity method, using a Leco RH2 hydrogen analyzer.

The reversibility of embrittlement of galvanized wire samples was tested by aging corroded wires at temperatures of 25°C and 150°C for up to 72 hours prior to tensile testing. The room-temperature aged samples were left hanging in the laboratory overnight, and the high-temperature aged samples were left in a tempering oven. The latter samples were water-cooled prior to tensile testing. Because the high-temperature aging process could possibly temper the wire samples, the material properties of uncorroded samples aged under identical conditions were also measured.

RESULTS OF EXPERIMENTAL PROGRAM

The experimental program consisted of 21 separate corrosion tests, some of which were repeated in order to increase the amount of available data. In each test, approximately 20 to 45 samples were placed in the cyclic accelerated corrosion chamber for a period of 400 to 800 hours. Looking at the amount of mass loss during each cycle and comparing the results with those obtained in previous studies (Eiselstein & Caligiuri (1988)), it was concluded that each chosen cycle (with a 0.05% NaCl solution) is equivalent to 0.09 year of service life. Based on these estimates, a typical test duration of 84 cycles may be considered equivalent to 7.3 years in service conditions (Barton et al. (1999)).

First, the hydrogen content in different types of corroded wires was analyzed. The results are presented in figure 2, where the hydrogen concentration for corroded galvanized and stripped galvanized wires are presented. The uncorroded samples are analyzed to serve as the baseline. They show minimal amount of hydrogen that could come from either the manufacturing process of the wire or from the cleaning process (partial/complete removal of zinc coating). By comparing the hydrogen contents for corroded and uncorroded wires, it is clear that a substantial amount of hydrogen is generated and absorbed into the steel core of the wires. This hydrogen comes in part from the corrosion products and in part from the action of the galvanic zinc-iron cell, as it can be seen by comparing the hydrogen concentration of corroded stripped galvanized wires and of corroded galvanized wires. Wires with only a small portion of zinc coating present the highest concentration of hydrogen, ranging from 30 ppm to 80 ppm. By removing the corrosion products from the wire surface, it is noteworthy that still a substantial amount of hydrogen is present in the wire, clear sign that part of the hydrogen is absorbed into the wire. However, it is interesting to notice that this amount seems to be independent by the applied load on the wires (up to 5,000 lbs), indicating a no-dependence of the hydrogen absorption with the external load. Such a relationship, instead, was noticed in previous studies on hydrogen embrittlement of prestressing wires (Toribio et al. (1997)).

One of the most interesting results of this laboratory program is represented in figure 3, where the reduction of the ultimate strain of galvanized wires is plotted as a function of time of exposure to the aggressive environment for different pH chloride solutions (pH values = 2, 3, 4.6). It is clear that, as the exposure time in the corrosion chamber increases, the ultimate strain decreases from approximately 6 percent (for the uncorroded wires) to a threshold value that depends on the aggressiveness of the solution. The more aggressive is the corrosive solution, the lower is the ultimate strain reached by the wire at fracture. In some cases, like the one corresponding to a solution of pH 2.0, the ultimate strain reaches a value of 2.5 percent, equivalent to a reduction of more than 50 percent the elongation of a virgin wire. However, it is very interesting to notice that, when the variation of the wire's ultimate strain is plotted as a function of the corrosion depth, the rate of loss of ultimate strain versus corrosion depth for all the different pH solutions is similar, as shown in figure 4. Here, the ultimate strains are plotted as a function of an average depth of the corroded area, as derived from the measurement of the mass loss. The curves corresponding to the different solutions are now quite close, indicating a clear dependence between the ultimate elongation of the wire and the depth of the corroded area. The same trend is shown in figures 5 and 6, where the variation of ultimate strain for all the types of wire tested in this program (ungalvanized, galvanized and stripped galvanized) is presented. It is noteworthy that also ungalvanized wires show an analogous behavior as galvanized wires, with an ultimate strain that decreases with time of exposure and with corrosion depth. The relationship between the reduction of wire's ultimate strain and its corrosion depth has been also tested looking at solutions less aggressive than the chloride solutions. Sulfate solutions, with equal pH as the chloride ones, were used in the cyclic corrosion chamber. Such solutions, being less corrosive than the chloride ones, provide much slower reductions in ultimate strain versus time when compared with those corresponding to the reductions induced by the equivalent chloride solution. However, when the ultimate strains are plotted versus the average corrosion depth, the two types of solutions show identical rate.

This trend can help us explain the embrittlement process in bridge wires as the result of corrosion pitting and surface irregularities. In fact, as the steel wire surface corrodes, corrosion products, pitting and surface irregularities are generated over time. Such events alter the cylindrical geometry of the wire, inducing complex three-dimensional stress distributions and local stress concentrations that lead to the fracture of the wire. For the case of uniform corrosion over a limited area of steel, the state of stress will change from one-dimensional to three-dimensional, analogously to the one at necking: the morphology of the fracture surface will show all the characteristics of a crack that starts from the center of the wire and moves outwards. Instead, after extensive corrosion, the presence of corrosion pitting and surface irregularities will act as stress raisers, inducing stress concentrations similar to the ones induced by notches. In this case, fracture will start from the surface of the wire, in correspondence of the pit or irregularity, and will propagate through the cross section of the wire. These fracture mechanisms have been validated by numerical FEM analyses as well as by microscopic analyses using the Scanning Electron Microscope (SEM), as shown in a following section.

In addition, the effects of various corrosion inhibitors on the variation of ultimate strength and strain of bridge wire have been investigated. In these tests, galvanized wires, with a portion of zinc coating removed and coated with various protective corrosion inhibitors (Pre-lub 19, raw linseed oil, and pre-lube 19-60 RLO), have been tested over 124 corrosion cycles (744 hours in the corrosion chamber). As expected, the results show that all the coated wire samples, having a protective barrier between the zinc or steel and the atmosphere that slows down the general corrosion process and thus the formation of pitting and surface irregularities, perform better than the uncoated wires. This conclusion is consistent with the assumption that the leading phenomenon controlling the embrittement of the wire is the one related to the corrosion pitting and surface irregularities.

Among the various experimental tests conducted during this research projects, tests 2, 3, 4, and 14 were devoted to studying the possible effects of pre-loading on the deterioration rate of the wire and on its ultimate strength and ductility. It is known that, if a steel specimen is subjected to an external loading during the corrosion process, a deterioration mechanism known as stress-corrosion cracking should occur, making the specimen under sustained load more brittle than an unloaded one. However, such a relationship could not be observed in the galvanized and ungalvanized high-strength bridge wires studied in this study. In addition, no effect of the pre-load on the amount of hydrogen absorbed into the inner steel core was noticed.

FEM ANALYSES OF THE FRACTURE MECHANISM

An in-depth numerical analysis was then conducted to validate the conclusions of the experimental testing program. Such a theoretical analysis was mainly focused on the effects of uniform corrosion and corrosion pitting on the reduction of the ultimate strain of the wire, separating the two contributions and highlighting the differences.

An FEM model of a 25.4 cm [10 in] long wire, with a 0.5 cm long zinc strip removed, was built using the finite element ANSYS software. The element chosen was the 8-node SOLID45 3-D structural solid element, with three degrees of freedom per node. This element has also the capability of handling plastic deformations and large strain conditions. The material properties used in this analysis were assumed to be isotropic (a satisfactory assumption for virgin steel) and were obtained from the previous experimental analysis of uncorroded wires. The elastic modulus is set equal to 206850 Mpa [30×10^6 psi] while the stress-strain relationship was loaded into the program as a multilinear curve using ten stress and strain data points. As a yielding criterion, the Von Mises criterion was chosen while the fracture criterion is based on reaching the maximum allowable longitudinal strain. Any finite element that experiences a longitudinal strain larger than 0.057 is deactivated so that the fracture can propagate through the wire.

The total number of finite elements in each wire is 4864. The cross-section of the wire is meshed so that the outer steel layers are modeled by 8 circumferential rings, each having a thickness of 0.00838 mm [0.00033 in]. The purpose of 8 rings on the outer perimeter of the wire is to simulate the effect of progressive corrosion. In fact, the thickness of each layer corresponds to the loss of thickness of a steel wire after 100 hours of exposure to an aggressive chloride solution with a pH of 3. In this way, uniform corrosion conditions, corresponding to a range of exposure time between 100 and 800 hours, can be simulated. Pitting is simulated by randomly removing a finite element on the external surface of the wire.

Figure 7 show the middle right section of a wire after 400 hours of uniform corrosion and for an elongation of 0.40 in. The strain distribution presents maximum strain occurring at the center of the cross-section, similar to the strain distribution at the necking. Analogously, the fracture mechanism will start from the center of the wire and propagate toward the surface of wire. It is interesting to notice how the strain distribution will change in the presence of a surface pit, as shown in figure 8 where the strain distribution for a wire after 400 hours of uniform corrosion and a surface pit is presented. Here, it is clear that the notch effect induced by the surface pit causes the maximum strain to move to the wire surface, in the vicinity of the pit, and that fracture initiates at the wire surface. To check the validity of the FEM analyses, it is also interesting to compare the experimental results with those obtained through numerical simulations. In figure 9, the ultimate strains from the wire FEM models (for both the case of uniform corrosion and of uniform corrosion plus pitting) and from the laboratory tests are presented as functions of the exposure time in the corrosion chamber. The majority of the experimental data falls within the envelope defined, as the upper boundary, by the cases of uniform corrosion and, as the lower boundary, by the data corresponding to uniform plus corrosion pitting. The excellent agreement between the numerical results and the experimental data confirms that, as also shown through the experimental testing, the reduction of ultimate strain is directly affected by the surface geometry of wire, as the result of corrosion and surface irregularities.

SEM ANALYSIS OF THE FRACTURE SURFACES

An SEM analysis performed at the E. Mindlin Structural Deterioration laboratory of Columbia University confirmed the conclusions of the experimental and numerical analyses. Wires with no or slight surface corrosion show a fracture morphology that is typical of a ductile material. Crack initiation will occur at the center where, because of the presence of high stresses, nucleation and growth of a large number of voids or holes will occur. In a ring area, usually called the shear-lip zone, the change in the rate of growth of the crack is clearly seen. When the wire instead undergoes substantial corrosion, the morphology of the fracture surface changes completely. Figure 10 shows a typical fracture surface that occurred in wires with extensive corrosion, showing the fracture pattern (mirror, mist, hackle and bifurcation) that can be seen in a tension test of a rod with a surface crack. This is also confirmed by the presence of "steps" on the fracture surface, which are only present when the fracture initiates from the surface of the material (Hull (1999)) and which are a characteristic feature of a brittle fracture. This fracture mechanism is quite similar to the strain distribution obtained through the numerical FEM analyses, as seen in figure 6.

CONCLUSIONS

From the analysis of the experimental data and of the FEM results, it can be concluded that, when a high-strength steel galvanized wire, with an exposed area of steel, is corroded, its ultimate ductility is controlled by the surface geometry of the wire. When the cylindrical geometry of the wire changes because of localized uniform corrosion, it can be shown that the ultimate ductility of that wire will decrease and that the fracture will initiate at the center midpoint of the wire. On the contrary, when the corrosion pits occur on the wire surface, the wire's ultimate ductility of the wire will decrease even further, and the fracture will initiate at the location of the surface pit.

ACKNOWLEDGEMENTS

This research has been sponsored through research grants by the National Science Foundation (CMS-9457305, CMS-9532082), whose support has been greatly appreciated.

REFERENCES

ASTM. (1994a). *Standard Practice for Modified Salt Spray (Fog) Testing*. G85-94, West Conshohocken, PA.

ASTM. (1994b). *Standard Practice for Preparing, Cleaning, and Evaluating Corrosion Test Specimens*. G1-90(1994)e1, West Conshohocken, PA.

ASTM. (1996). *Standard Test Methods for Tension Testing of Metallic Materials*. E8-96a, West Conshohocken, PA.

Barton, S., Vermaas, G., Duby, P., West, A., and Betti, R. (1999). Accelerated Corrosion and Embrittlement of High-Strength Bridge Wire. *Journal of Materials in Civil Engineering,*

Betti, R., and Bieniek, M.P. (1998). *Conditions of NYC Suspension Bridge Cables*. Report, Columbia University.

Eiselstein, L.E., and Caligiuri, R.D. (1988). Atmospheric Corrosion of the Suspension Cables on the Williamsburg Bridge. *Degradation of Metals in the Atmosphere, ASTM STP 965*, American Society for Testing and Materials, Philadelphia, 78-95.

Hartt, W.H., Kumria, C.C., and Kessler, R.J., (1993). Influence of Potential, Chlorides, pH, and Precharging Time on Embrittlement of Cathodically Polarized Prestressing Steel. *Corrosion,* **49(5),** 377-385.

Hull, D. (1999). Fractography – Observing, Measuring and Interpreting Fracture Surface Topography. Cambridge University Press, Cambridge, United Kingdom.

Seabrook, J. B., Grant, N. J., and Carney, D., (1950). Hydrogen Embrittlement of SAE 1020 Steel. *Trans. AIME Journal of Metals*, **188**, 1317-1321.

Stahl, F.L., and Gagnon, C.P. (1996). *Cable Corrosion in Bridges and Other Structures: Causes and Solutions*, ASCE Press, New York.

Toribio, J. (1997). "Hydrogen Embrittlement of Prestressing steels: the Concept of Effective Stress in Design." *Materials and Design,* **18(2)**, 81-85.

Figure 1: Accelerated Corrosion Chamber with Preloaded Wires

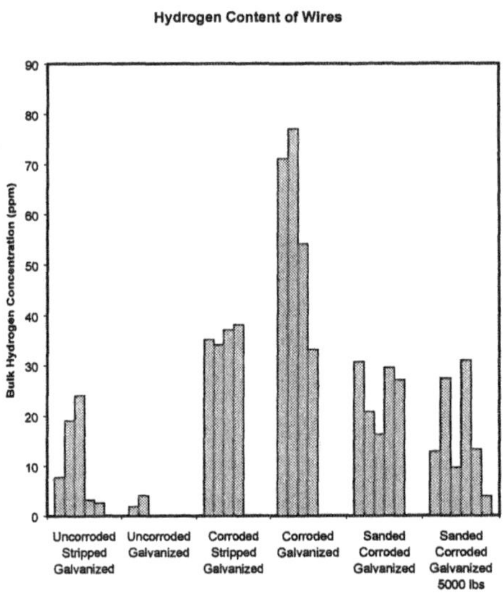

Figure 2: Hydrogen Concentration in Corroded Wires

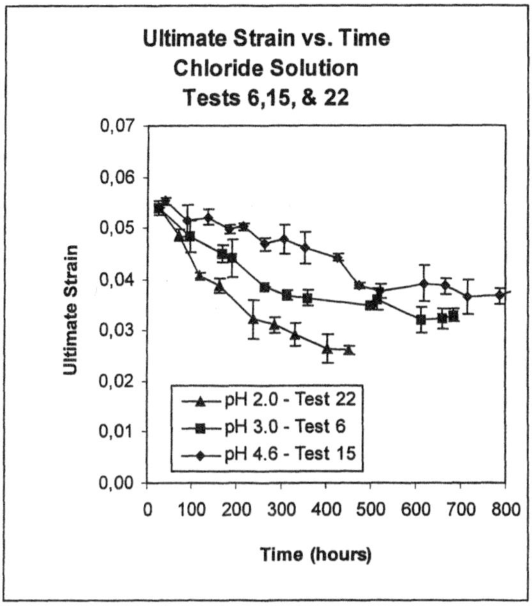

Figure 3: Ultimate Strain versus Time of Exposure in Corrosion Chamber

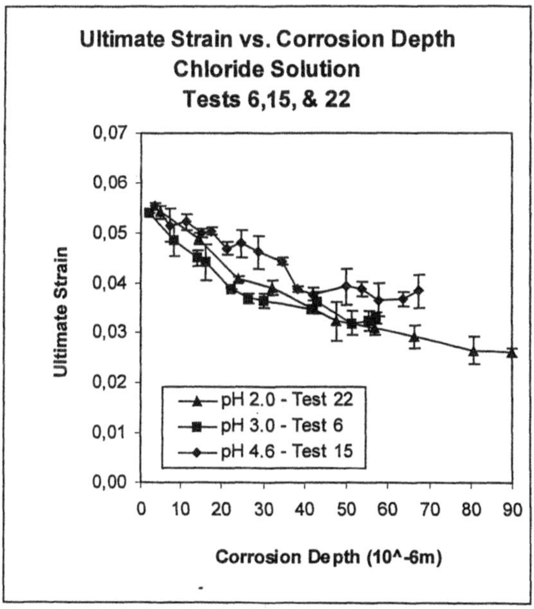

Figure 4: Ultimate Strain versus Average Corrosion Depth

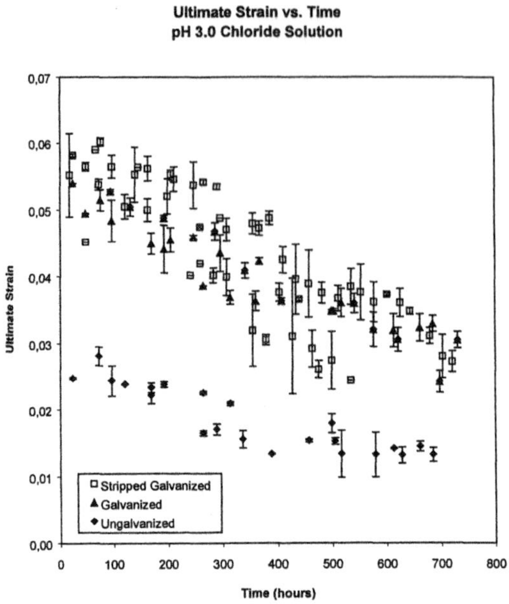

Figure 5: Ultimate Strain of Galvanized and Ungalvanized Wires versus Time of Exposure to Aggressive Environment

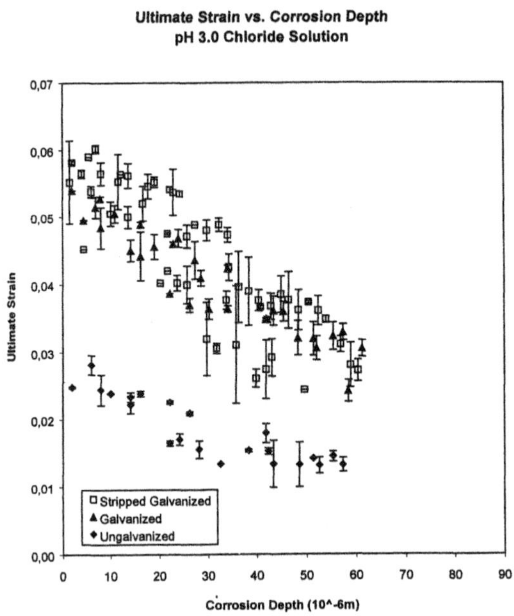

Figure 6: Ultimate Strain for Galvanized and Ungalvanized Wires versus Corrosion Depth

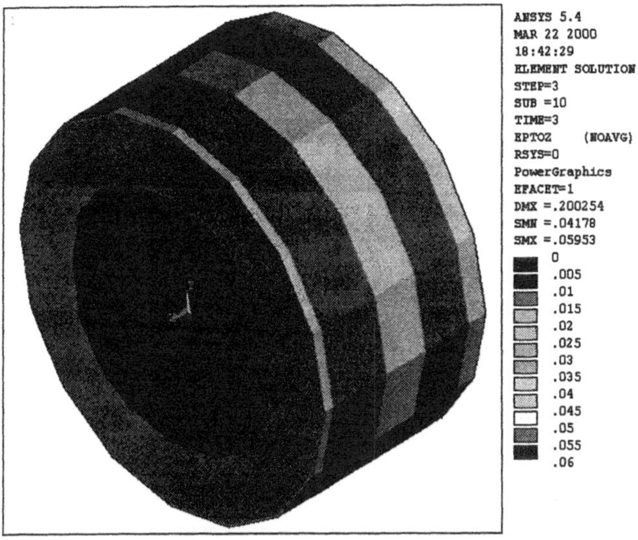

Figure 7: Strain Distribution at Midpoint after 400 Hours of Uniform Corrosion

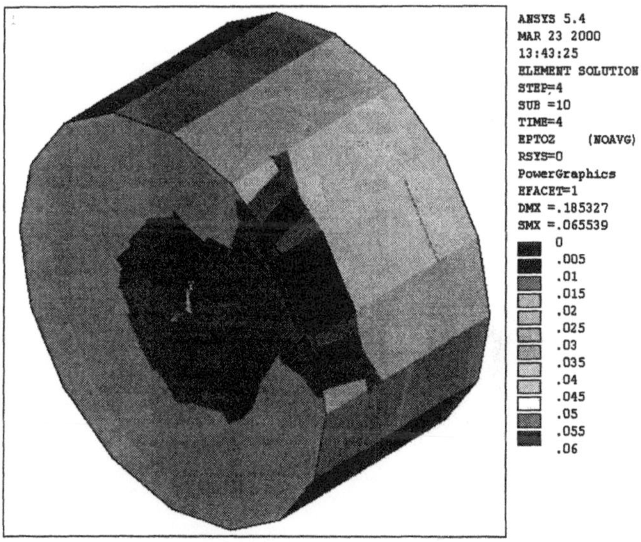

Figure 8: Strain Distribution at Midpoint after 400 Hours of Uniform Corrosion and Pitting

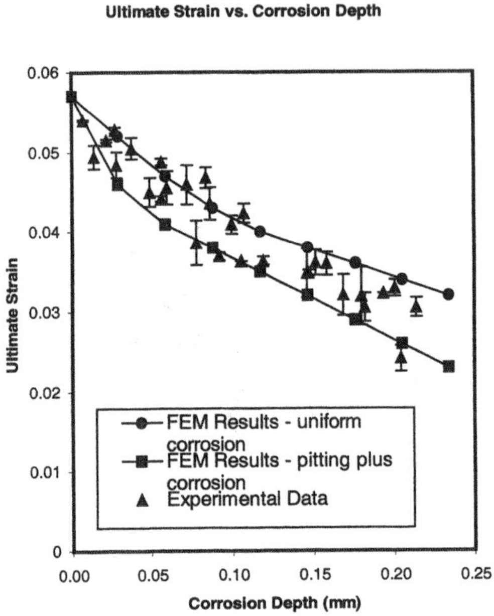

Figure 9: Comparison between Numerical and Experimental Results

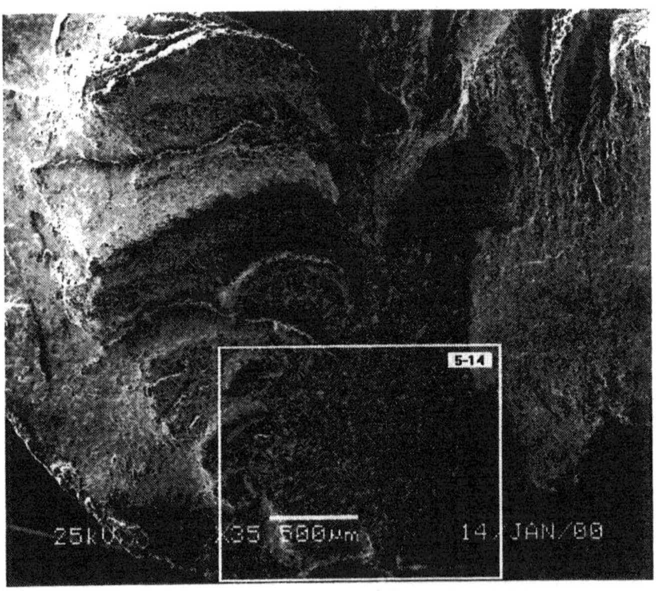

Figure 10: Fracture Surface after 648 Hours of Exposure to Aggressive Corrosive Environment

ACCELERATED TESTING FOR CONCRETE REINFORCING BAR CORROSION PROTECTION SYSTEMS

D. Darwin,[1] J. Balma,[1] C. E. Locke, Jr.[2] and T. V. Nguyen[2]

[1]Department of Civil and Environmental Engineering, University of Kansas
Lawrence, KS 66045, U.S.A.
[2]Department of Chemical and Petroleum Engineering, University of Kansas
Lawrence, KS 66045, U.S.A.

ABSTRACT

The corrosion of reinforcing steel is a major problem in reinforced concrete bridges. The problem, due principally to chlorides in deicing salts and sea water, is the main durability concern for most transportation structures. As a result, methods that can significantly reduce or halt chloride-induced corrosion have been aggressively pursued for well over 30 years. Laboratory tests represent the principal method used to evaluate corrosion protection systems, but there is little correlation between the performance of the systems in the laboratory and in the field. The current study is aimed at providing such a correlation. Key observations to date include the indication that corrosion of reinforcing steel is a cathode-limited process; some (currently) standardized specimens provide little early indication of corrosion performance; and some conventional corrosion evaluation techniques, such as linear polarization resistance, do not consistently represent the corrosion behavior of reinforcing steel cast in concrete.

KEYWORDS

accelerated tests, chlorides, concrete, corrosion, corrosion inhibitors, corrosion testing, microalloys, reinforcing bars.

INTRODUCTION

The corrosion of reinforcing steel in highway structures results in maintenance and replacement costs in the United States that are measured in billions of dollars. The use of deicing salts has resulted in the steady deterioration of roadway bridge decks. The deicers penetrate the decks and attack the reinforcing steel, causing corrosion. Bridge substructures in marine environments are likewise attacked by the chlorides in sea water. Due to these problems, the cost of maintaining highway structures in the U.S. continues to increase. In 1979, the cost of bridge repairs in the federal-aid system due to corrosion damage was estimated to be 6.3 billion dollars (Locke 1986). By 1986, the estimated cost was $20 billion and was forecast to increase at the rate of $500 million per year (Cady & Gannon 1992). A 1991 National Research Council report estimated annual expenditures on repair and "corrosion proofing" of bridges and parking structures to be between $200 and $500 million. As a result, methods that can significantly reduce or halt chloride induced corrosion have been aggressively pursued for well over 30 years.

The methods used to reduce corrosion of reinforcing steel may be divided into two categories. The first includes methods that slow the initiation of corrosion, that is the time it takes the chlorides to reach the reinforcing steel in the concrete. The second category includes methods that lengthen the corrosion period, the time between initiation of corrosion and the end of service life. In most cases, the effectiveness of the methods used to reduce corrosion are evaluated using one or more laboratory tests. The techniques are then applied in practice.

The costs required to implement these techniques vary, with ratios as large as six to one (W. R. Grace 1997), and there is little evidence of a quantitative correlation between the performance of the techniques in the laboratory and in the field. The lack of a direct correlation is due largely to the evolving nature of the laboratory tests and the need to innovate in practice without waiting for the development of a correlation or, in some cases, waiting for lab results that may take several years. This strategy is often justified because, even with the large differences in cost, measures to prevent damage are invariably less expensive than the total replacement of a bridge deck or bridge substructure.

The goal of research at the University of Kansas is to develop a detailed correlation between accelerated laboratory tests and field performance for a broad range of corrosion protection systems. The study includes ongoing state and university field surveys in northeast Kansas evaluating the performance of bridge decks (Schmitt & Darwin 1998, Miller & Darwin 2000), and is being carried out in partnership with the National Science Foundation, the Kansas Department of Transportation, and the manufacturers of several corrosion protection systems.

BACKGROUND

The Problem

Reinforcing steel embedded in concrete is normally in a passive or noncorrosive condition due to the high pH of the concrete pore solution. Passivation involves the formation of a tightly adhering iron-oxide layer on the reinforcing bar surface, which protects the iron from corrosion. To obtain a passive condition, the pH must be between 12.5 and 13.8 (Jones 1992). If the pH of the concrete pore solution is lowered, the iron-oxide layer becomes unstable and corrosion will occur.

The pH of concrete can be lowered in two ways: by carbonation, due to the penetration of CO_2 into the concrete; or indirectly, by the presence of aggressive ions, like Cl^-, found in deicing salts and sea water. Chloride ions also weaken areas of the iron-oxide layer, allowing chloride ions to react with available iron cations on the bar surface to form an iron-chloride complex. In the presence of hydroxyl ions, the iron-chloride reacts to form ferrous hydroxide and releases the chloride ions, which, in turn, again react with available iron cations. As a result, the passive iron-oxide layer is dissolved, initiating corrosion.

Laboratory Tests

During the past two decades, a number of laboratory test techniques have been developed to provide a realistic model for the corrosion of reinforcing steel in both uncracked and cracked concrete. Bench-scale tests, such as the Southern Exposure, ASTM G 109, and Cracked Beam tests, are used most often. Although these tests typically require one to two years for completion, they qualify as accelerated tests, considering that the service life of actual structures should be 30+ times as long. More recently, modified versions of bench-scale tests, in which chlorides are diffused into the concrete in the presence of an electrical potential, have also been used. In addition to the bench-scale tests, a number of truly accelerated tests have been developed over the past decade that show promise of indicating the viability of corrosion protection systems in much shorter time periods. The main aspects of the bench-scale and rapid corrosion tests are described next.

Bench-Scale Tests

The specimen used in the Southern Exposure, or SE, test (Pfeifer & Scali 1981) consists of a small slab containing two mats of reinforcing steel (Figure 1a). The concrete is wet cured for three days and then air cured until the test begins at 28 days. The top mat consists of two bars; the bottom mat consists of four bars. The mats are electrically connected across a resistor (typically 10 ohms), a dam is placed around the edge of top surface, and the sides of the concrete are sealed with epoxy. A 15 percent sodium chloride solution is placed inside the dam, allowing the chlorides to penetrate into the concrete. The slabs are subjected to a seven day alternate ponding and drying regime, with ponding at 22°C (72°F) for four days and drying at 38°C (100°F) for three days. Corrosion current and the corresponding corrosion rate are determined by measuring the voltage drop across the resistor. The test provides a very severe corrosion environment and is generally believed to simulate 15 to 20 years of exposure for marine structures and 30 to 40 years of exposure for bridges within a 48-week period (Perenchio 1992). In recent practice, the period of exposure has been extended to 96 weeks to obtain an improved picture of the performance of the corrosion reducing technique under evaluation.

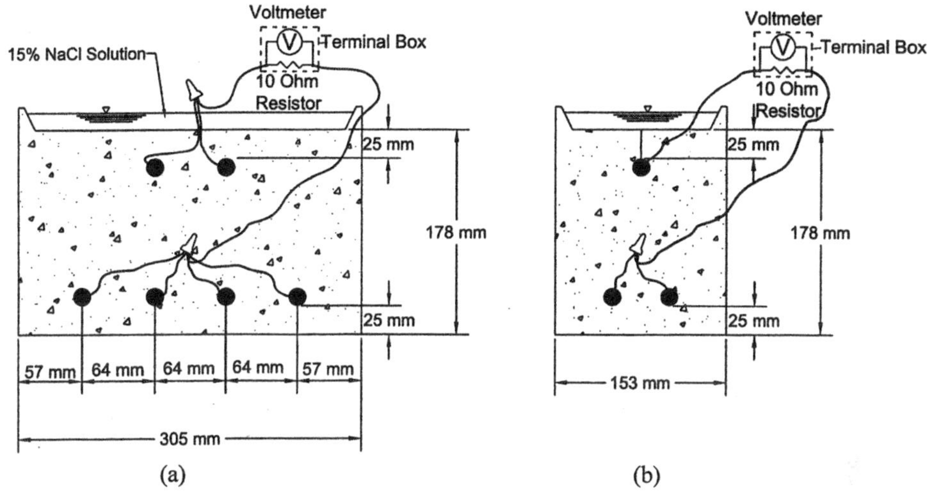

Figure 1 Test specimen for (a) Southern Exposure Test and (b) Cracked Beam Test

A similar specimen (ASTM G 109) was developed to evaluate the effect of chemical admixtures on the corrosion of reinforcing steel. The G 109 specimen is a little less than half the width and 25 mm shorter than the SE specimen, with one reinforcing bar on the top and two reinforcing bars on the bottom. A 3 percent, rather than 15 percent, solution is used, and the 10 ohm resistor is replaced by a 100 ohm resistor. The ponding and drying periods are two weeks each, and temperature is maintained at 23°C (73°F) throughout the evaluation period, which often lasts up to 96 weeks (Berke et al. 1990).

Two versions of the cracked beam specimen are used to model the corrosion of reinforcing steel in cracked concrete. The sizes of the specimens vary, but are, in general, similar to that illustrated in Figure 1b, half the width of the SE specimen, with one bar on the top and two bars on the bottom. "Cracks" are placed in the concrete either perpendicular or parallel to the reinforcing steel, often using a thin stainless steel spacer that is inserted during specimen fabrication. The spacer is removed after the concrete stiffens, leaving a direct path to the reinforcing steel. A dam is placed around the specimen in a manner similar to that used for the SE and G 109 specimens. Like the SE specimen, the cracked beam specimen is subjected to cycles of wetting and drying with a 15 percent chloride solution, continuing up to 96 weeks.

The SE, cracked beam and, to a lesser degree, the G 109 specimens have a drawback in that the wetting-drying cycles can result in high salt concentrations throughout a specimen. High chloride concentrations near the bottom steel will reduce the measured macrocell current (the principal output of these specimens), thus reducing the usefulness of the results. At the same time, microcell corrosion may be quite high, but go unmeasured. This drawback can be overcome using linear polarization and electrochemical impedance techniques. The test methods can also be modified by using an electrical potential to diffuse chlorides into the specimen until the concentration reaches the desired level, without wetting-drying cycles.

Rapid Tests

With the goal of developing more rapid tests for early evaluation of corrosion protection techniques, "lollipop" style specimens have been in use for a number of years. These specimens, such as illustrated in Figure 2, simulate the cementitious environment of reinforcing steel by casting a thin layer of concrete or, in most cases, mortar around a short reinforcing bar. While "lollipop" specimens have been used for many years, work began in 1989 at the University of Kansas under the SHRP program (Martinez et al. 1990, Chappelow et al. 1992) and continued under the NCHRP-IDEA Program (Smith et al. 1995, Senecal et al. 1995, Schwensen et al. 1995, Darwin et al. 1996) with the goal of using the technique to obtain a realistic measure of the performance of corrosion protection systems in a short period of time. [Note, the goal of the NCHRP-IDEA study was to evaluate a new corrosion-resistant steel.] The tests allow very rapid evaluation of both the corrosion potential and the formation of a corrosion macrocell. The basic test specimen is illustrated in Figure 2. The contact surface between the mortar and the bar simulates the contact obtained between concrete and reinforcing bars in actual structures through the use of realistic water-cement and sand-cement ratios.

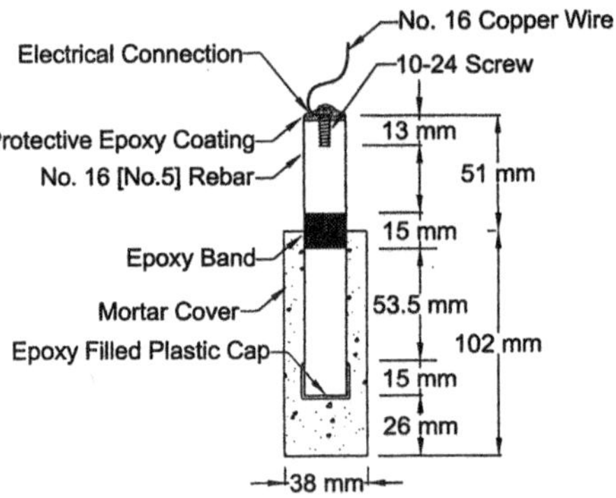

Figure 2 Cross-Section of Test Specimen Used for Rapid Corrosion Potential and Macrocell Tests

The corrosion potential test (Figure 3) requires two containers. The test specimen is placed in a 5 liter container, along with crushed mortar fill and a simulated concrete pore solution containing a preselected concentration of sodium chloride. A standard Calomel reference electrode is placed in a separate container, along with a saturated potassium chloride solution. The two containers are connected by a salt bridge and the potential (voltage) of the steel with respect to the Calomel electrode is measured at selected time intervals using a digital voltmeter. This voltage is called the corrosion potential of the steel.

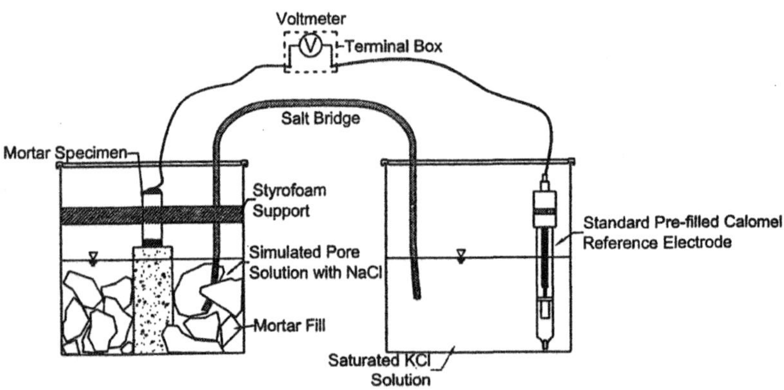

Figure 3 Schematic of Corrosion Potential Test

The mortar fill consists of the same mixture as used in the test specimen. The fill is used primarily to serve as a buffer and to help simulate the relative amount of cementitious material that exists in the actual structure. The simulated pore solution represents the liquid in the saturated pores and capillaries in concrete (Farzammerhr 1985, Farzammerhr et al. 1987). Together with the mortar fill, it helps establish a realistic environment to measure the progress of corrosion of reinforcing steel. The salt bridge allows for the completion of the corrosion cell at the time the corrosion potential is measured.

To obtain a rapid measure of the degree of corrosion that occurs through the formation of a macrocell, the corrosion potential test is modified as shown in Figure 4. The container with the Calomel electrode is replaced by another container with two standard specimens surrounded by mortar fill and immersed in a simulated pore solution (with no chlorides added). The test specimen in the pore solution containing sodium chloride (anode) is electrically connected through a single 10-ohm resistor to the two specimens in the simulated pore solution (cathode). The macrocell test specimen is completed by a salt bridge that connects the liquid in the two containers. Air (scrubbed to remove CO_2) is bubbled into the liquid surrounding the cathode to insure an adequate supply of oxygen. The air causes some evaporation, which is countered by adding deionized water to the container to maintain a constant volume of the solution. The corrosion current and the rate of corrosion can be determined by measuring the voltage drop across the resistor. Different concentrations of sodium chloride can be used at the anode.

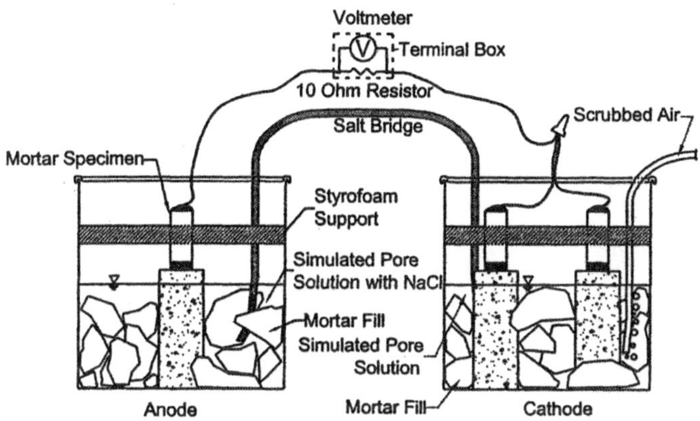

Figure 4 Schematic of Macrocell Test

Other than in the corrosion potential test shown in Figure 3, corrosion performance is normally evaluated by measuring the voltage drop across the resistor in the test specimens (Figures 1a, 1b, and 4). The voltage drop is then converted to a corrosion current, which is, in turn, used to determine the metal loss or corrosion rate. The techniques simulate the corrosion that occurs in a reinforced concrete structure that is being penetrated by chlorides. Some reinforcing bars (the top mat in a bridge deck) are exposed to higher chloride concentrations than others (bottom mat in the deck). The difference in chloride concentration not only results in a loss of passivity by the reinforcing bar but modifies the local environment of the reinforcing bars to increase the difference in electrical potential between the two layers of steel, increasing the rate of corrosion.

A comparison of results for Southern Exposure tests to 48 weeks and macrocell tests to 100 days is shown in Figures 5a and 5b (Senecal et al. 1995, Darwin et al. 1996). The figures show the evaluation of two conventional steels, H and T, and two corrosion-resistant steels, CRSH and CRST. The two figures show that both tests indicate that the CRST steel is significantly more corrosion-resistant than the other three steels.

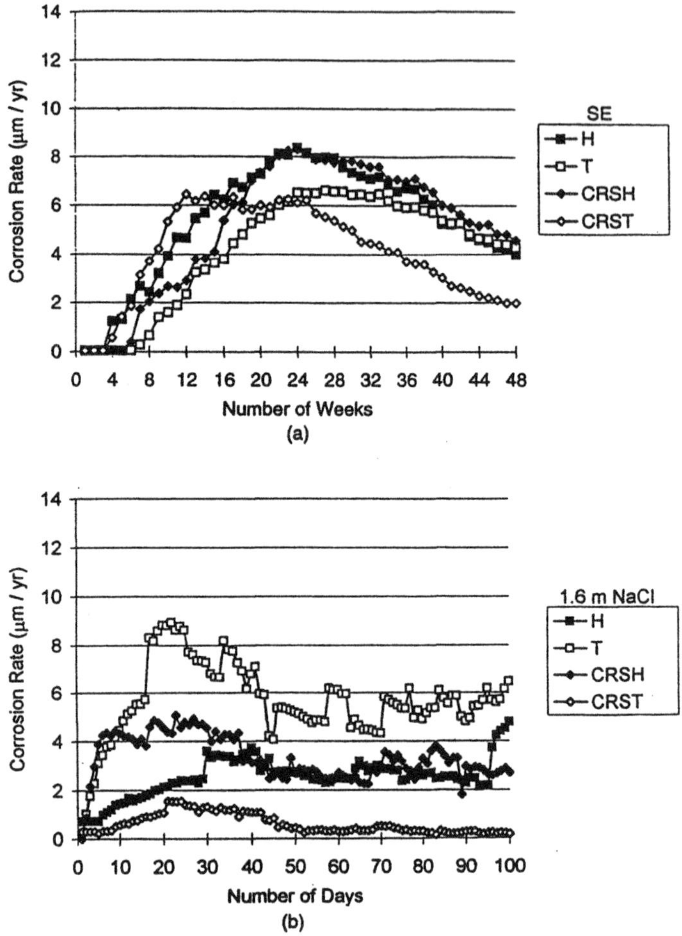

Figure 5 Corrosion Rate versus Time for (a) Southern Exposure Test and (b) Macrocell Test Comparing Conventional (H and T) and Corrosion Resistant (CRSH and CRST) steels (Senecal et al. 1995, Darwin et al. 1996)

Other Techniques

The key drawback of using corrosion potential to evaluate to corrosion performance is that it only measures the tendency of a metal to corrode for a particular exposure condition, not the actual corrosion rate. A good example is provided by Krauss and Nmai (1996) who observed that steel embedded in concrete containing a corrosion inhibiting admixture exhibited a high negative corrosion potential, but no measurable corrosion. As a result, macrocell current is generally considered to be a superior measure of corrosion performance. However, using macrocell current as the sole measure of corrosion has its own drawbacks, since the procedure does not detect microcell corrosion and is reduced in bench-scale tests as chlorides penetrate to the bottom mat of steel, equalizing the corrosion potential of the two layers of bars.

A better measure of total corrosion can be obtained using polarization-resistance measurements, which use a noncorroding counter electrode and a reference electrode (shown schematically in Figure 6) to establish a polarization curve by imposing a range of potentials on the metal and measuring the corresponding corrosion currents using a potentiostat. A portion of the polarization curve will exhibit a linear relationship. The slope of the linear region is proportional to the resistance of the metal. The corrosion current density is then obtained using the relationship

$$i_{corrosion} = B/R_p \tag{1}$$

in which i is the corrosion current density (amps/cm^2), B is a constant (determined to be 26 mV for reinforced concrete), and R_p is the slope determined from the polarization curve (kΩ·cm^2). Although polarization-resistance measurements are not often used with the bench-scale tests shown in Figures 1a, 1b, and 4, they have been used for laboratory studies of basic corrosion mechanisms and are used with increasing frequency in the field (Escalante et al. 1986, Escalante & Ito 1990, Feliú et al. 1990, Broomfield 1996), suggesting that the bench-scale tests can be improved through application of the technique.

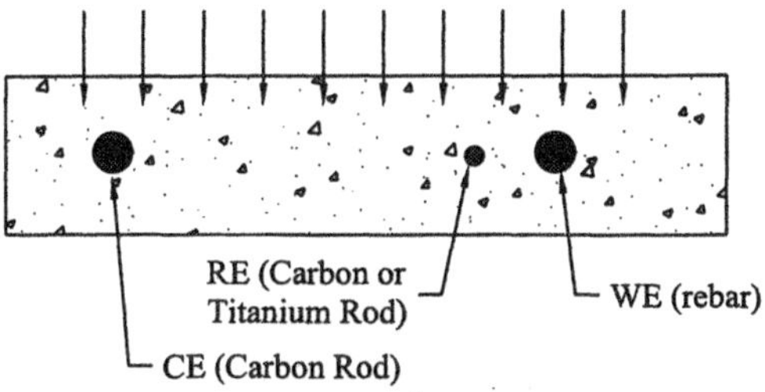

Figure 6 Schematic Showing Additional Electrodes for Polarization-Resistant and Electrochemical Impedance Spectroscopy Tests

Another particularly useful technique is that of electrochemical impedance spectroscopy. The test method uses a potentiostat, like the polarization resistance method, but instead of using direct current, alternating current is applied to the system. This is done to obtain more mechanistic information about the system. The different constituents that make up reinforced concrete may be thought of as a network of capacitances and resistances (Berke & Hicks 1990). By applying a variable current to the reinforced concrete, various constituents within the concrete may be isolated and quantified for their respective effect on resistance to corrosion. For example, the resistance due to an epoxy coating on a reinforcing bar or the reinforcing bar itself may be directly measured, whereas, with the polarization resistance method, the resistance of the coating and the concrete are measured together and are indistinguishable. The technique allows the various contributing components that occur simultaneously, such as the electrolyte resistivity, the corrosion active surface area, and the corrosion rate, to be measured separately.

Other, simpler, techniques are used in the evaluation of specimens such as those in Figures 1 and 2, which include determination of the corrosion potential of individual layers of bars using a copper-copper sulfate electrode (CSE). This allows determination of the relative degree of corrosion in the two layers of bars. Measurement of the resistance between the two bar layers (the mat-to-mat resistance) provides additional data on the progression of the corrosion process.

Field Evaluations

The corrosion of reinforcing bars in structures has historically been evaluated in several ways. Most commonly, the observation of concrete delamination has provided a measure of the extent of the corrosion of reinforcing bars. This technique, unfortunately, involves the observation of a severely damaged structure, which usually entails significant rehabilitation.

The next (improved) level of evaluation involves the determination of the corrosion potential of the reinforcing bars. Using the techniques described in ASTM C 876, a copper-copper sulfate reference electrode is used to determine the variation of corrosion potential by connecting to a reinforcing bar within the structure and moving the electrode across the surface. A difference in potential of –350 mV with respect to the CSE indicates that active corrosion may be under way. As described earlier, the key drawback of this widely-used technique is that corrosion potential alone does not provide a measure of the rate of corrosion.

Because of this drawback, polarization resistance and electrochemical impedance techniques have seen expanding field application (Escalante & Ito 1990, Broomfield 1996). A major concern with polarization resistance measurements in the field is to limit the portion of the corroding (working) electrode being measured. This is generally handled through the use of an external counter electrode or "guard ring," as shown in Figure 7. With application of a guard ring, useful comparisons have been obtained (Feliú et al. 1990). It should be noted, however, that corrosion rate measurements are highly dependent upon the conditions existing at the test time and can be dependent upon changing moisture, temperature, and chloride contents (Berke 1996, Feliú et al. 1996). This requires that readings be taken on a regular basis under varying ambient conditions. Application of the galvanostatic pulse technique (Newton & Sykes 1988, Elsener & Böhni 1990) provides additional information that is useful in interpreting the corrosion state of reinforcing steel. The technique, which involves the application of a short anodic pulse, applied through a small counter electrode, can determine the corrosion state of the bars as well as provide a measure of the concrete resistivity (a function of the concrete humidity and salt content).

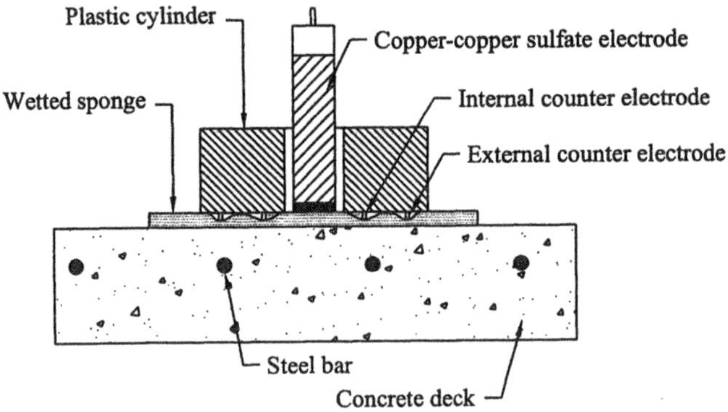

Figure 7 Schematic of Corrosion Rate Device with Guard Ring

Field and Laboratory Correlation

While a number of test specimens and evaluation techniques are available, direct correlations between the performance of corrosion protection systems in the laboratory and in the field are rare. Even the measures of performance used most often (corrosion rate in the lab, corrosion potential in the field) differ markedly. The current study aims to compare the performance of corrosion protection systems as evaluated in the laboratory and in the field and to establish those accelerated testing techniques that give the best match with field performance.

CURRENT RESEARCH

The principal goal of the current study is to establish accelerated corrosion testing techniques for use in the evaluation of concrete reinforcing steel protection systems that provide comparable results to the long-term performance of bridge structures. The study includes the evaluation and modification of laboratory test procedures to establish those that provide the best match with the corrosion behavior of reinforced concrete bridge decks subjected to normal and accelerated exposure. The corrosion evaluation techniques address the protection provided by epoxy-coated bars, corrosion inhibiting admixtures, corrosion-resistant steel, the effects of various deicers, and modifications in concrete mix proportions. Field performance is compared based on bridges selected from a group of 69 that have been evaluated under two research studies by the University of Kansas (Schmitt & Darwin 1998, Miller & Darwin 2000), five bridges scheduled for construction during the period of the proposed work, and field test specimens constructed in conjunction with the five new bridges. A series of laboratory test specimens will also be fabricated on the job sites using the materials used for the new bridge construction. The study is being carried out in concert with the Kansas Department of Transportation and the producers of several corrosion protection systems.

Laboratory Tests

The study includes "standard" versions of the Southern Exposure test (Figure 1a), the cracked beam test (Figure 1b) and G 109 test, as well as bench-scale specimens with electrically diffused chlorides and the rapid corrosion potential (Figure 3) and macrocell (Figure 4) tests. Although the Southern Exposure and G 109 test specimens differ principally in size, the tests differ markedly in the exposure conditions (chloride concentration and temperature history). As described earlier, these bench-scale test specimens evaluate the performance of corrosion protection systems principally based on the macrocell current between the top and bottom layers of steel. Additional information can be obtained from these specimens using linear polarization and electrochemical impedance techniques.

Field Tests

Field tests in the study involve work on several levels. First, five existing bridges, selected from a group of 69 being monitored by the University of Kansas, are being evaluated for current corrosion performance. The data base on these bridges includes material properties, construction history, crack profiles, and overall condition. Concrete cores have been taken to determine the permeability of the bridge decks, measured using ASTM C 1202/AASHTO T-277, the so-called Rapid Chloride Permeability Test. Additional samples are used to establish the chloride concentrations through the deck, with special emphasis on the values at the levels of the top and bottom mats of reinforcing steel. Corrosion performance is measured based on both half-cell potentials (ASTM C 876) and polarization resistance/electrochemical impedance measurements. Measurements are taken with the goal of capturing corrosion response under the full range of moisture and temperature conditions for each structure. The results of the field evaluations will be used to develop matching laboratory specimens (permeability and chloride concentration) subjected to both simulated field and accelerated corrosion conditions.

At a second level, field tests involve the instrumentation of newly constructed bridge decks in which electrical connections for obtaining corrosion measurements are preinstalled on the reinforcing steel. The bridge decks will be constructed with different corrosion protection systems. Parallel with construction of the bridge decks, "field test platforms" will be fabricated using the same geometry, reinforcing steel, concrete, and corrosion protection systems as used in the bridge decks. These 1.3 x 2.6 m (4 x 8 ft) sections will be subjected to both realistic corrosion environments, matching those of the bridge, and aggressive corrosion environments matching, to the degree possible, those obtained in a laboratory. Bench-scale test specimens will also be fabricated in the field using the same materials as used in the actual structures and field test platforms. Laboratory results will be compared with field performance for a minimum of eight years.

Observations

The following observations are based on the work completed during the first 18 months of the study.

1. The corrosion of reinforcing steel appears to be a cathode-controlled process.

2. Some microalloyed steels appear to provide improved corrosion resistance when compared to conventional reinforcing steel.

3. To date, the alternate deicing chemical, calcium magnesium acetate, is significantly less corrosive than sodium chloride.

4. The G 109 specimen shows little corrosion response during the first year, while the Southern Exposure and cracked beam specimens provide response starting in the first few weeks and major response by 26 weeks.

5. Linear polarization resistance techniques applied to bare steel in a simulated pore solution and specimens with mortar coatings do not consistently predict the corrosion performance of reinforcing steel cast in bench-scale laboratory specimens. Some microalloyed steel provide improved corrosion

resistance when cast in concrete, a resistance that is not indicated by polarization resistance measurements.

SUMMARY

Corrosion of reinforcing steel represents the major durability problem in reinforced concrete bridge structures in the U.S. The need for improving the corrosion resistance of reinforced concrete, especially in conditions of chloride exposure, has been apparent for many years. In response, a number of protection methods have been developed and are currently in use. The protection systems were developed in the laboratory without a direct correlation to actual field performance. Thus, full-scale structures become the real test specimens. Upon completion, the current research will provide direct correlation between the performance of a variety of corrosion protection systems in the laboratory and in the field (and as a "side" benefit, provide a large-scale comparison between the systems). That direct correlation will allow more effective and rational evaluation techniques to be instituted. The final result will be more efficient selection procedures, more rapid movement of corrosion technology into practice, the extension of structure service life and the overall reduction in the repair and rehabilitation costs associated with chloride induced corrosion in reinforced concrete structures.

REFERENCES

ASTM C 876-91-(1991). Standard Test Method for Half-Cell Potentials of Uncoated Reinforcing Steel in Concrete. *2000 Annual Book of ASTM Standards*, **4.02**, American Society for Testing and Materials, West Conshohoken, PA.

ASTM G 109-92. (1992). Standard Test Method for Determining the Effects of Chemical Admixtures on the Corrosion of Embedded Steel Reinforcement in Concrete Exposed to Chloride Environments. *2000 Annual Book of ASTM Standards*, **3.02**, American Society for Testing and Materials, Philadelphia, PA.

Berke, N. S., Shen, D. F., and Sundberg, K. M. (1990). Comparison of the Polarization Resistance Technique to the Macrocell Corrosion Technique. *Corrosion Rates of Steel in Concrete*, Berke/Chaker/Whiting, Eds., ASTM STP 1065, American Society for Testing and Materials, Philadelphia, PA, 38 51.

Berke, N. S., and Hicks, M. C. (1990). Electrochemical Methods of Determining the Corrosivity of Steel in Concrete. *Corrosion Testing and Evaluation: Silver Anniversary Volume*, Babraiam/ Deam, Eds., ASTM STP 1000, 425-440.

Berke, N. S. (1996). Overview, *Techniques to Assess the Corrosion Activity of Steel Reinforced Concrete Structures*. ASTM STP 1276, Berke/Escalante/Nmai/Whiting, Eds., American Society for Testing and Materials, vii-ix.

Broomfield, J. P. (1996). Field Measurement of the Corrosion Rate of Steel in Concrete Using a Microprocessor Controlled Unit with a Monitored Guard Ring for Signal Confinement. *Techniques to Assess the Corrosion Activity of Steel Reinforced Concrete Structures*, Berke/Escalante/Nmai/ Whiting, Eds., American Society for Testing and Materials, West Conshohoken, PA, 91-106.

Cady, P. D., and Gannon, E. J. (1992a). *Condition Evaluation of Concrete Bridges Relative to Reinforcement in Concrete,* **1**, State of the Art of Mixing Methods, SHRP-S/FR-92-103; Strategic Highway Research Program, National Research Council, Washington, D.C.

Chappelow, C. C., McElroy, A. D., Blackburn, R. R., Darwin, D., deNoyelles, F. G., and Locke, C. E. (1992). *Handbook of Test Methods for Evaluating Chemical Deicers*, Strategic Highway Research Program, Nat. Res. Council, Washington, D.C.

Darwin, D., Locke, C. E., Senecal, M. R., Schwensen, S. M., Smith, J. L. (1996). Corrosion-Resistant Steel Reinforcing Bars, *Materials for the New Millennium*, K. P. Chong, Ed., ASCE, Reston, VA, 482 491.

Elsener, B., and Böhni, H. (1990). Potential Mapping and Corrosion of Steel in Concrete. *Corrosion Rates of Steel in Concrete*, Berke/Chaker/Whiting, Eds., ASTM STP 1065, American Society for Testing and Materials, Philadelphia, PA, 143-156.

Escalante, E., Whitenton, E., and Qiu, F. (1986). *Measuring the Rate of Corrosion of Reinforcing Steel in Concrete – Final Report*, NBSIR 86-3456, National Bureau of Standards, Gaithersburg, MD, Oct.

Escalante, E., and Ito, S. (1990). Measuring the Rate of Corrosion of Steel in Concrete. *Corrosion Rates of Steel in Concrete,* Berke/Chaker/Whiting, Eds., ASTM STP 1065, American Society for Testing and Materials, Philadelphia, 86-106.

Farzammehr, H. (1985). Pore Solution Analysis of Sodium Chloride and Calcium Chloride Containing Cement Pastes. *Master of Science Thesis,* University of Oklahoma, Norman, OK.

Farzammehr, H., Dehghanian, C., and Locke, C. E. (1987). Study of the Effects of Cations on Chloride Caused Corrosion of Steel in Concrete. *Revista Técnica de la Facultad de Ingeniería*, Univ. Zulia, Venezuela, **10:1**, 33-40.

Feliú, S., González, J. A., Feliú, D., Jr., and Andrade, C. (1990). Confinement of the Electrical Signal for In Situ Measurement of Polarization Resistance in Reinforced Concrete Structures. *ACI Materials Journal*, **87:5**, 457-460.

Feliú, S., González, J. A., and Andrade, C. (1996). Electrochemical Methods for On-Site Determinations of Corrosion Rates of Rebars. *Techniques to Assess the Corrosion Activity of Steel Reinforced Concrete Structures*, ASTM STP 1276, Berke/Escalante/Nmai/Whiting, Eds., American Society for Testing and Materials, West Conshohoken, PA, 107-118.

Jones, D. A. (1992). *Principles and Prevention of Corrosion,* Macmillan Publishing Company, New York.

Locke, C. E. (1986). Corrosion of Steel in Portland Cement Concrete: Fundamental Studies. *Corrosion Effects of Stray Currents and the Techniques for Evaluating Corrosion of Rebars in Concrete,* ASTM STP 906, American Society for Testing and Materials, Philadelphia, PA, 5-14.

Martinez, S. L., Darwin, D., McCabe, S. L., and Locke, C. E. (1990). Rapid Test for Corrosion Effects of Deicing Chemicals in Reinforced Concrete. *SL Report* 90-4, University of Kansas Center for Research, Inc., Lawrence, KS.

Miller, G. G. and Darwin, D., (2000). Performance and Constructability of Silica Fume Bridge Deck Overlays. *SM Report* No. 57, University of Kansas Center for Research, Inc., Lawrence, KS.

Newton, C. J, and Sykes, J. M. (1988). A Galvanostatic Pulse Technique for Investigation of Steel Corrosion in Concrete. *Corrosion Science*, **28:11**, 1051-1074.

Perenchio, William F. (1992). Corrosion of Reinforcing Bars in Concrete. Annual Seminar, Master Builders Technology, Cleveland, OH.

Pfeifer, Donald W., and Scali, Mauro J. (1981). Concrete Sealers for Protection of Bridge Structures. *NCHRP Report* No. 244, National Cooperative Highway Research Program, Transportation Research Board, Washington, D.C.

Schmitt, T. R. and Darwin, D. (1998). Effect of Material Properties on Cracking in Bridge Decks. *Journal of Bridge Engineering*, ASCE, **4:1**, 8-13.

Schwensen, S. M., Darwin, D., and Locke, C. E., Jr. (1995). Rapid Evaluation of Corrosion-Resistant Concrete Reinforcing Steel in the Presence of Deicers. *SL Report* 95-6, University of Kansas Center for Research, Inc., Lawrence, KS.

Senecal, M. R., Darwin, D., and Locke, C. E., Jr. (1995). Evaluation of Corrosion-Resistant Steel Reinforcing Bars. *SM Report* No. 40, University of Kansas Center for Research, Inc., Lawrence, KS.

Smith, J. L., Darwin, D., and Locke, C. E., Jr. (1995). Corrosion-Resistant Steel Reinforcing Bars Initial Tests. *SL Report* 95-1, University of Kansas Center for Research, Inc., Lawrence, KS.

W. R. Grace (1997). Engineering and Economic Performance Summary. *Report*, W. R. Grace & Co. Conn., Cambridge, MA.

IN-CORE LEACHING OF CHLORIDE FOR PREDICTION OF CORROSION OF STEEL IN CONCRETE

A.A. Sagüés[1], S.C. Kranc[1], L. Caseres[1], L. Li [1,3] and R.E. Weyers[2]

[1] Department of Civil and Environmental Engineering, University of South Florida, Tampa, FL 33620, U.S.A.
[2] Department of Civil and Environmental Engineering, Virginia Polytechnic Institute and State University, Blacksburg, VA 24061, U.S.A.
[3] Present address: Department of Ocean Engineering, Florida Atlantic University, SeaTech Campus, Dania Beach, FL 33004, U.S.A.

ABSTRACT

Leaching into small cavities in field-extracted concrete cores is explored as a method to sample the free chloride concentration profile in field-extracted concrete cores. If successful, this method will permit more accurate evaluation of durability of civil infrastructure subject to reinforcement corrosion. In the present phase of the work, a model for evolution of cavity water concentration with time was expanded to include linear binding of the chloride in concrete. Under the assumptions used, for a given value of apparent concrete diffusivity linear binding significantly accelerates the initial approach to equilibration between cavity and pore water. Exploratory tests with moderate permeability concrete provided encouraging indication that the in-situ leaching method was feasible for chloride ions.

KEYWORDS

Chloride, concrete, binding, leaching, corrosion, reinforcement, durability, forecasting.

INTRODUCTION

Chloride-induced corrosion of reinforcing steel in concrete is responsible for widespread premature deterioration of the national infrastructure, including, in particular, coastal structures (Ahlskog 1990). Steel is initially protected from corrosion by the highly alkaline (pH >12.5) concrete pore solution, which promotes passivity. However, in time chloride ions (from externally applied deicing salts or seawater) penetrate the concrete cover until a critical concentration builds at the steel surface and passivity breakdown ensues. Expansive corrosion products form on the steel surface, cracking and spalling the cover concrete.

Because of the magnitude of this problem, considerable attention has been given to developing accurate diagnostic techniques for forecasting the development of corrosion damage in existing structures. Particularly desirable are short-term, simple tests of field-extracted samples of concrete to accurately forecast the development of corrosion with time in existing structures.

Corrosion durability predictions for reinforced concrete are made using variations of the two-stage mechanism initially proposed by Tuutti (1982). The first or *corrosion initiation* stage corresponds to the period of time in which the concentration of chloride ions at the rebar surface builds up toward the critical value for passivity breakdown. The initiation stage ends when passivity breaks down, at which time the *corrosion propagation* stage begins and significant corrosion starts. The amount of corrosion products needed to create visible cracks in concrete is quite small, corresponding typically to the loss of ~50 μm of steel at the perimeter of the reinforcing bar (Andrade et al 1996). Since corrosion rates during the propagation stage often exceed 10 μm/year (Andrade & Alonso 1992), the propagation stage commonly lasts for only a few years before significant structural damage is observed (Cady & Weyers 1984; Weyers et al 1993; Liu & Weyers 1996).

As a consequence, design improvements for durability have focused on extending the length of the initiation stage of corrosion (Berke & Hicks 1992), which can be achieved by the use of thicker concrete cover and concrete resistant to chloride penetration. As present U.S. highway design service life goals call for 75-year repair free design (AASHTO 1992), the initiation stage will encompass much of the lifetime of the structure. The project described in this paper concerns the development of methodology to aid in the accurate determination of the length of the initiation stage in existing structures.

Most present methods to estimate the duration of the initiation stage are based on simplifying assumptions of the mechanism of transport of chloride ions in concrete and on nominal estimates of the critical chloride concentration for passivity breakdown, C_C (unless otherwise indicated all concentrations in the following are expressed as mass per unit volume of concrete).

Chloride transport is usually assumed to proceed by simple diffusion so that:

$$\partial C_T/\partial t = \partial (D_A \partial C_T/\partial x)/ \partial x \tag{1}$$

where C_T is the total chloride concentration, x the distance from the concrete surface, and D_A the apparent diffusivity of chloride ions in concrete. For cases where D_A and chloride concentration C_S at the concrete surface are approximately constant in time, the concentration at any position and time approaches the relationship (Weyers et al 1993; Berke & Hicks 1992):

$$C_T(x,t) = C_s\,(1\text{-erf}(x/2(D_A\, t)^{1/2})) \tag{2}$$

If a concrete core is extracted from an existing structure of age t_A, and cut into slices corresponding to various midpoint distances x_i from the external surface, analysis of the chloride concentration from each slice yields a value of $C_T(x_i,t_A)$ for each x_i. By fitting the set of value pairs to Eqn. 2 one obtains estimates of D_A and C_S. If the value of C_C is known, use of Eqn. 2 together with D_A and C_S yields the length t_I of the initiation stage, so that the residual initiation period is t_I-t_A. Extensions of the same method account also for cases in which there is an initial chloride concentration in the concrete.

The simplified method to determine t_I described above fails to take into consideration a number of important complicating factors and uncertainties that affect actual structures (Saetta et al 1993). Among those complications, chloride ion binding (Nilsson et al 1994; Rasheduzzafar et al 1991; Sagüés & Kranc 1996; Mangat & Molloy 1995; Jensen & Pratt 1989; Tritthart 1989) is particularly significant because it retards chloride penetration and can cause the actual chloride ion content of the pore solution to be very different from the total content determined by analysis of the concrete slices. Chloride ions in concrete (with a total concentration C_T) can be considered to be partitioned into a free chloride portion C_F that corresponds to ions present in the pore solution, and a bound chloride portion C_B sorbed or chemically combined in the paste and aggregates. The isotherm relationship between C_F and C_B is usually non linear (Sagüés & Kranc 1996), and as a result of binding the chloride transport equation takes the more complicated form (Nilsson et al 1994; Sagüés & Kranc 1996):

$$\partial C_F/\partial t = \partial \left(D_F \left(1+dC_B/dC_F\right)^{-1} \partial C_F/\partial x\right)/\ \partial x \quad (3)$$

where D_F is the diffusion coefficient for free chloride and the concrete has been assumed to have uniform porosity and all properties invariant with time. The value of D_F is equal to the D_A in Eqn. 1 only in the absence of binding. In the ideal case of linear binding ($C_B = k\ C_F$) it can be shown that $D_A = D_F/(1+k)$. In all other cases there is not a simply defined value for D_A and Eqn. 1 is only roughly satisfied. Likewise, solution of Eqn. 3 no longer takes the simple form of Eqn. 2 even under the other assumptions used there. The next generation of durability forecasting approaches will necessitate an accounting for these issues.

Knowledge of the distribution of total and free chloride in concrete in a structure under actual field conditions would yield the operating binding isotherm, and significantly advance the ability to forecast chloride penetration and consequently service life. However, determination of a binding isotherm is usually a complicated experimental process, especially when using actual concrete samples instead of more easily characterized cement paste or mortar systems (Nilsson et al 1994). Moreover, information from laboratory systems may not apply directly to the conditions determining transport in the more complex field environment. This investigation explores the application of a novel straightforward method for analysis of field extracted concrete cores, to obtain chloride distribution and binding information directly applicable to durability forecasting.

CONCEPT INVESTIGATED

The basis of the test approach is the application of the technique of in-situ leaching of concrete, which has been recently explored by Sagüés et al (1998) and more recently by Li et al (1999). A small diameter hole (or similar small volume cavity) is opened in the concrete by drilling, cutting or a comparable technique. A small amount of water is introduced in the cavity and allowed to equilibrate with the surrounding pore solution over an extended period of time (days to weeks), taking adequate precautions to avoid carbonation or other artifact reactions. The composition of the cavity water is easily evaluated in-situ or ex-situ by well established micro analytical techniques. Analysis of pore water solution extracted with conventional pore water expression (Sagüés et al 1998; Li et al 1999) is in reasonable agreement with the in-situ leachate results. Preliminary mathematical modeling of the system yielded projections of the time for equilibration approach in agreement with the observed behavior (Sagüés et al 1998). This modeling also indicates that useful tests can be conducted in the time frame of interest with practicl cavity dimensions and liquid contents.

In the in-core leaching sampling procedure being developed, a concrete core is extracted from the structure for which a durability forecast is to be made. The core is allowed to approach water saturation in a 100% relative humidity chamber. Cavities of appropriate dimensions are drilled or otherwise made along the side at regularly spaced distances x from the external surface end of the core, and partially filled with distilled water. After a satisfactory approach to equilibrium has been obtained, the water in the cavities is analyzed to determine chloride content (proportional to C_F) and pH as a function of x. A matching core is sliced and analyzed to obtain total chloride (C_T) also as function of x. The C_T and C_F

values are correlated to obtain C_B and the effective isotherm C_B (C_F). Numerical solution of Eqn. 3, with the appropriate boundary conditions yields the $C_T(x,t)$ projection for future times, including the concentration evolution at the depth where the reinforcing steel is located. The results can then be used in conjunction with chloride ion transport models and a statistical treatment of the incidence of steel depassivation to obtain a more accurate estimate of t_I, thus providing the basis for a rational damage function development model.

The objective of the present exploratory investigation was to address the feasibility of the in-situ leaching method for chloride ions in concrete, by conducting (1) detailed modeling and experimental confirmation of the kinetics of leaching of the relevant species into small concrete cavities and (2) initiate application of the procedure to concrete cores extracted from actual structures in service. Progress achieved during the early phases of this work is described below.

KINETICS OF LEACHING IN CONCRETE CAVITIES

The success of the concept depends on the development within the cavity, in a reasonable time, of a chemical composition representative of that of the pore solution. Early modeling work to estimate the length of time needed to approach equilibration addressed only the one-dimensional case of cylindrical systems with no binding. Those results (Sagüés et al 1998) indicated that the time needed to achieve near equilibrium is proportional to $D^{-1}\ r_C^{\ 2}$ and roughly proportional to $\varepsilon^{-2}\ V_r^{-2}$, where r_C is the cavity radius, D the diffusion coefficient (binding absent) of the leached species, ε the porosity of the concrete and V_r the ratio of volume of the cavity to volume of water inside it. This analysis permitted semiquantitative projections of the time for equilibration in the more complex systems of interest. Those projections together with experimental results showed encouragingly short times for equilibration for ions responsible for pH development. Additional work with nitrite ions showed also reasonable short times to approach equilibrium (Li et al 1999). The present exploratory work addressed the application of the method to chloride ions, for which transport in concrete can be very slow (especially in high quality concrete) and complicated by binding.

The first phase of activities has addressed the effect of linear binding on the approach to equilibrium of the concentration in the cavity water. Subsequent work will concern non linear binding effects. Following the derivation by Sagüés et al (1998), it will be assumed that the pores are filled with pore water and that the porosity ϵ is constant in space and time. Furthermore, it will be assumed that equilibrium between free and bound chloride is nearly instantaneous everywhere on the time scale of the experiment. Therefore, the concentration C_F in moles per cm^3 of concrete is given by:

$$C_F = \epsilon\ C_P \tag{4}$$

where C_P is the chloride concentration of the pore water.

For simplicity, the cavity will be treated as a long cylinder of length L centered in a cylindrical specimen also of length L and radius $r_e >> r_c$ (thus ignoring transport beneath the cavity). In such one-dimensional cylindrical problem Eqn. 3 for the linear binding case can be written as:

$$\partial C_F / \partial t = [D_F /(1+k)]\ [\ \partial^2 C_F / \partial r^2\ +\ r^{-1}\ \partial C_F / \partial r] \tag{5}$$

where r is the distance from the center of the cavity. The initial concrete free chloride concentration in the specimen is C_{F0}, (corresponding to an initial total concentration C_{T0} = (1+k) C_{F0}). The cavity is assumed to be only partially filled with water, but with walls efficiently wetted by capillarity.

Transport inside the cavity is assumed to be fast enough so as to have the same chloride concentration in the water at the bottom and on the moist cavity walls.

The number of moles of chloride flowing per unit time into the cavity, dm/dt, is given by Fick's first law as:

$$dm/dt = 2\, D_F\, \pi\, r_c\, L\, (\partial C_F / \partial r)_{rc} \quad (6)$$

Calling the volume of the cavity $V_c = \pi r_c^2 L$, V_{wc} the volume of the water in the cavity and defining $V_r = V_c/V_{wc}$, then the value of the chloride concentration in the cavity water (denoted by C_W) varies with time as:

$$dC_W / dt = 2\, D_F\, r_c^{-1}\, V_r\, (\partial C_F / \partial r)_{rc} \quad (7)$$

Applying Eqn. 4 and assuming that the water in the cavity is in equilibrium with the pore water at the walls at all times:

$$(dC_F / dt)_{rc} = 2\, \epsilon\, D_F\, r_c^{-1}\, V_r\, (\partial C_F / \partial r)_{rc} \quad (8)$$

Eqn. 3 can then be solved to obtain $C_F(r,t)$ for the boundary conditions in Eqn. 8 as well as in the following:

$C_F\, (r_c, 0) = \epsilon\, C_{W0}$	(C_{W0} is the initial chloride content in the cavity)	(9a)
$C_F\, (r, 0) = C_{F0}$	($r_e > r > r_c$)	(9b)
$C_F\, (r_e, t) = C_{F0}$	(Since $r_e >> r_c$, C_F at r_e is considered to be unaffected by leaching within the time frame of the experiment)	(9c)

Equations (5) to (9) can be expressed in terms of D_A and C_T by substituting $D_A = D_F /(1+k)$ and $C_T = C_F\, (1+k)$ respectively. Substitution shows that the resulting system of equations and conditions is identical to that developed (Sagüés et al 1998) for the no-binding case, except that now ε is multiplied everywhere by a factor (1+k). Thus for a given value of D_A the presence of linear binding has the same effect, on the dynamics of approach to terminal concentration in the cavity, as increasing the porosity by a factor (1+k).

Figure 1 shows solutions to Eqns. 5 to 9 numerically calculated using finite differences. The model equations and the results can be stated in non-dimensional terms using (for the usual case of $C_{W0}=0$) the generalized concentration $C_g = C_w\, C_{F0}^{-1}\, \epsilon = C_w\, C_{T0}^{-1}\, (1+k)\, \epsilon$, the generalized time $T = t\, D_F\, r_c^{-2}\, (1+k)^{-1} = t\, D_A\, r_c^{-2}$, and the parameter $P = \epsilon\, (1+k)\, V_r$. For the no-binding case (k=0) the results correspond to the conditions addressed by Sagüés et al (1998) (the present set of curves represent a refined numerical calculation scheme and supersede those given there). Figure 1 also shows that, since P includes the term (1+k), binding accelerates the process of equilibration as measured by the evolution of C_g toward unity. This finding could be viewed as being inconsistent with the retarding effect of binding on chloride penetration in concrete (Nilsson et at 1994). The apparent discrepancy is explained by recalling that binding reduces the free chloride concentration (and consequently the terminal value of C_w) and that binding permits maintaining a high gradient of C_F near the cavity wall for a longer time than otherwise.

At low leaching times diffusional transport has significantly depleted only a very thin region around the cavity surface, so the behavior closely resembles that of diffusional leaching from a flat wall. Adapting the well known solution for that case (Crank 1975) to the present system yields $C_g = 4\ \pi^{-1/2}\ P\ T^{1/2}$. That limit is in agreement with the vertical offset and straight line behavior (at low T) with slope 1/2 of the curves in the log-log representation used in Figure 1. The accelerating effect of binding is manifested at early times as a reduction in the time needed to reach a given C_g value by a factor $(1+k)^2$, when comparing concretes with and without linear binding but having the same value of D_A.

The model results suggest that binding plays an important role in facilitating the determination of free chloride content in concrete with this method. Typical values in practice for ε, r_c and V_r are 0.1, 0.25 cm and 3 respectively. Under those conditions but in the absence of binding, a sizable fraction (e.g. C_g=0.8) of the terminal chloride content in the cavity water would only be reached only after a long time (e.g. ~1 year) if $D_A < 10^{-8}$ cm²/sec as is in most low permeability concretes (Cady & Weyers 1984; Bamforth 1994). However, chloride binding does take place normally in concrete and improves the prognosis considerably. If conditions roughly approximated linear binding with k~2.5 (a situation that could be encountered in a typical concrete when chloride content is moderate (Nilsson et at 1994; Rasheeduzzafar et al 1991), then C_g=0.8 could be achieved after only ~ 1 month for $D_A = 10^{-8}$ cm²/sec. This shorter time frame has allowed useful results in the recent work using this method with pH and nitrite ion measurements (Sagüés et al 1998; Li et al 1999). The model results indicate that (as in the no-binding case) additional acceleration can be obtained by reducing the cavity size as much as possible, since the estimated time to reach a given C_g value decreases with square of r_c.

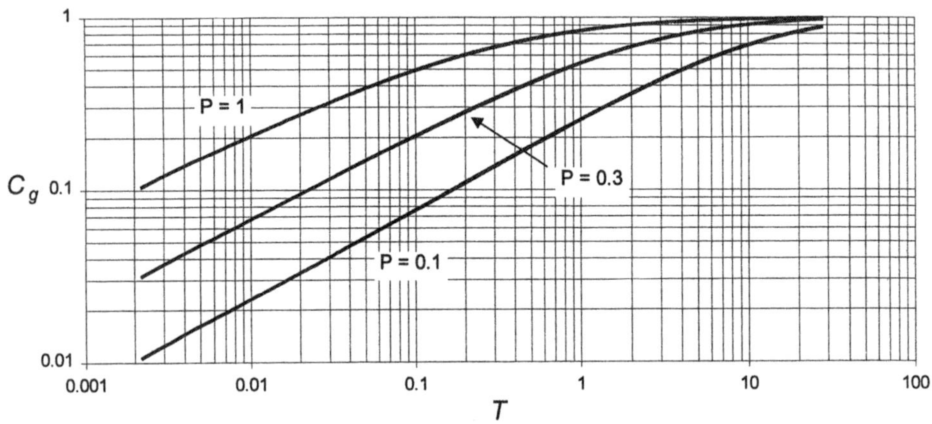

Figure 1: Non-dimensional presentation of the model results.

$$C_g = C_W\, C_{F0}^{-1}\, \epsilon = C_W\, C_{T0}^{-1}\,(1+k)\,\epsilon$$
$$T = t\, D_F\, r_c^{-2}\,(1+k)^{-1} = t\, D_A\, r_c^{-2}$$
$$P = \epsilon\,(1+k)\,V_r$$

Work is in progress at present to obtain terminal concentration estimates from limited time sequences, by extrapolation guided by the trends obtained in the model results. Modeling of the behavior with non-linear chloride binding will be addressed in the next phase of the project execution.

EXPLORATORY IN-SITU CHLORIDE LEACHING EXPERIMENTS

Experimental work in this program is in an early phase. Exploratory tests have been conducted with available specimens of concrete and mortar. In addition, measurements toward obtaining free chloride profiles have been initiated with cores extracted from the substructure of a marine bridge.

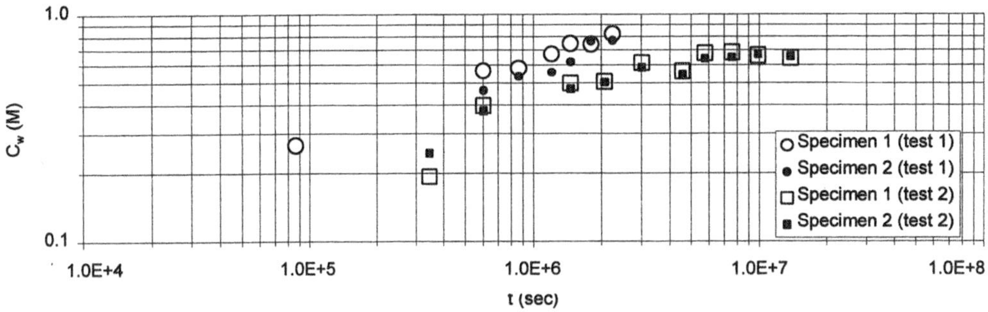

Figure 2: Results of two sequences of in-situ leaching experiments with duplicate concrete specimens.

In all cases the procedure was to place the test specimen (usually a cylinder 40-100 mm in diameter and 50-100 mm long) in a 100% relative humidity (RH) chamber for a period ranging from 2 to 6 weeks. The surface of the specimen was periodically lightly mist-sprayed. 3 to 4 small holes (3 to 6 mm diameter, 30 mm deep) were then drilled on the top of the sample, and cleaned of dust. A plexiglass cap with a rubber stopper was then attached with epoxy to the concrete around the top of each hole. Typically 0.2 to 0.4 cm^3 of distilled water were introduced in each hole. The pH of the solution in the holes was measured periodically with a microelectrode using the procedure given by Sagüés et al (1998). To determine chloride content, ~ 0.01 to 0.03 cm^3 of solution (sampling in sequence each time another hole of a given specimen) were periodically extracted with a disposable pipette and diluted into 50 cm^3 of distilled water. Sample mass was accurately determined by weighing. After acidifying with 0.5 cm^3 of 1:1 nitric acid solution, the chloride content was determined by potentiometric titration with silver nitrate with appropriate blank controls.

As an example, Figure 2 shows results from duplicate 10 cm diameter, 7 cm long cylindrical specimens of concrete made with Type I cement (388 kg/m^3), nominal water-to-cement ratio w/c=0.5, 9 mm limestone coarse aggregate and sand fine aggregate. NaCl was added to the mix to attain a total Cl^- content C_T=10.7 kg/m^3 (acid-soluble). The concrete was wet cured for ~1 month and had matured afterwards for ~ 2 years at room ambient conditions. Tests of companion concrete samples per ASTM C-642-90 indicated a porosity ε=0.14. The specimens were placed in the 100% RH conditioning chamber for 1 month, after which three test cavities with rc=0.25 cm were drilled on one of the ends of each cylinder. In the first test sequence the cavities were filled each with 0.4 cm^3 of water (V_r=1.5) and monitored for 27 days. Monitoring was discontinued for 120 days after which the cavities were found to be nearly empty, and replenished with 0.4 cm^3 of water to start the second test sequence which lasted for 170 days. The resistivity of the concrete at the end of the tests was ρ ~ 7 kΩ cm, as determined with a Wenner-array probe and appropriate correction for shape (Morris et al 1996).

Both test sequences showed similar trends in the duplicate specimens (composite of results from all three cavities in each specimen), with relatively fast initial buildup of the chloride content in the cavity water (C_W). In the second, longer sequence C_W appeared after ~ 2 months of exposure to approach a terminal value comparable to the highest values recorded during the first sequence (~ 0.8M). If the results are treated as if only linear chloride binding were taking place, the same data plotted as C_W versus $t^{1/2}$ have initial slopes which yield D_A values on the order of ~ 2 10^{-8} cm^2/sec. Although the binding isotherm is expected to be more complicated than linear, this value of D_A is consistent with the low ρ observed, per the empirical correlation between ρ and D_A reported by Berke & Hicks (1992). The results provide initial information applicable to determining the binding isotherm of this system. If all the chloride were present as free chloride in water saturated pores, the terminal value of C_W would have been expected to be ~ 2.2 M. Since C_W seemed to approach a terminal value ~ 0.8 M, the data indicate that only about 1/3 of the chloride in those samples was present as free chloride while 2/3 was bound. This partition is on the order of that noted elsewhere for concretes with comparable total chloride and cement content (Li & Sagüés 2001). Similar experiments conducted with other specimens of the same concrete but with different chloride content would have yielded complementary information to construct the operating isotherm. Experiments are now in progress to that effect using a series of concrete specimens of varying chloride content prepared at the beginning of this project.

The preliminary tests have provided encouraging indication that the method, which was previously used for pH and nitrite ion determination, is feasible also for chloride ions at least for concrete with a moderate value of D_A. The dynamics of cavity concentration evolution were also generally in agreement with the expectations from the model developed to account for binding.

Experiments have begun to assess the applicability of the method to high quality concrete having D_A values near the low end of normal practice. Preliminary tests have produced, as expected, lower rates of increase in C_W with time. Schemes including reducing r_c and increasing V_r as much as practicable are being examined to accelerate the approach to a terminal value. In addition, it is sought to develop an appropriate extrapolation model to evaluate the terminal value from a truncated data sequence. The most important experimental difficulty encountered so far is relatively rapid loss of cavity water to the surrounding concrete in some of the specimens evaluated. That loss is indicative of incomplete saturation of the specimen, which is usually unavoidable but less of a problem if C_W stabilization times are short. For concrete with low values of D_A the cavity water may require one or more replacements before a terminal value of C_W is approached. A C_W evolution model including a convective transport term is being developed to account for the effect of water loss and to obtain an estimate of C_{F0} from the data sequence.

Concurrent with the other tests, experiments are also in progress with multiple cavities drilled in the same concrete core (extracted from an actual structure in service) at different distances from the external surface, toward achieving the main objective of determining binding isotherm and comprehensive chloride transport information from the field-extracted specimen.

SUMMARY OF PROGRESS

1. A previously developed model for evolution of cavity water concentration with time was expanded to include linear binding. Early in the process and for a given value of D_A, linear binding (C_B=k C_F) reduces the time needed to reach a given fraction of the terminal pore water concentration by a factor $(1+k)^2$.

2. Exploratory tests with concrete with a moderate value of D_A provided encouraging indication that the in-situ leaching method was feasible for chloride ions. The dynamics of cavity concentration evolution were also generally in agreement with the expectations from the simplified model used.

3. Further experiments are in progress to evaluate applicability of the method to low permeability concrete and to demonstrate performance in field extracted cores.

ACKNOWLEDGEMENT

This investigation was supported by the National Science Foundation, Grant No. CMS-9872323.

REFERENCES

AASHTO-AGC-ARTBA Joint Committee, Subcommittee on New Highway Materials (1992), Manual for Corrosion Protection of Concrete Components in Bridges. Task Force 32 Report.

Ahlskog J.J. (1990). Bridge Management - The Answer to the Challenge. *Bridge Evaluation, Repair and Rehabilitation*, Ed. By A. S. Nowak, Proceedings of the NATO Advanced Workshop in Bridge Evaluation, Repair and Rehabilitation, Baltimore, MD, Kluwer Academic Publishers, Boston, MA

Andrade C.and Alonso C. (1992). Values of Corrosion Rate of Steel in Concrete in Order to Predict Service Life of Concrete Structures. *Application of Accelerated Corrosion Tests to Service Life Prediction of Materials*, STP 1194, G. Cragnolino, Ed., ASTM, Philadelphia

Andrade C., Alonso C., Rodriguez J. and Garcia M. (1996). Cover Cracking and Amount of Rebar Corrosion: Importance of the Current Applied Accelerated Tests. *Concrete Repair, Rehabilitation and Protection*, R. Dihr and M. Jones, Eds, E&FN Spon, London, pp. 263-273

Bamforth P. (1994). Admitting that Chlorides are Admitted . *Concrete Magazine*, November/December , p.18

Berke N.S.and Hicks M.C. (1992). Estimating the Life Cycle of Reinforced Concrete Decks and Marine Piles Using Laboratory Diffusion and Corrosion Data. *Corrosion Forms and Control for Infrastructure*, ASTM STP 1137, Victor Chacker, Ed., American Society for Testing and Materials, Philadelphia, p.207

Cady P.D. and Weyers R.E. (1984). Deterioration Rules of Concrete Bridge Decks. *Journal of Transportation Engineering, American Society of Civil Engineers* **110: 1**, 35 - 44.

Crank J. (1975). *The Mathematics of Diffusion*. 2nd. Ed., Oxford University Press, Oxford

Jensen H. and Pratt P. (1989). The Binding Of Chloride Ions By Pozzolanic Product In Fly Ash Cement Blends. *Adv. Cem. Rsch.* **2**, 121-129

Li L., Sagüés A.A., and Poor N. (1999). In-Situ Leaching Investigation of pH and Nitrite Concentration in Concrete Pore Solution. *Cement and Concrete Research* **29**, 315

Li L. and Sagüés A.A. (2001). Chloride Corrosion Threshold of Reinforcing Steel in Alkaline Solutions—Open-Circuit Immersion Tests. *Corrosion* **57** (in press)

Liu Y.and Weyers R. E. (1996). Time to Cracking for Chloride - Induced Corrosion in Reinforced Concrete. *4th International Symposium on Corrosion of Reinforcement in Concrete Construction: Corrosion of Reinforcement in Concrete Construction*, Cambridge, England, pp. 88-107

Mangat P. and Molloy B. (1995). Chloride Binding in Concrete Containing PFA, GBS or Silica Fume under Sear Water Exposure. *Mag. of Concr. Rsch.* **47**, 129-141

Morris W., Moreno E.I., and Sagüés A.A. (1996). Practical Evaluation of Resistivity of Concrete in Test Cylinders Using a Wenner Array Probe. *Cement and Concrete Research* **26:12,** 1779-1787

Nilsson O.L., Massa M.t, and Tang L. (1994). The Effect of Non-linear Chloride Binding on the Prediction of Chloride Penetration into Concrete Structures. *Durability of Concrete*, ACI Publication SP-145, V.M. Malhotra, Ed., American Concrete Institute, Detroit, p. 469

Rasheeduzzafar, Hussain S. and Al-Saadoun S. (1991). Effect of Cement Composition on Chloride Binding and Corrosion of Reinforcing Steel in Concrete. *Cement and Concrete Research* **21**, 777-794

Saetta A., Scotta R.and Vitaliani (1993) R.. Analysis of Chloride Diffusion into Partially Saturated Concrete. *ACI Materials Journal*, **47**, 441-451

Sagüés A. A. and Kranc S.C. (1996). Effect of Structural Shape and Chloride Binding on Time to Corrosion of Steel in Concrete in Marine Service. *Corrosion of Reinforcement in Concrete Construction,* C.L. Page, P.B. Bamforth and J.W. Figg, Eds, The Royal Society of Chemistry, Cambridge, p. 105-114

Sagüés A.A., Moreno E.I. and. Andrade C. (1998). Evolution of pH During In-Situ Leaching in Small Concrete Cavities. *Cement and Concrete Research* **27**, 1747

Tritthart J. (1989). Chloride Binding in Cement, II. *Cement and Concrete Research* **19**, 683-691

Tuutti K. (1982). *Corrosion of Steel in Concrete* (ISSN 0346-6906), Swedish Cement and Concrete Research Institute, Stockholm

Weyers R. et al (1993). *Concrete Bridge Protection, Repair and Rehabilitation Relative to Reinforcement Corrosion: A Methods Application Manual*, SHRP-S-360, National Research Council, Washington, DC

Polymeric and Composite Materials

Long Term Durability of Structural Materials
P.J.M. Monteiro, K.P. Chong, J. Larsen-Basse, K. Komvopoulos (Eds)

Enviro-Mechanical Durability of Polymer Composites

K. Verghese[1], J. Haramis[2], S. Patel[1], J. Senne[1], S. Case[1], and J. Lesko[1]

[1] Department of Engineering Science and Mechanics
[2] Department of Civil and Environmental Engineering
Virginia Tech
Blacksburg, VA 24061, USA

ABSTRACT

The goal of this effort was to investigate the effect of temperature and environment (particularly temperature cycling) on the behavior of polymer composite materials typical of that used in infrastructure applications as well as in aerospace applications. To this end, freeze-thaw tests were conducted on a glass/vinyl ester composite using a DSC to verify the presences of freezable water within the composite material. Additionally, freeze-thaw conditioning of composites was conducted according to the protocol specified in ASTM C 666. Tensile test data encompassing strength, stiffness, and strain-to-failure for "as-received", moisture saturated, and freeze-thaw conditioned material is presented as well as saturation moisture uptake data.

To examine the role of temperature cycling on an aerospace composite, durability studies were carried out on a woven (5 harness satin) graphite/epoxy system targeted at applications in areas of the aircraft industry where operating environment is of concern. Fatigue and residual strength testing was carried out on the composite material in question with the objective of developing a residual strength based life prediction model in which the effects of environment are considered. In this case, the approach called for the evaluation of tensile strength degradation during fatigue cycling in specified environments. The results could then be combined to predict the fatigue life of the material under periodically changing environments. The study was duplicated with material previously aged (unstressed) under hygrothermal cycling to study the effects of environment alone on the composite.

KEYWORDS

Hygro-thermal aging, freeze-thaw, thermal cycling, residual strength, residual stiffness, lifetime prediction

INTRODUCTION

The prediction of fatigue damage and fatigue life for composite materials has been the subject of many investigations during recent years. In fact, a recent conference (Second International Conference on

Fatigue of Composites) was devoted entirely to the subject. Despite these efforts, Curtis (2000) pointed out at the recent ECCM-9 meeting that

> "at present there are major deficiencies in life prediction methodologies for composite materials, which often force large factors of safety to be adopted. That is, composite structures used in high cycle fatigue applications are often over designed and are therefore somewhat heavier and more costly than necessary. Increasing design strains further will exacerbate this problem. Improved life prediction methodologies are thus essential and would result in more efficient use of these materials and the added benefits of lower weight and lower cost structures."

In fact, little has changed since a review paper by Schutte (1994) in which she stated that

> "the state of the art in predictive methodologies for life-time behavior of composites lacks assimilation of the large base of information available (on glass/epoxy and polyester composites). Furthermore, some phenomenological approaches accompanied by experimental work exist for this purpose; ***however, it is most desirable that this methodology be based upon a more fundamental understanding of the processes that are responsible for the mechanisms of failure.*** " (emphasis added)

It is the goal of this proposed research to develop some of that understanding and to incorporate it into a framework that readily admits the addition of additional understanding as it is developed. The following sections describe examples of the manner in which this goal is accomplished. In particular, we will consider the effect of freeze-thaw cycling on a glass/vinyl ester composite and the influence of temperature cycling and moisture cycling on the behavior of a graphite/epoxy composite. We will first consider the freeze-thaw behavior of the glass/vinyl ester composite.

FUNDAMENTAL CHARACTERIZATION OF WATER DURING FREE-THAW AGING

A commercial vinylester produced by the Dow Chemical Co. and marketed under the brand name Derakane 441-400 was used. Styrene is added as the co-monomer and helps reduce the room temperature viscosity of the resin. Benzoyl peroxide is added to the system and the resin undergoes a rapid, free radically initiated addition polymerization to form the crosslinked network. A Perkin Elmer, Pyris 1 differential scanning calorimeter (DSC) was used to perform the freeze-thaw experiments on both the resin and the composite. Liquid nitrogen was used to achieve the cryogenic temperatures needed for the experiment. Regular aluminum pans were used for the single cycle experiments whereas special Perkin Elmer, pressure pans had to be used in the cyclic experiments. These are pans with special threaded lids to prevent leakage due to pressure build up inside the pans leading to subsequent opening of the seal. In the case of the cyclic experiments a few drops of water was dispensed into the pan in addition to the saturated specimen in order to investigate the role of this free-water. A temperature cycle of -150°C to +50°C was used for the single cycle experiment and –18°C to +4°C for the cyclic experiment. In both cases a rate of 5°C/min was used and the melt, endotherm was monitored owing to its greater reliability as explained by McKenna (1995). This thaw rate is clearly much greater than that observed in our terrestrial environment (5°C/hr).

Heat flow measurements as obtained by DSC for an unreinforced resin sample subjected to a single cycle showed no endotherm. Again, this resin has been saturated in a water bath at 65°C and has an equilibrium moisture content of 0.9% by weight (Table 1). The lack of a melt endotherm indicates the absence of freezable water in the pure resin. This is not surprising at all considering the fact that the water essentially resides in the free volume of the resin and this free volume size is of the order of about 6-20 Å, which according to the Thompson's equation is too small (minimum size of 40 Å) interactions in the form of hydrogen bonding exist in the presence of water for the vinyl ester resin. These specific interactions bind the water to the backbone of the resin and prevent it from freezing.

Table I: Saturation and DSC data obtained on both the neat resin as well as composite

Sample Type	Maximum Moisture (wt.%)	Absorbed water (mg)	Area under Endotherm (mJ)	T_m (°C)	Amount of Freezable Water (mg)
(resin) Derakane 441-400	0.90	0.2	None	-	None
(composite) E-Glass/Derakane 441-400 (v_f = 52%)	0.55	0.1	1.3	-6.5	0.04

The melt endotherm region in a similar heat flow experiment that was carried out on a saturated (at 65°C) E-glass/Derakane 441-400 composite sample is shown in Figure 1. The existence of a weak melt endotherm is clear, indicating the presence of freezable water. The peak position is suppressed due to the impurities in the water as well as the small geometry available for freezing. From this melt endotherm region the enthalpy under the peak can be estimated. Using the enthalpy of fusion of water (333 J/gm) and safely associating the peak to be caused due to the phase transition of water alone, the amount of freezable can be ascertained (Table 1), corresponding to 40% of the absorbed moisture in the composite sample. The remaining water is bound.

Now we consider the question of change in damage state for the composite studies in Table 1 under cyclic freeze-thaw conditions. This question can be answered by reviewing a series of melt endotherm curves of the cyclic freeze-thaw experiment shown in Figure 2. Again, the composite specimen was saturated in a 65°C water bath before testing. In this experiment a few drops of water was dispensed into the pan (free water) in addition to the saturated specimen (bound water) in order to investigate the role of free versus bound water. It is clear that certain features of the curve change as freeze thaw cycles proceed. Initially, a shoulder appears at the lower temperatures and is attributed to the water contained inside the saturated specimen. The position of the peak seems to be shifted to around -4°C as opposed to –6°C and this is because these samples were saturated in distilled water as opposed to tap water. The second more prominent peak is associated due to its shear magnitude to the melting of the free water. After 5 cycles this low-end shoulder disappears. We speculate that this is due to the accumulation of damage in the composite that then open up the size of these spaces and allows the water to freeze in an increased free-water like manner. This however needs to be inspected with further testing. The question then is "where does the water reside in a composite?"

To answer this question we review the scanning electron micrograph of a different glass/vinyl ester composite (EXTREN™) (McBagonluri et al. 2000) shown in Figure 3a. It is very clear that interfacial cracks tend to form with the saturation of these materials and are of a size (600 nm) to sustain freezable moisture. More apparent are the interply transverse matrix cracks of a typical pultruded composite shown in Figure 3b. This crack is formed in the resin rich regions of the "as received" material. These transverse cracks are due to the large cure shrinkage associated with the vinyl ester resin (~6% by volume) and the mismatch in coefficient of thermal expansion between glass fibers and vinyl ester (Phifer 1999). Similar cracks have also been seen on structural composites like bridge beams. The authors therefore have sufficient reason to believe the idea that water can reside in these areas in the composite and that these regions are large enough to facilitate the freezing of water during aging.

It is clear from the data shown above that it is virtually impossible to freeze water in a highly crosslinked amorphous polymer like vinyl ester. This is in part due to the geometric space constraints and due to hydrogen bonding that further impedes the process. In the composite system however, the crack dimensions are large enough to facilitate the freezability of water. The authors believe that it is this mechanism then of freezing and the associated volume increase during the transition that leads to the propagation of cracks and the accumulation of damage. Presently, work is underway to understand the effects of fatigue, moisture and freeze-thaw on the remaining tensile strength of cross-ply glass/vinyl ester composites. Such work is described in the next section.

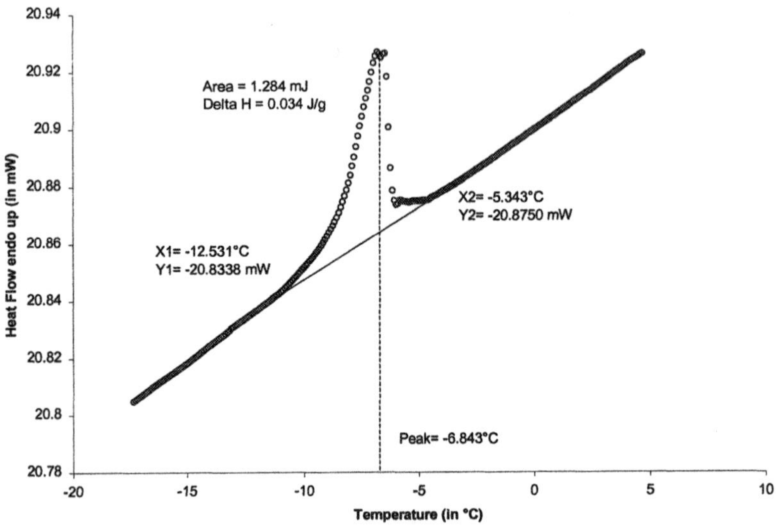

Figure 1: Melt endotherm region for the composite specimen. Indicated here are both the position of the melt endotherm at –6.5°C and the area under the peak.

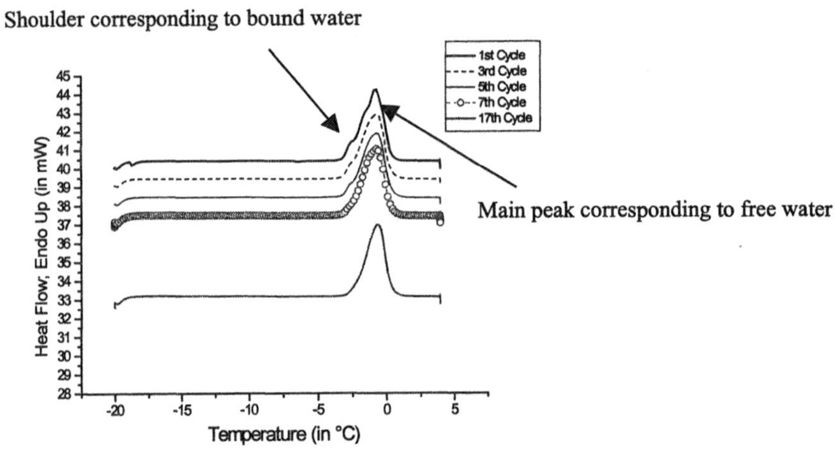

Figure 2: Cyclic DSC data on the composite specimen. The pan contained both the saturated specimen as well as a few drops of free-water.

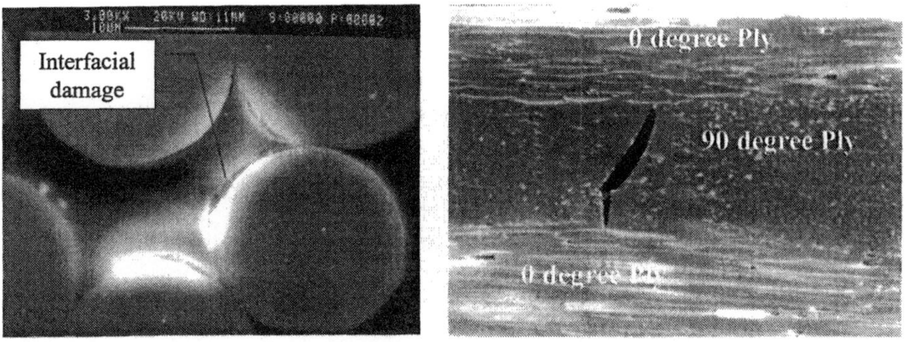

Figure 3: a) Right - Scanning electron micrographs of fresh water aged material at 3K magnification, b) Left - Optical micrograph of the pultruded cross ply laminate (E-glass/Derakane 441-400) viewed from the transverse direction; microcrack size (300μm long by 40μm wide)

FREEZE-THAW EFFECTS ON COMPOSITE LAMINATES

Both saturated and dry fiber reinforced polymeric composite samples were be placed in an accelerated freeze-thaw environment and tested for mechanical property degradation, changes in crack density, and moisture uptake at specified intervals to assess damage. A group of saturated controls were held at a constant temperature above freezing and tested in the same manner as the freeze-thaw samples. All samples are pultruded glass reinforced cross-ply (0/90) laminates with polymeric matrix materials. Ultimate tensile strength, stiffness, and strain-to-failure were determined quasi-statically for each class of as-received material in accordance with ASTM D 3039 using a displacement rate of 2.5 mm/min (0.10 inches/min). A total of thirty samples were tested, ten of each material type. In addition, two samples of each material were set aside for crack density analysis using optical microscopy techniques. Also, 324 samples were fully saturated in a 65°C (149°F) water bath with moisture uptake measured throughout the saturation process. Weight measurements were taken hourly on the first day the samples were placed in the saturation tank, every three hours on the second day, every four hours the third day, every six hours the fourth day, and once everyday thereafter. The samples reached saturation within 45 days.

The freeze-thaw conditioning parameters chosen for this research study were based on ASTM C 666. This test protocol calls for a ramp down from 4.4°C (40°F) to -17.8°C (0°F) followed by a hold at -17.8°C (0°F), a ramp up to 4.4°C (40°F) and a hold at -17.8°C (0°F). There may be a minimum of 4.8 and a maximum of 12 conditioning cycles per day with 75% of the cycle time set aside for freezing and 25% for thawing. Two high performance cascading refrigeration freeze-thaw conditioning chambers were be used to achieve a ten cycle per day rate.

A series of trays was fabricated to hold each sample in accordance with ASTM C 666, namely to surround the samples with between 0.8 mm (1/32 inch) to 3.2 mm (1/8 inch) of water. Additional design goals for the trays were to minimize the volume of water and maximize convective heat transfer with the air inside chamber. A finished tray is shown in Figure 6. Each tray can hold eight samples and every other slot was machined all the way through to allow airflow vertically through the trays. A second type of tray was also fabricated to allow the application of four point bending loads to each sample capable of causing 0.55% strain at the centerline on the surface of the tension face as shown in Figure 6. This strain level was chosen because it is beyond the "knee" in the stress-strain curve for each material in this research study and it likely opens up cracks that might be large enough to allow additional freezing to take place. This tray can also hold up to eight samples.

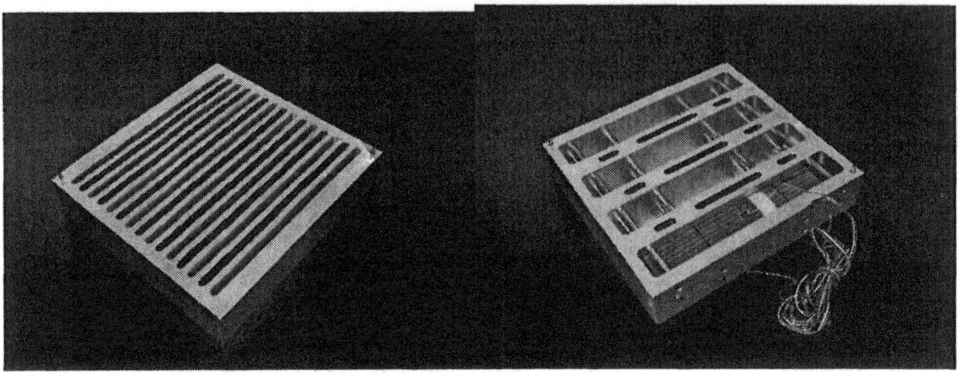

Figure 4: Unloaded sample freeze-thaw tray (left) and loaded sample freeze-thaw tray (right).

Three levels of freeze-thaw exposure were included in this study: 100, 300, and 1000 cycles. A set of control samples was also placed in a constant 4.4°C (40°F) bath for the duration of each freeze-thaw exposure level. Subsequent to the cycling, ultimate tensile strength, stiffness, and strain-to-failure were determined quasi-statically in accordance with ASTM D 3039 for each class of material after saturation. A total of thirty samples were tested, ten of each material type. In addition, one sample of each material was placed in a desiccator after conditioning to examine the recoverability of mechanical properties once moisture content returned to the as-received state.

Strength results for a glass/toughened vinylester composite are summarized in Figure 5. The initial strength was 389 MPa in the as-received condition versus 237 MPa for the post-saturation condition. These figures represent a 40-50% reduction in strength after saturation, although freeze-thaw conditioning does not appear to have any effect on strength at less than 300 cycles. There was a slight increase in strength for the dry samples, but this is most likely due to desiccation experienced during conditioning.

HYGRO-THERMAL CYCLING EFFECTS ON GRAPHITE/EPOXY COMPOSITES

In addition to the study of composites typical of those used in infrastructure applications, we have also been investigating the effect of moisture and thermal cycling on the behavior of graphite/epoxy composites typical of those used in aerospace applications. One of the main objectives of this project was to use a residual strength based approach for life prediction of the specified material; therefore the testing program was designed to obtain residual strength data. Since relating the damage state to residual strength was also of interest, several intermediate steps were taken to track damage progression.

The composite material of interest in this study was a $[(0/90)]_{4s}$ graphite/epoxy five harness satin weave to be used in a subsonic aircraft engine. For this particular application, not only are the individual environments of concern, but the alternation of environments as the component goes between storage and operating conditions must also be considered. In the course of this study, the effects of the individual and combined (alternating) environments given in the mission profile on the durability of the material system were characterized.

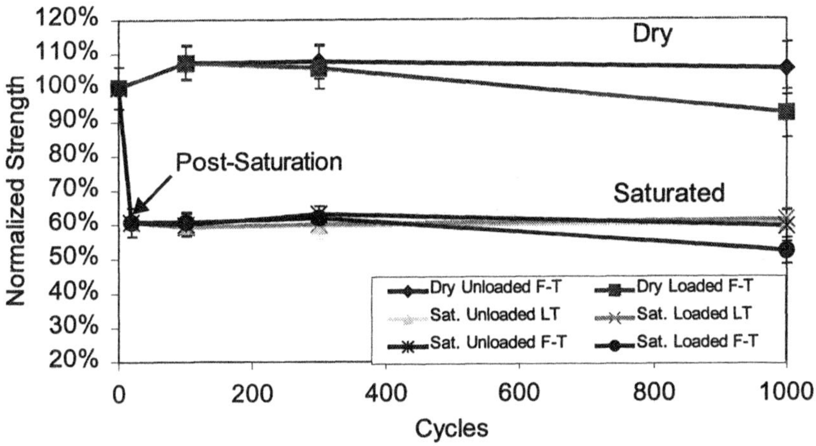

Figure 5: Influence of freeze-thaw cycling for toughened glass/vinyl ester composites.

One of the main objectives of this project was to use a residual strength based approach for life prediction of the specified material, so the testing program was designed to obtain residual strength data. Since relating damage state to residual strength was also of interest, several intermediate steps were taken to track damage progression. Finally, as the change in remaining strength and damage state with different environmental conditions was of particular concern, the tests on unaged and aged specimens were conducted under each of four conditions:

1. Room temperature (to provide baseline behavior)
2. Elevated temperature (120°C, engine operating condition)
3. Wet (saturated and then tested at 85% relative humidity at 30°C, storage condition)
4. Hygrothermal cycling – alternation between temperature and humidity condition during fatigue as shown in

The aging process involved exposure of the material to an environment which alternated in moisture/temperature conditions according to the mission profile shown in Figure 6. The results given in this paper come from material aged in this fashion for 12000 hours. One of the major observations that can be made about the aged material is that microcracks were present in the material prior to mechanical loading. Cracks are clearly present at the crossover points of aged material and absent from the unaged material. In addition, material aged for 6000 hours only contained transverse microcracks, but material aged for 12000 hours had a higher level of transverse micocracks, and it also had longitudinal microcracks. This preferred orientation of microcracks is surprising since the material is nominally orthotropic. Manufacturing may have played a role in producing this result. In keeping with the trend of the cracking, the material aged for 12000 hours had darker resin than the material aged for 6000 hours, and both had darker resin than unaged material. It is important to note, however, that optical microscope images of polished cross-sections of the aged material revealed that the cracks exist predominately at the surface layers of the laminate.

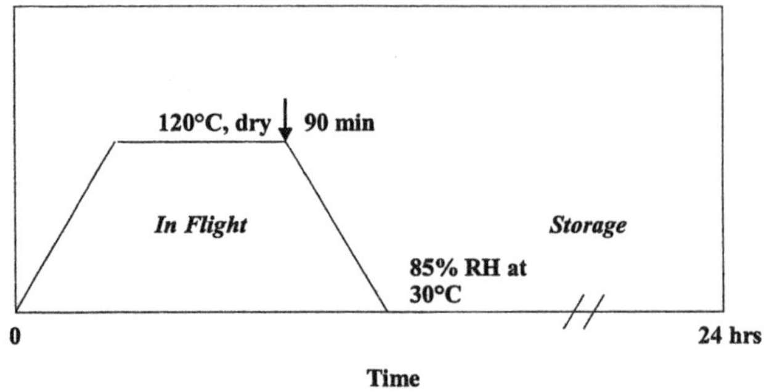

Figure 6: Mission profile for aircraft engine

Since specimens tested under the "wet" condition were first saturated in the humid environment, it was possible to compare moisture uptake behavior of the aged and unaged material. The results of the moisture saturation process are shown in Figure 7. The diffusivity and maximum moisture content of the aged material ($4.02 \times 10^{-9} cm^2/sec$ and 0.5%, respectively) were significantly higher than that found for the unaged material ($2.23 \times 10^{-9} cm^2/sec$ and 0.36%). Cracking in the aged material is the most obvious answer to these differences. Moisture can be drawn into cracks by capillary action, so the cracks provide an enhanced transport system for the moisture. While this behavior may explain the higher slope of the moisture uptake curve, the higher saturation moisture content is still somewhat questionable since the cracks are small and do not necessarily hold water.

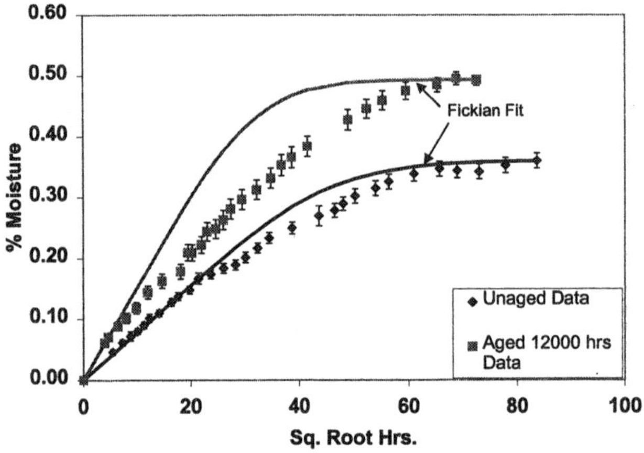

Figure 7: Moisture uptake curves (85% rh at 30°C) for aged and unaged material

One possible explanation explored for the higher moisture content in this case was aging induced degradation of the matrix and the consequential formation of voids. However, DMA and TGA data, given in Table II, indicated no signs of resin degradation. The DMA suggests that the T_g increases slightly in going from unaged material to material aged for 6000 hours, and it increases again in going to material aged for 12000 hours. While the increases are minimal, the tests were repeatable. The increase in T_g is most likely due to additional crosslinking during aging. The TGA data supports this claim, as the resin degradation temperature was found to increase significantly. Additional crosslinking, however, would imply that the material should have a lower moisture saturation content since the density of the material would be higher. Another explanation for the increase in maximum moisture content that deserves further investigation is the possibility of moisture residing in the fiber/matrix interface if the interface degraded due to aging.

Table II: DMA and TGA Results

Hygrothermal Aging	T_g (°C)	Degradation Temperature (°C)
None	200	375
6000 hrs	203	397
12000 hrs	208	406

The life prediction methodology used in this study is one originally developed by Reifsnider and Stinchcomb (1986) and further developed by Reifsnider and co-workers (1995, 1996). According to the modeling philosophy, failure of the laminate is assumed to occur when its residual strength degrades to the value of the maximum applied load. In addition, the residual strength is assumed to be a damage metric such that equivalent damage states are assumed to be represented by equivalent residual strengths. Based on kinetics arguments, Reifsnider (1995) developed a damage evolution integral

$$\Delta Fr = -\int_{\tau_1}^{\tau_2} (1 - Fa) j\tau^{j-1} d\tau \tag{1}$$

by which the change in normalized residual strength (ΔFr) of a material may be calculated by implementing an appropriate scalar failure criterion, Fa. The material parameter j is an experimentally determined quantity which may vary with factors such as failure modes. In the case of mechanical fatigue, the characteristic time,τ, can be expressed as n/N where n is the number of fatigue cycles the material has undergone out of a total of N (usually fatigue life) cycles.

Because this study was meant to examine the behavior of the material under a service condition which involved alternating periods of specific moisture and temperature conditions, it was thought that a similar paradigm to the critical element ideas originally presented by Reifsnider and Stinchcomb (1986) above could be used to account for the different environments. The effects of temperature and moisture on the residual strength of the material during fatigue were to be considered individually and then combined to predict the fatigue behavior of the material under periodically changing environments. Since the experimental data for the material reflected no significant and quantifiable effects of environment on its residual strength, it was found that a single j value could be used in Eqn. (1) along with the maximum stress failure criterion

$$Fa = \left(\frac{\sigma_{\max}^{applied}}{X_t} \right) \tag{2}$$

to model the strength degradation of the material during fatigue in any of the specified environments.

The values for τ were estimated based on the fatigue-life data, and the value of *j* was determined to be 1.3. As can be seen from the room temperature and combined fatigue/hygrothermal cycling residual strength data and predicted residual strength curves shown in Figure 8, reasonable agreement was found between experimental and predicted residual strength values.

STRUCTURAL DURABILITY ANALYSIS

The ultimate goal in performing these characterization tests is to develop an understanding of the material behavior that may be incorporated into a structural durability analysis code. In particular, we will consider the fatigue response of a hybrid pultruded structural section presently employed in the Tom's Creek Bridge in Blacksburg, Virginia (Hayes and Lesko, 1999). The double web I beam structural shape, a 20.3 cm (8 in) deep section is serving as a sub-scale prototype for a 91.4 cm (36 in)

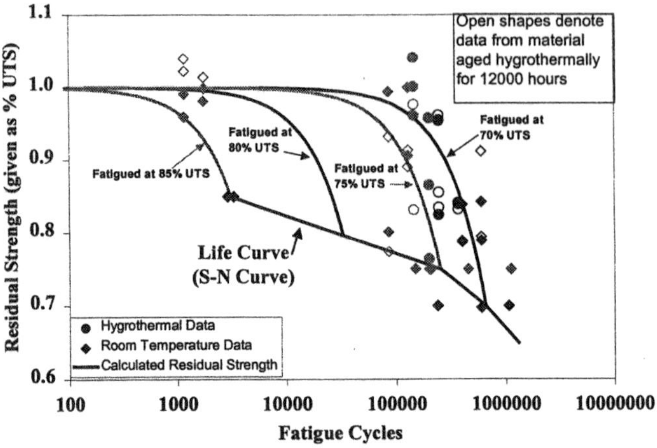

Figure 8: Residual strength curves for aged and unaged material

beam being developed for 10 to 18 meter span bridges. The beam is a pultruded section composed of both E-glass and carbon fiber in a vinyl ester resin. The approximate fiber volume fraction (both glass and carbon) for the structure is 55%. The carbon is located in the flanges to increase the section's bending stiffness. Glass fiber is present in the pultruded structure primarily in the form of stitched angle ply mats, roving and continuous strand mat. Tests of individual beams at a span of 5.74 m have shown an average bending modulus of 44 to 48 GPa depending on the type of carbon fiber used. The moment of inertia for this section is 5328 cm^4 (128 in^4). A beam was tested to failure under a four-point loading and failed at an applied load of 129 kN (29 kips). This represents a 152 kN•m (112 kip-ft) moment capacity at a 5.3m (17.5 ft.) span (the span of the Tom's Creek bridge). The shear contribution to deflection is approximately 5% at this span. The failure of this beam, and those under investigation for fatigue performance, is dominated by delamination at the hybridized region within the top flange as shown in Figure 9. As limited fatigue data is and will be available for the structural FRP beam, a life prediction model for this beam has been developed (Senne, 2000) based on coupon test data. This model, although far from complete in describing the full combined enviro-mechanical fatigue response of the structural member, forms the basis for future efforts which consider other pertinent degradation mechanisms. As an illustration of the results from this model, we may consider the resulting lifetime predictions as shown in Figure 10. For details on this preliminary model development, the reader is referred to Senne (2000).

SUMMARY

In this paper we have examined the role of temperature cycling on the behavior of two different classes of material systems: glass/vinylester under freeze-thaw conditions and graphite/epoxy under humidity and temperature cycling conditions. We have been able to isolate the location of the freezeable water in the glass/vinylester composites and have conducted experimental tests to measure residual properties after freeze-thaw cycling. In addition, we have measured residual properties for the glass/epoxy composites after hygro-thermal aging. Residual properties of these types form the basis for residual strength based life prediction models, and we have considered one specific example of the application of such a model.

Figure 9: Failure of the 20.3 cm deep double web I-beam under four point bending.

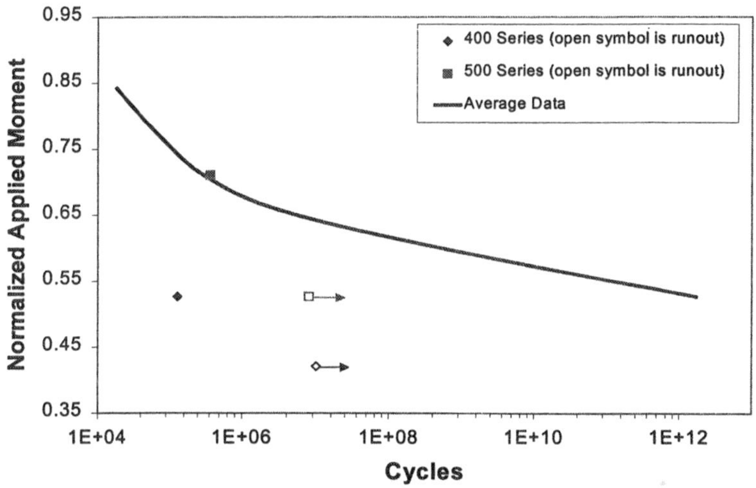

Figure 10: Life prediction results for the 20.3 cm deep double web I-beam under four point bending.

REFERENCES

Broutman, L. J. and Sahu, S. (1972). A new damage theory to predict cumulative fatigue damage in fiberglass reinforced plastics, *Composite Materials: Testing and Design (Second Conference), ASTM STP 497,* American Society for Testing and Materials, pp. 170-188.

Curtis, P. T. and Davies, A. J. (2000). Fatigue Life Prediction of Polymer Composite Materials, *Proceedings of ECCM9: Composites—From Fundamentals to Exploitation,* Brighton, U. K., , published electronically by IOM Communications.

Diao, X, Mai, Y., and Ye, L. (1999). Statistical analysis and experiments on fatigue of fiber composite laminates, (keynote address), *Proceedings of the Seventh International Fatigue Congress (FATIGUE'99)*, Beijing, P.R. China,

Hashin, Z. and Rotem, A. (1978). A cumulative damage theory of fatigue failure, *Materials Science and Engineering,* **34**, pp. 147-160.

Hayes, M. D. and Lesko, J. J. (1999). The Tom's Creek Bridge Rehabilitation Project, http://filebox.vt.edu/eng/esm/jlesko/tcb/tcb.html

Huston, R. J. (1994). Fatigue life prediction in composites, *International Journal of Pressure Vessels and Piping,* **59**, pp. 131-140.

Liu, B. and Lessard, L. B. (1994). Fatigue and damage tolerance analysis of composite laminates: stiffness loss, damage modelling, and life prediction, *Composites Science and Technology,* **51**, pp. 43-51.

McBagonluri, F., Garcia, K., Hayes, M., Verghese, N. & Lesko, J. J. (2000). Characterization of Fatigue and Combined Environment on Durability Performance of Glass/Vinyl Ester Composite for Infrastructure Applications, *International Journal of Fatigue*, **22:1**, pp. 53-64.

McKenna, G. and Catheryn, L. J. (1990) On the Anomalous Freezing and Melting of Solvent Crystals in Swollen Gels of Natural Rubber, Rubber Chemistry and Technology, **64**

Poursartip, A., Ashby, M. F., and Beaumont, P. W. R. (1982). Damage accumulation during fatigue of composites, *Scripta Metallurgica,* **16**, pp. 601-606.

Reifsnider, K L, and Stinchcomb, W W. (1986). A Critical-Element Model of the Residual Strength and Life of Fatigue-Loaded Composite Coupons, *Composite Materials: Fatigue and Fracture*, ASTM STP 907, H. T. Hahn, ed., American Society for Testing and Materials, Philadelphia, pp. 298-313.

Reifsnider, K., Lesko, J., and Case, S. (1995). Kinetic Methods for Prediction of Damage Tolerance of High Temperature Polymer Composites, *Composites '95: Recent Advances in Japan and the United States*, I. Kimpara, H. Miyairi, and N. Takeda, Ed., Proceedings Japan-U.S. CCM-VII, Kyoto, pp. 49-55.

Reifsnider, K. L., Iyengar, N., Case, S. W. and Xu, Y. L. (1996). Damage Tolerance And Durability of Fibrous Material Systems: A Micro-Kinetic Approach, *Durability Analysis of Structural Composite Systems*, A. H. Cardon (ed). A. A. Balkema (Rotterdam), pp. 123-144.

Reifsnider, K., Case, S. and Iyengar, N. (1996). Recent Advances in Composite Damage Mechanics, *Proceedings of Conference on Spacecraft Structures, Materials & Mechanical Testing* (ESA SP-386), Noordwijk, The Netherlands, pp. 483-490.

Schutte, C. L., (1994). Environmental Durability of Glass-Fiber Composites, *Materials Science and Engineering*, R13, pp. 265-324.

Senne, J. L., (2000). *Fatigue Life of Hybrid FRP Composite Beams*, M.S. Thesis, Virginia Polytechnic Insititute and State University, available electronically at http://scholar.lib.vt.edu/theses/available/etd-07132000-14530012/

Verghese, K. N. E., Hayes, M. D, Garcia, K., Wood, J., Riffle, J. S., and Lesko, J. J. (1999) Influence of Matrix Chemistry on the Short Term, Hydrothermal Aging of Vinyl Ester Matrix and Composites Under Both Isothermal and Thermal Spiking Conditions, *Journal of Composite Materials*, **33:20**, pp. 1918-1938.

Long Term Durability of Structural Materials
P.J.M. Monteiro, K.P. Chong, J. Larsen-Basse, K. Komvopoulos (Eds)

Long – Term Material Characterization of a Cured In Place Plastic (CIPP) Sewer Rehabilitation Liner Material

Clifton Vining, William Jordan and David Hall

Mechanical Engineering Program, Louisiana Tech University
Ruston, Louisiana 71272, USA

ABSTRACT

Cured-In-Place Plastic (CIPP) pipeline rehabilitation liners must be designed to prevent creep-induced collapse caused by external groundwater pressure. Current liner design models account for the time dependent response of liner materials using estimates of the creep modulus at 50 years. Room and elevated temperature creep testing of a CIPP material are reported here in an attempt to more accurately determine the value of the creep modulus at 50 years. The compressive creep modulus predicted at 50 years is found to be 275 ksi. These results are not significantly inconsistent with the industry practice of assuming that the 50 year modulus is ½ of the short-term elastic modulus.

KEYWORDS

time-temperature superposition, aging, creep, liner buckling, polymeric materials

BACKGROUND

Cured-In-Place Plastic (CIPP) pipe liners are commonly employed to rehabilitate deteriorated sewer and water lines (Figure 1). Such liners are often installed below the water table and are consequently subjected to external hydrostatic pressure which may lead to creep-induced buckling (Figure 2) of the liner within the host pipe before the anticipated service life of 50+ years is achieved. Insufficient understanding of this buckling phenomenon is regarded as a significant limitation of the technology and a barrier to more cost – effective sewer rehabilitation.

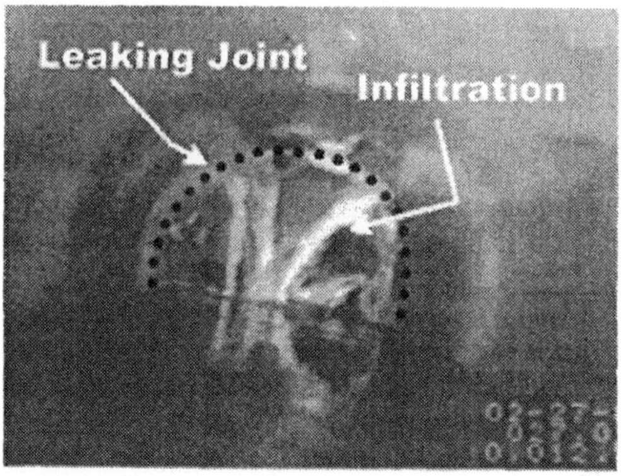

Figure 1 - Deteriorated Sewer Line with Infiltration of Groundwater.

Figure 2 - Buckled CIPP Liner Test Specimen.

The ongoing liner buckling research at the Trenchless Technology Center (TTC) at Louisiana Tech University centers around experimental and computational analysis of liner buckling with the objective of developing improved liner design models (Hall, Zhu). The experimental research involves room temperature and elevated temperature material characterization tests of CIPP liner materials as well as buckling tests of liners at these same temperatures. The softening of the CIPP with increasing temperature is used to establish a relationship between time and temperature such that tests performed on a convenient time scale (less than one year) can be extended to much longer time periods (decades). This work is the result of a grant provided by the Division of Civil and Mechanical Systems of the National Science Foundation.

The objective of the material characterization research discussed here is to develop a material model to accurately predict the behavior of CIPP materials over a 50+ year period. Such a material model is useful because it can be embedded into computational or analytical buckling models to predict the performance of the pipe liner / host pipe systems over long time periods. Liner systems are currently designed according to ASTM F1216 (ASTM) which is based on Timoshenko's buckling equation (Timoshenko) for a free standing ring. The elastic modulus in the ASTM design equation is commonly replaced with the estimated creep modulus of the liner material at 50 years. The standard practice is to take the 50 year modulus as ½ of the short-term flexural modulus of the material. Consequently, the long-term design is based solely on the short-term properties of the material. New models to predict liner buckling are being designed at the TTC which are based on the creep properties of the material. The accelerated material characterization testing discussed in this paper will allow for accurate determination of these long-term properties using relatively short-term (less than one year) tests.

PROBLEM DESCRIPTION

The CIPP liners used in this study are constructed of a nonwoven polyester felt impregnated with a polyester resin. The liner is inverted into a deteriorated host pipe and cured with hot, circulating water. Short-term tensile testing, elevated and room temperature creep testing in bending, tension, and compression, and aging tests in compression have been conducted on specimens cut from liner segments to provide the data necessary to characterize this material. Although the variation of the response of the material as a function of time is important in models that are currently being developed at the TTC, the work in this paper will focus on determining the creep modulus at 50 years. Accurate determination of this modulus is of great interest to the industry since it is the only material parameter in the ASTM design equation that accounts for the time dependent behavior of the material.

NOTATION

The notation used in this paper is given below.

σ	=	stress
ϵ	=	strain
ϵ_i	=	initial elastic and plastic strain
ϵ_c	=	creep strain
E	=	Young's Modulus
E_c	=	Creep Modulus
t	=	time
λ	=	effective time
μ	=	aging shift factor rate of change
α	=	$1 - \mu$

EXPERIMENTAL SETUP

The equipment described in this section was used to characterize the short-term and long-term mechanical properties of the CIPP material. The loading mechanisms, specimen fixtures, data collection method and other important characteristics of the testing system are described below.

The time-dependent deformation characteristics of the CIPP material were determined for tensile, compressive, and flexural loading using five SATEC creep/stress rupture devices. Each of these five SATEC devices has two loading mechanisms such that ten specimens can be tested simultaneously. The tensile and compressive load mechanisms provide the load through a lever arm with a magnification factor of 16, as shown in Figure 3. The lever arm must be balanced prior to testing to eliminate errors in the loading magnitude. The flexural loading mechanism applies the load directly to the specimen with no magnification, as shown in Figure 3. Being a three-point bending mode of loading, far less load is needed to induce the desired stress levels in the specimen. The loading mechanisms in all of these systems provides a constant stress over time.

The testing devices are equipped with environmental chambers which allow both room temperature and elevated temperature creep rupture tests to be performed. The ovens have a temperature control unit which maintains the temperature at the set point. The specimens and the fixtures which hold the specimens are enclosed in the ovens. The fixture for flexural loading can be seen in Figure 3, while the fixture for compressive loading is seen in Figure 4. For compressive testing, a rectangular sample is placed in a cylindrical cavity such that the corners of the specimen just touch the sides of the cavity. A plunger applies the load to the top of the specimen.

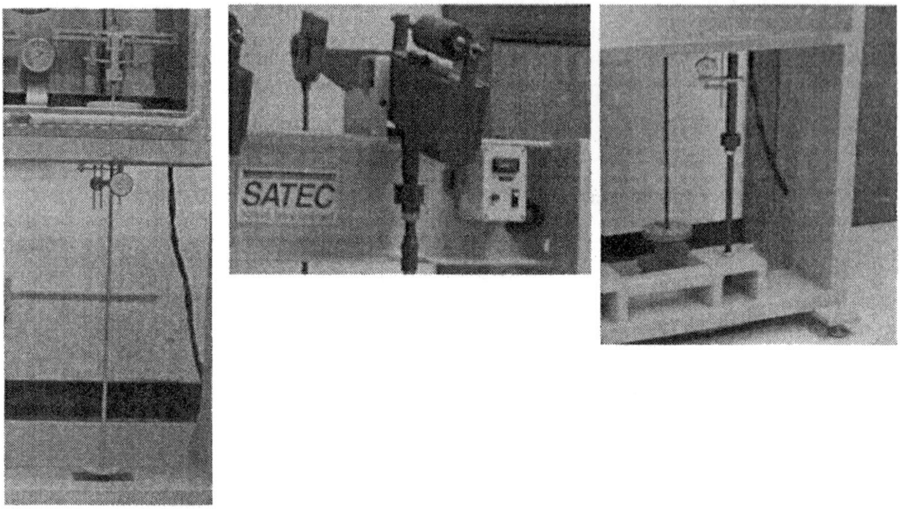

Figure 3 - Left to Right: Flexural Loading Mechanism, Lever Arm for Tensile and Compressive Loading Mechanisms, Lower Portion of Loading Mechanism for Tensile and Compressive Loading Mechanism with Dial Indicator.

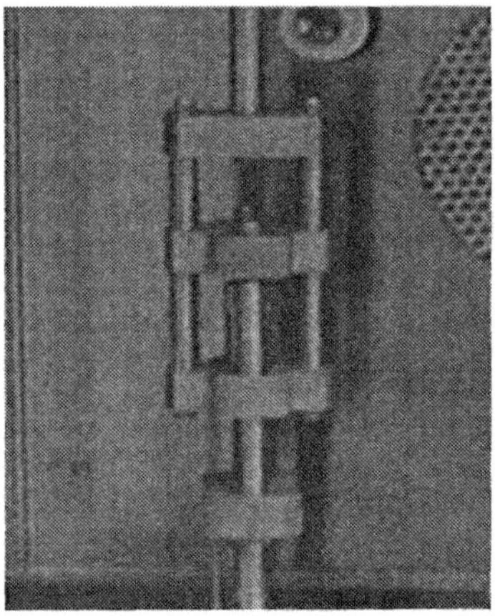

Figure 4 - Compressive Fixture.

Table 1 shows the matrix of test temperatures and stresses for which creep deformation testing was completed. Additional tests were conducted to determine the short-term elastic modulus as a function of time.

Table 1 - Matrix for Creep Deformation Testing.

Stress State	Nominal Stress Levels (psi)	Temperature (°F)
Tension	600, 1600, 2650, 3200	80, 120, 150, 180, and 210
Compression	2300, 2900, 4600, 5800	80, 120, 150, 160, 170, 180
Flexure	1000, 1750	80, 120, 150, 180

The SATEC creep testers described use dial indicators to measure specimen deflection. The indicators were graduated in 0.001 of an inch, giving a resolution 0.0005 inches. The indicators were quite at acceptable high temperatures, but at room – level temperatures, more finely graduated indicators would be desirable.

The MTS TestStar IIs was used to perform the short - term tensile tests on the current liner material. The load cell has a capacity of 20,000 pounds and is accurate to ± 3 lbs. The machine has a massive steel frame, with an adjustable table.

EXPERIMENTAL PROCEDURES

The experimental testing done on this project was conducted primarily by ASTM standards, and by recommendations of selected authors (Struik). Four types of tests were conducted to quantify the information needed to perform the necessary calculations. The short – term tensile testing gave the Young's Modulus. The Modulus vs. Temperature testing was conducted to determine the initial Young's Modulus at the start of the high – temperature creep tests. The creep tests provided the momentary creep data at selected temperatures. Finally, the aging tests provide information on the necessary information on aging in this material.

Short – Term Tensile Testing

The short term tensile tests were conducted in accordance with ASTM 638 – 89. The tests were conducted using a constant displacement rate, and the strain of the specimen was recorded electronically. From this data, the breaking strength and modulus of elasticity could be found, as well as whether the material failure was brittle or elastic. Once Young's Modulus is found, a base for the calculation of the long term material properties has been determined.

Modulus as a function of Temperature

Testing the material at different temperatures necessitates the use of different moduli, as the initial modulus changes with temperature. The actual test consisted of heating the specimen in steps, and performing the necessary tests. The change in displacement at different loads and temperatures was used to calculate the E as a function of temperature.

A byproduct of E vs. T data was the approximate location of the glass transition temperature. The glass transition temperature is defined as the region with the greatest slope on the E vs. T curve. The glass transition region appears to be from 150 °F to 160 °F, which is consistent with tabulated data in other references for polyester (Callister).

The data from this test provided an estimation of the modulus at various temperatures, used in computing the initial elastic strain from the creep tests. The set – up of the SATEC oven makes the measurement of initial strain difficult in some cases. The calculation of the strain from the modulus and stress was deemed an appropriate alternative to measuring it.

Creep

The room & high temperature creep tests were conducted as per ASTM D2690. Creep testing gives the actual material behavior under the assigned conditions for the duration of the test. Using creep data from tests under varying levels of temperature and stress can provide information needed to identify trends in a materials behavior at long times.

Aging

As a polymer ages, its rest state is changing (Matsumoto), working its way back to equilibrium. One deformation mechanism of amorphous polymers is chain motion on the molecular level. Aging is the thermoreverisible phenomenon of the polymer reverting back to the lowest energy state available. The polymer stiffens itself with this tendency to go toward the lowest energy state. The aging of the polymer must be taken into consideration when predicting the long – term behavior of the material.

The aging procedure consisted of an equalization period, during which the specimen is held above the glass transition temperature. Following that time, the specimen is quenched to the temperature of interest. Once quenched, the specimen is aged until time for the first creep test. After the creep test, the specimen was unloaded, and allowed to age to the next test time. This cycle of age, creep test, age, creep test was repeated 4 times for aging time of 1, 3, 10, and 30 hours. The above test cycle is based on suggestions in (Struik). The specimen was tested in compression, since the least amount of scatter is found in the compression creep data.

In the performance of the test, two issues stand out. If the test is conducted at room temperature with a stiff material, the displacements will be small. A dial indicator or other measurement device of appropriate resolution should be used. Also, the loading and unloading of the system could damage the specimen if care is not taken.

DATA REDUCTION & ANALYSIS

The short term tests are relatively routine and their results are presented directly. The creep and aging tests are more complex and how their data was analyzed will be described in more detail.

Short-Term Testing

The material was subjected to short term tensile loading. The material was tested in bending and tension. There is some debate on the actual failure mode of the liner, and this work will help in that analysis. Studies have not shown conclusively what stress state dominates during buckling: tension, compression, or flexure. The most appropriate value of the Modulus is likely the Modulus of the dominating stress state, prior to buckling. The material failed in a brittle manner at room – temperature.

The E vs. temperature curve can be seen in the figure below. A third degree polynomial was found to describe the material behavior quite well in the temperature range of interest.

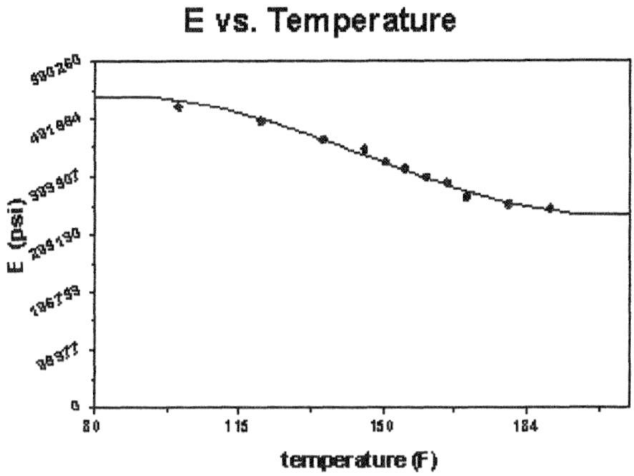

Figure 5 - Young's Modulus as a function of temperature.

Using the specimen dimensions and the load, the initial stress and strain were calculated to be:

$$\sigma = 3072 psi \qquad \varepsilon_i = 0.0079$$

The creep strain of the specimen is defined as the change instantaneous length divided by the initial length.

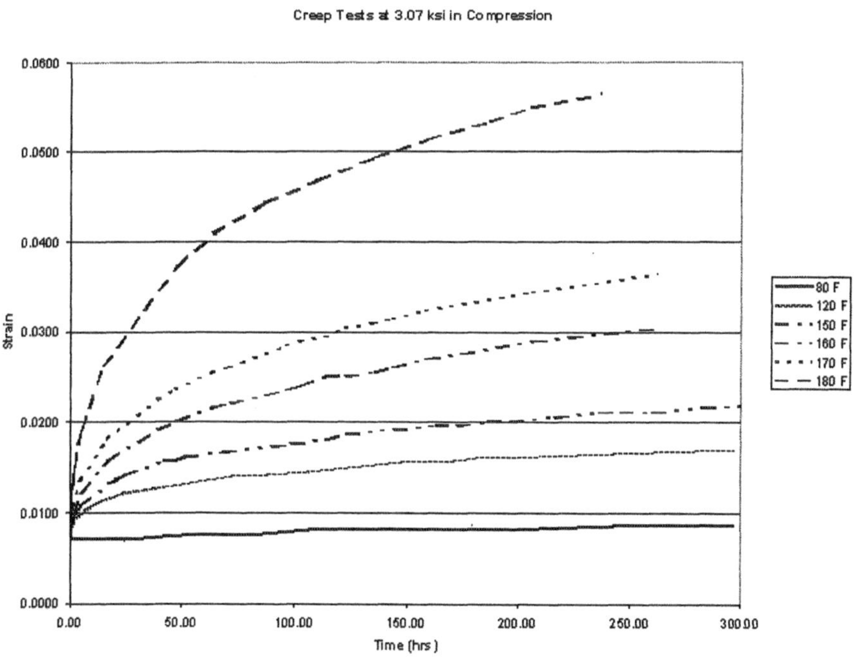

Figure 6 - Creep strain versus time curves for various temperature.

Figure 6 shows the creep curves for 3072 psi and selected temperatures:
During creep, a material deforms continuously, but if one were to unload the material, and perform a tensile test on it, the Young's modulus would be about the same as before the creep test. The creep modulus is not the instantaneous modulus, but allows us to handle the phenomenon of creep efficiently in analytical and empirical terms.

The creep modulus of a material is defined by:

$$E_c = \frac{\sigma}{\varepsilon_{total}} \qquad (1)$$

$$where: \quad \sigma = stress$$

$$\varepsilon_{total} = \varepsilon_c + \varepsilon_i$$

The figure below shows creep E vs. time plots on a log – log scale.

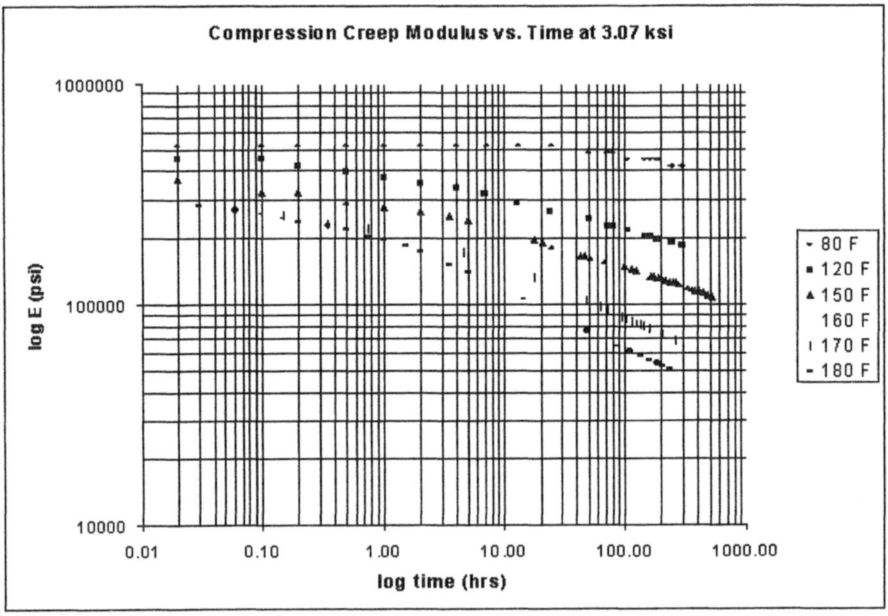

Figure 7 - Creep modulus versus time.

In Ferry's text, instruction is provided on the fitting of the WLF equation to dynamic mechanical analysis (DMA) data and the subsequent application of the WLF equation to creep data. Lacking DMA data for this material, time – temperature superposition was applied to the creep data. However, the effects of aging should be considered (Struik, Bradshaw [1997], Matsumoto, Bradshaw [1994], Janas), and in fact, they cannot be ignored. The aging of a polymer invalidates the time – temperature superposition technique (Matsumoto). Several references (Struik, Bradshaw [1997], Janas) provide a formulation of the known techniques to used perform the aging work on this data, which is Time – Aging Time Superposition. Struik's text is classical for this type of work and the work here is based on that text.

The data from the aging tests yields results similar to the creep test, modulus vs. time curves. The difference is that one aging tests yields a family of modulus curves at one temperature, each one at different aging times. Figure 8 below shows the 2 of the compressive modulus vs. time curves taken from the aging test:

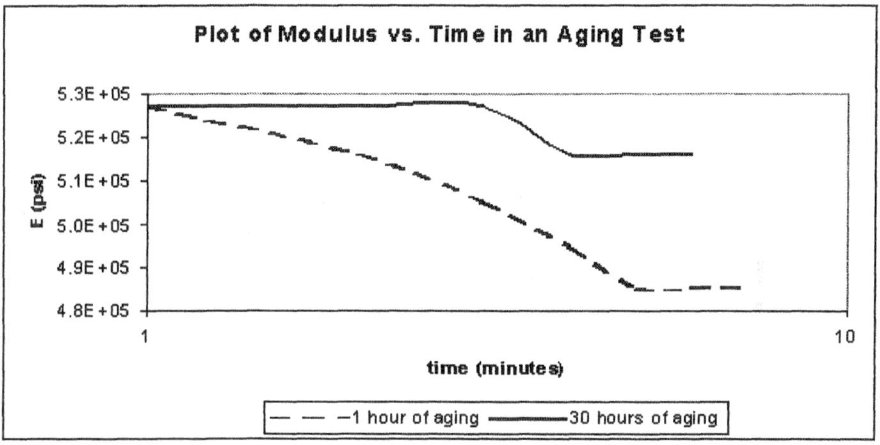

Figure 8- Plot of Young's modulus vs. log time to demonstrate the stiffening of the material with increasing aging time.

By inspection, one notices modulus decreases less with an increase in aging time. The increase in stiffness is from the aging of the specimen.

Since the Momentary Master Curve (MMC) has no information about aging, one introduces aging into the MMC via the effective time concept proposed by Struik. The effective time concept says that a test at a long aging time may be used to predict the behavior of the material far into the future at a shorter aging time. The effective time concept can be expressed mathematically by the following for small strain tests:

$$\lambda = t_e \cdot \ln(1 + \frac{t}{t_e}) \qquad if\ \mu = 1$$

$$\lambda = \frac{t_e}{\alpha} \cdot \left[\left(1 + \frac{t}{t_e}\right)^{\alpha} - 1 \right] \qquad if\ \mu < 1 \tag{2}$$

Struik provides commentary and examples about the use of the effective time technique, prior to giving a worked example followed by a straightforward method in his text. One example provided by the author has an error of about 10% at an extrapolation factor of 2000.

The method Struik formulates has 3 main steps:

1. Determination of the momentary moduli
2. Determination of the shift rate
3. Determination of the long – time moduli via effective time calculation

The momentary compliances were determined in the creep tests. Per the ASTM standards, the samples were aged 24 hours prior to testing. Using Struik's guidelines that the test time for momentary data shouldn't exceed 10 – 20% of aging time, the first 6 hours of the creep tests results provide the momentary creep data. The MMC for this material at 80 °F is below.

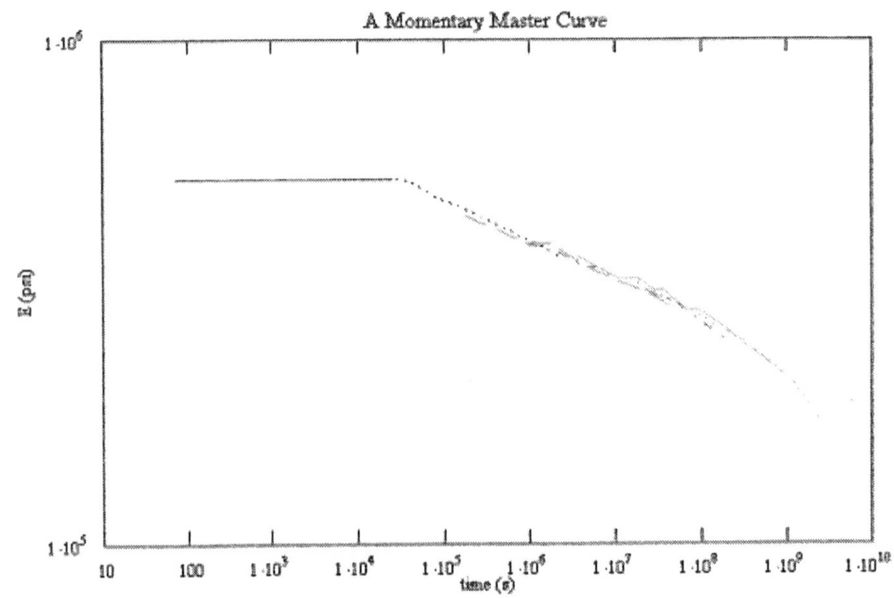

Figure 9 - Momentary Master for the current liner material with no aging information contained in the curve.

The shift rate was determined by conducting an aging test, and shifting the aging E vs. time curves until a reasonable overlap existed. The curves are somewhat stepped and not smooth, making a good superposition quite difficult. Even in semi – log coordinates, the smoothness of the curves leaves much to be desired.

The determination of the long – time moduli is the ultimate goal. The first two steps provide the necessary data to perform the calculations in this step. Prior to using the effective time equation to predict the long – term modulus, some additional data manipulation occurred. Struik gives a table of correction factors to use on the momentary creep data. The changes, the author claims, are small but important. Once the data correction is done, the effective time equation is used to predict the following curve.

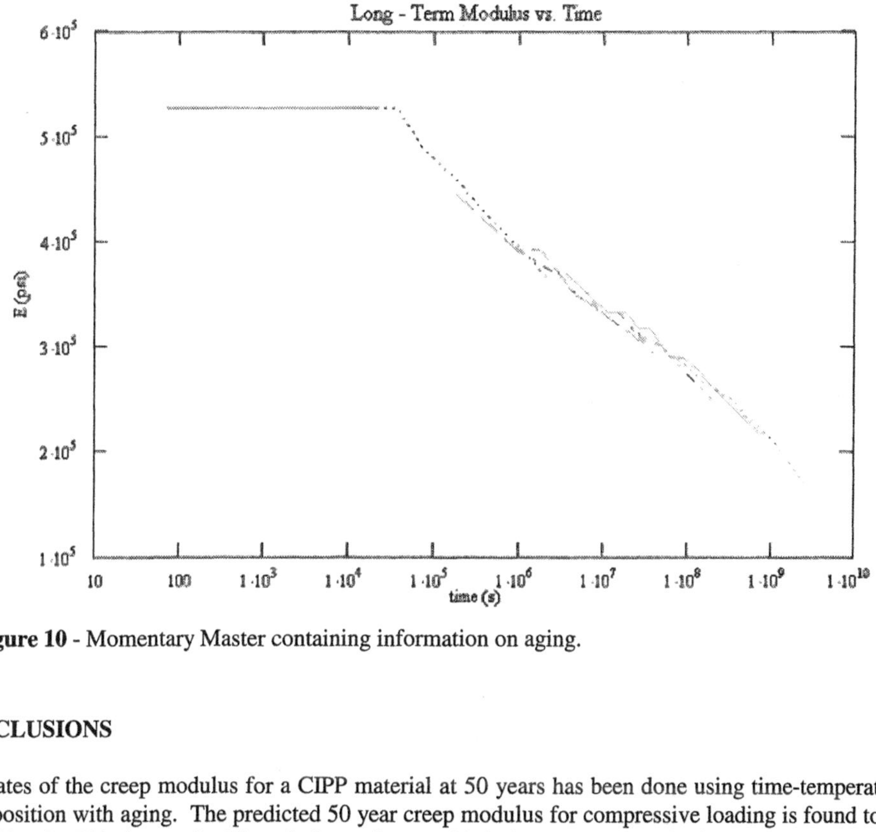

Figure 10 - Momentary Master containing information on aging.

CONCLUSIONS

Estimates of the creep modulus for a CIPP material at 50 years has been done using time-temperature superposition with aging. The predicted 50 year creep modulus for compressive loading is found to be 275,000 psi. This is not significantly inconsistent with industry practice of using 1/2 the short term modulus as the 50 year creep modulus. Work to determine the flexural and tensile moduli at 50 years is currently underway at Louisiana Tech University. These modulii will help designers of pipeline rehabilitation systems to more accurately select the thickness of liner systems.

ACKNOWLEDGEMENTS

This research is based on the generous support of the National Science Foundation under grant number CMS-9872378. Special thanks goes to Mr. S. Venkatraghavan for his work in conducting most of the elevated temperature creep deformation tests.

REFERENCES

ASTM. (1995). Standard Practice for Rehabilitation of Existing Pipelines and Conduits by the Inversion and Curing of a Resin – Impregnated Tube. ASTM Designation F1216 – 95.

Bradshaw, R. D. and Brinson, L. C. (1997). Physical Aging in Polymers and Polymer Composites: An Analysis and Method for Time – Aging Time Superposition. *Polymer Engineering and Science*, **37:1**, 31 – 44.

Bradshaw, R. D. and Gates, T. S. (1994). Effects of Physical Aging on Long – Term Creep of Polymers and Polymer Matrix Composites, NASA Technical Memorandum #109081.

Callister, W. D. (1997). *Materials Science and Engineering: An Introduction.* John Wiley & Sons, New York.

Ferry, John D. (1980). *Viscoelastic Properties of Polymers,* John Wiley & Sons, New York.

Hall, D. E. (1998). *Accelerated Testing of Cured in Place Pipe Liners*, NSF Grant # CMS-9872378

Janas, V. F. and McCullogh, R. L. (1987). The Effects of Physical Aging on the Viscoelastic Behavior of a Thermoset Polymer. *Composite Science and Technology*, 30, 99 – 118.

Matsumoto, D. S. (1988). Time – Temperature Superposition and Aging in Amorphous Polymers. *Polymer Engineering and Science*, **28:20**, 1313 – 1317.

Struik, L. C. E. (1978). *Physical Aging in Amorphous Polymers,* Elsevier Scientific Publishing Co., New York.

Timonshenko, S. and Gere, J. M. (1961). *Theory of Elastic Stability*, McGraw – Hill, New York.

Zhu, Meihan. (2000). *Evaluation of Long – Term Pipe Liner Buckling Models*, Louisiana Tech University.

Long Term Durability of Structural Materials
P.J.M. Monteiro, K.P. Chong, J. Larsen-Basse, K. Komvopoulos (Eds)

LIFETIME PREDICTION OF POLYOLEFIN GEOSYNTHETICS UTILIZING ACCELERATION TESTS BASED ON TEMPERATURE

Y. G. Hsuan[1] and R. M. Koerner[1]

[1]Geosynthetic Research Institute
Civil and Architectural Engineering
Drexel University
Philadelphia, PA 19104

ABSTRACT

A major issue in the use of geosynthetic products is to estimate the material's lifetime under various application environments. The degradation mechanism of the geosynthetic is governed by the polymer type and formulation used in manufacturing the product. This paper focuses on the oxidation degradation mechanisms of polyolefin geosynthetics, which are made from either polyethylene or polypropylene. The test data obtained from a durability study based on a high-density polyethylene (HDPE) geomembrane are used to illustrate the lifetime prediction methodology. The study utilized elevated temperatures alone to accelerate the oxidation degradation. The results of the study indicated that the antioxidant package plays a significant role in preserving the mechanical properties of the HDPE geomembrane. Such phenomenon should also apply to other polyolefin geosynthetic products. In addition, a brief description of the on-going high-pressure durability study on three different oriented polyolefin geosynthetics is presented.

KEYWORDS

Geosynthetic, geomembrane, geotextile, geogrid, polyolefin, polyethylene, polypropylene, antioxidant, lifetime, oxidative induction time, temperature, laboratory testing.

INTRODUCTION

In the last 30 years, a variety of polymer products have gradually been introduced into civil engineering field. The significant increase in the use of polymer probably started with the application of geosynthetics, which have been integrated into many geoenvironmental, geotechnical and transportation related projects. Nowadays, geosynthetics have been recognized as an engineering material. In some cases, they have been used as a structural component to replace conventional civil engineering materials, such as steel and concrete.

Polyethylene (PE) and polypropylene (PP) are two most widely used polymers in the production of geosynthetics. These two polymers consist of only carbon and hydrogen atoms, and belong to the polyolefin family. All polyolefines have similar chemical properties and degradation mechanisms, and they are known to have greater susceptibility to oxidation than other polymers (Grassie and Scott, 1985). The oxidation degradation can be initiated by energy arriving from thermal, photo, or radiation actives etc. Oxidation involves a series of chemical reactions that cause polymer degradation via chain scission, eventually leading to a decrease in engineering properties. Although quite arbitrary, the limit of service life to polymeric materials is often selected as a 50% reduction in a specific functional property. From an engineer perspective, the lifetime of a product should be longer than the required service life of the engineering system. In waste containment applications, the required service life can be few hundred years, whereas 75 to 100 years are generally specified for highway systems.

In the majority of commercial polyolefin products, antioxidants are added to protect the polymer and prolong the lifetime of the product. Conceptually, the lifetime of a stabilized product can be considered in three distinct stages, as illustrated in Figure 1. These stages are (a) depletion time of antioxidants, (b) induction time to the onset of polymer degradation, and (c) degradation of the polymer to decrease a meaningful engineering property to 50% of its original value.

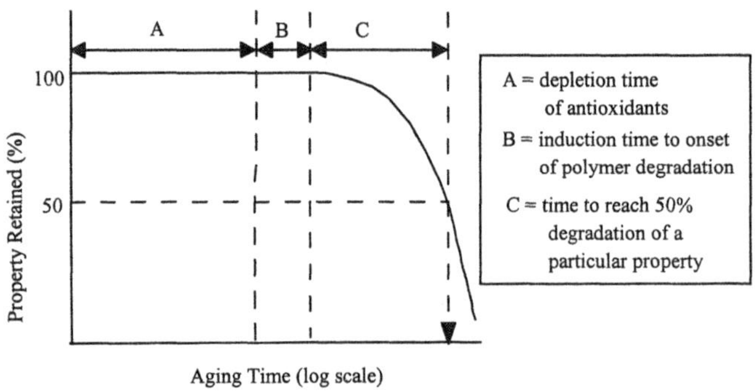

Figure 1: The three conceptual stages illustrating the lifetime of polyolefines

Stage A – Depletion of Antioxidants

The purposes of antioxidant are (1) to prevent polymer degradation during processing, and (2) to prevent oxidation reactions from taking place during the service life of the product. Obviously, there can only be a given amount of antioxidants in any formulation. Once the antioxidants are completely depleted, oxygen will begin to attack the polymer leading to Stages B and C in Figure 1. The duration of Stage A depends on the types and amounts of antioxidants in addition to the ambient conditions of the application.

Stage B – Induction Time

In a pure polyolefin resin, oxidation occurs extremely slowly at the beginning. Eventually, oxidation occurs rapidly. The reaction then decelerates and once again becomes very slow. This progression is illustrated by the curve of Figure 2(a). The initial portion of the curve is called the induction period. In the induction period, the polymer reacts with oxygen forming alkylperoxides (ROOH), as indicated in reactions (1) to (3). However, the amount of ROOH in this stage is very small and the peroxide does not further decompose into other free radicals. Thus, the acceleration stage oxidation cannot be achieved. In a stabilized polymer such as one with antioxidants, the accelerated oxidation stage takes an even longer time to reach. The antioxidants create an additional depletion time stage prior to the onset of the induction time, as shown in Figure 2(b).

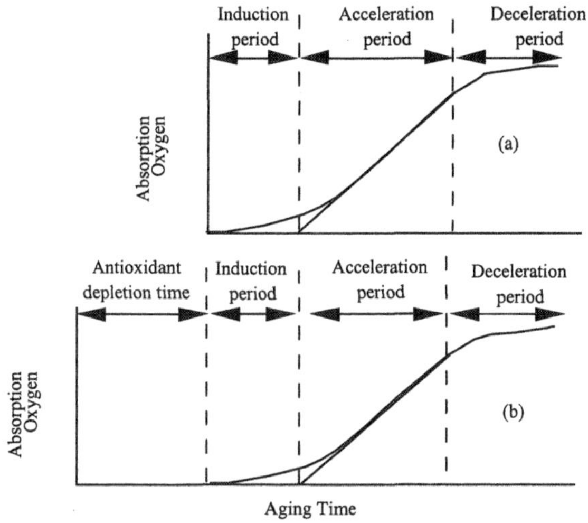

Figure 2: Curves illustrating various stage of oxidation for (a) pure polyolefines and (b) stabilized polyolefines

$$RH \rightarrow R\bullet + H\bullet \quad \text{(aided by energy)} \tag{1}$$
$$R\bullet + O_2 \rightarrow ROO\bullet \tag{2}$$
$$ROO\bullet + RH \rightarrow ROOH + R\bullet \tag{3}$$

where: RH represents the polyethylene chains, and the symbol "•" represents free radicals, which are highly reactive molecules.

Stage C – Polymer Degradation

As oxidation continues, additional ROOH molecules are formed. Once the ROOH concentration reaches a critical level, decomposition of ROOH begins, leading to a substantial increase in the amount of free radicals, as indicated in reactions (4) to (6). The free radicals attack the polymer chain readily, resulting an accelerated chain reaction, signifying the end of the induction period (Rapoport and Zaikov 1986).

$$ROOH \rightarrow RO\bullet + OH\bullet \quad \text{(aided by energy)} \tag{4}$$
$$RO\bullet + RH \rightarrow ROH + R\bullet \tag{5}$$
$$OH\bullet + RH \rightarrow H_2O + R\bullet \tag{6}$$

The oxidation produces a substantial amount of free radicals (R•), which can proceed further reactions, leading to either cross-linking or chain scission in the polymer. Subsequently the physical and mechanical properties of the polymer start to change. For the mechanical properties, both tensile break stress and break strain decrease, whereas to a lesser extent the yield stress increases and yield strain decreases. Ultimately, the degradation becomes so severe that all tensile properties decrease and the engineering performance is jeopardized. This signifies the lifetime of the product. As mentioned in the previous section, 50% change in engineering property was used to define as lifetime of the material. However, the product is still in place and may be meaningful for a considerable time thereafter.

In this paper, the oxidation degradation of polyolefin geosynthetics is illustration based on a HDPE geomembrane. The incubation environment is designed to be as close to the actual field condition as possible. The degradation mechanisms were accelerated using elevated temperatures. This results in a degradation of the tested sample in a relatively short period of time. The data from such elevated temperature testing can be extrapolated to predict the lifetime at a site-specific ambient temperature by using the Arrhenius method. The paper clearly demonstrates the importance of antioxidants in the overall lifetime of polyolefin geosynthetic products.

COMPOSITION OF HDPE GEOMEMBRANES

The components of an HDPE geomembrane consist of 96 to 97.5% polyethylene (PE) resin, 2 to 3 % carbon black and 0.5 to 1.0% antioxidants. It should be recognized that HDPE geomembranes are actually manufactured using polyethylene resin with a density between 0.932 and 0.940 g/cc. This resin density is classified as medium density polyethylene according to ASTM D 883. The addition of carbon black and antioxidants, however, increases the formulated density of the product to a range between 0.941 and 0.950 g/cc, which is defined as HDPE in ASTM D883. Therefore, the conventional term "HDPE" will be used.

- Polyethylene (PE) - The resin used for HDPE geomembranes is a linear copolymer, which is polymerized by using ethylene and α-olefin as comonomer. A greater amount of α-olefin added in the polymerization yields a lower density polyethylene polymer, which consists of higher free radical sites and lower crystallinity. Subsequently, PE of a lower density is more susceptible to oxidation than one with higher density. The same concept is also applied to PP, which consists of a lower percentage of crystallinity and higher number of tertiary carbon atoms than PE.
- Antioxidants - There are many types of antioxidants and each of them has unique functional characteristics and temperature ranges (Fay and King, 1994). The basic functions of the antioxidants are to provide stabilization by trapping (or deactivating) free radical species, or unstable species and converting them to stable molecules. Usually, synergistic mixtures of antioxidants of more than one type are used to prevent degradation occurring in processing as well as in service. Although the total amount of antioxidants in the geomembrane is relatively small, less than 1%, their existence is vital to the longevity of the product.
- Carbon black - Carbon black is added into a HDPE geomembrane formulation mainly for ultraviolet light stabilization. The loading range of carbon black in geomembranes is typically 2 to 3%, which is opacity level, above which additional of carbon black will not further improve ultraviolet resistance (Accorsi and Romero, 1995).

ARRHENIUS MODEL AND EQUATION

It is well established that chemical reactions of all types proceed more rapidly at higher temperatures that at lower temperatures. Sometimes the increase in the rate of reaction can be very dramatic for even a modest rise in temperature. With the introduction of the ideas involved in the kinetic theory of gases, in the mid-nineteenth century, a "theory" became possible to explain this phenomenon. The reaction rate can be presented conceptually as follows (Morrison and Boyd, 1992):

$$\text{Rate of reaction} = X * Y * Z \tag{7}$$

where: X = collision frequency; depending upon the density of particles (i.e, concentration or pressure).

Y = energy factor; determined by (a) the energy distribution of the gas molecules (i.e., velocity distribution, because in an ideal gas the energy is completely kinetic), and (b) the activation energy "E_{act}" for the reaction.
Z = probability factor; determined by how many of the colliding particles are properly oriented for the reaction.

Of these three terms, the energy factor is by far the most important in determining reaction rates as a function of temperature.

Figure 3 is a schematic diagram of the potential energy involved during the course of a reaction. The reactions must surmount an energy barrier, E_{act}, before going over into a change, or reacted, state. The heat of reaction, "ΔH", is also shown. It is the net consumption of energy required for a chemical reaction. If positive, the reaction requires additional heat progress and is endothermic. If negative, it dissipates heat as the reaction proceeds and is exothermic. Usually, little is known about the exact nature of the transition state in the HDPE geomembrane degradation, but E_{act} can be determined experimentally without exact details as to the transition state.

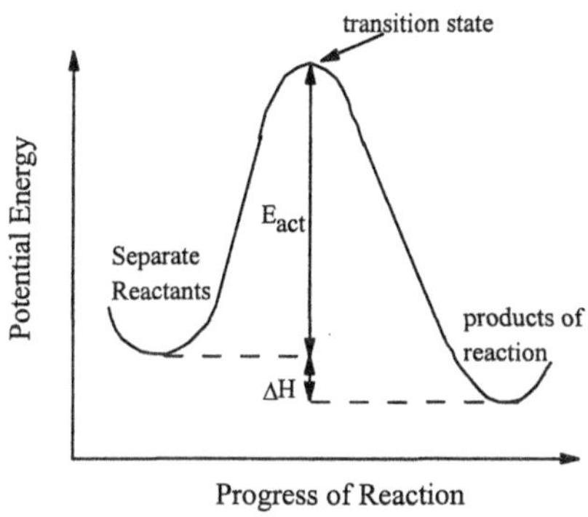

Figure 3: Potential energy involved in the progress of an assumed reaction

The distribution of velocities (i.e, energies) in an ideal gas was derived by Maxwell in 1852, and it is the usual bell-shaped (or Gaussian) curve common in statistics. This curve is shown in Figure 4. The fraction of particles in Figure 4 having energies greater than some value "E_{act}" is given by the relationship in Equation (8):

$$e^{-\frac{E_{act}}{RT}} \tag{8}$$

where:

E_{act} = activation energy for the particular reaction(s) (J/mol)
R = the gas constant (8.31 J/mole-K)
T = absolute temperature (K)

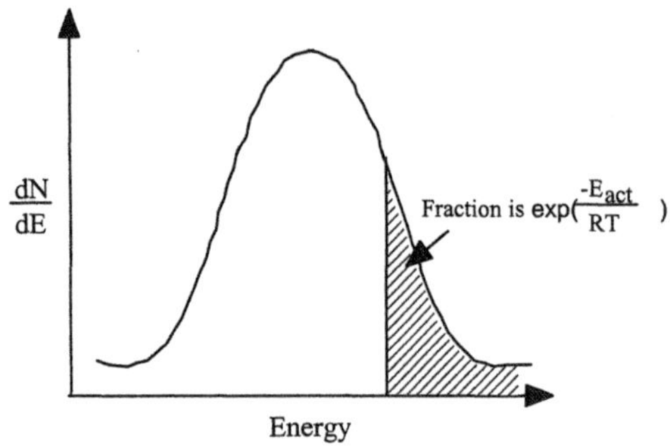

Figure 4: Distribution of energy in an ideal gas

The term "E_{act}" is fundamental to any particular reaction. Its value has been the focus of a significant amount of research and experimental investigation and will be seen to be critical in the work to follow.

By writing equation (7), using the relationship of equation (8) for the "Y" term gives equation (9):

$$R_r = (X)(e^{-\frac{E_{act}}{RT}})(Z) \tag{9}$$

For simple gas phase reactions, the constants "X" and "Z" can be estimated reasonable well. The activation energy term "Eact" remains for experimental determination. Assuming "X" and "Z" are independent of temperature and can be bracketed as a constant term "A", as shown in Equation (10).

$$R_r = (A)(e^{-\frac{E_{act}}{RT}}) \tag{10}$$

Equation (10) is the most widely referenced form of the so-called Arrhenius equation.

Taking the natural logarithm of both sides of Equation (10), a linear equation can be generated, as indicated in Equation (11).

$$\ln R_r = \ln A - \frac{E_{act}}{RT} \tag{11}$$

Plotting $\ln R_r$ against inverse temperature as shown in Figure 5, the slope of the resulting line will be "E_{act}/R" and the intercept on the vertical axis will be the constant "lnA". Figure 5 is called the "Arrhenius Plot", from which reaction rates at lower temperatures can be predicted by extrapolation from higher temperature experimental data.

For reactions involving more than the gas reactions used in the theory just discussed, there can be significant complications. This is probably the situation in the degradation of polymers. The main problem is that instead of single activation energy, there is often distribution of activation energies. This reflects the multiplicity of "reactions" that can lead to the final degraded product.

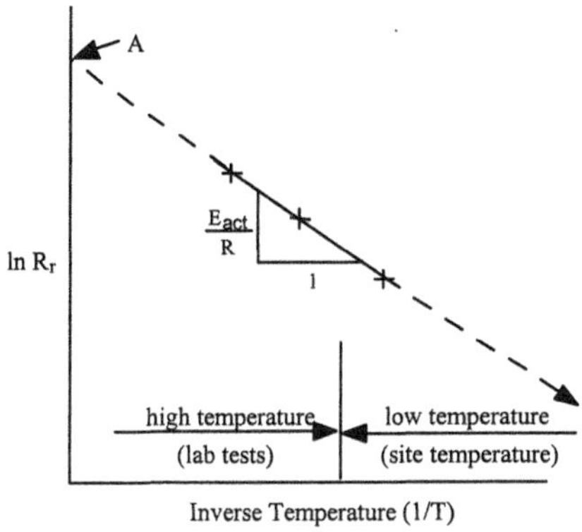

Figure 5: Generalized Arrhenius plot used for low temperature prediction from high temperature laboratory data

EXPERIMENTAL DESIGNS

It is most important in laboratory accelerated aging tests is that site conditions must be simulated as close as possible. This study attempts to simulate an HDPE geomembrane used as a landfill beneath 30 m of solid waste. Twenty (20) identical incubation devices were made for this study. The schematic diagram of the incubation cell can be seen in Figure 6. The cell was modifications of similar devices suggested by Mitchell and Spanner (1985). Using a 10 to 1 mechanical advantage, a static compressive stress of 260 kPa was applied to each sample. The stress was transmitted via a perforated load plate to 100 mm of sand and then to the geomembrane sample. A 300 mm head of water was maintained above the geomembrane. Beneath each sample was dry soil with small vent to the atmosphere. Five devices were maintained at each of four constant temperatures: 55, 65, 75 and 85°C. The incubated samples were retrieved at various time intervals and evaluated by a number of tests to monitor possibly properties changes.

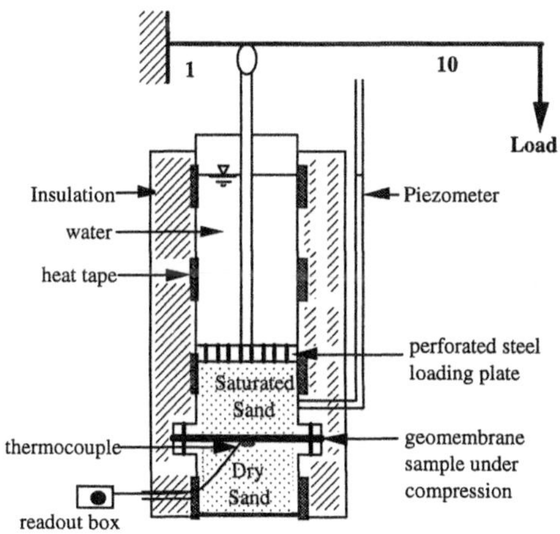

Figure 6: A schematic diagram of the compression column for incubation

EVALUATION TESTS ON INCUBATED SAMPLES

The geomembrane samples in the incubation devices were retrieved after predetermined lengths of time. The progression of the aging process was monitored by the results of a set of tests to track the behavior of the incubated geomembrane samples. The following tests were utilized:

Oxidative Induction Time (OIT)

OIT is the time required for a geomembrane specimen to be oxidized under a specific pressure and temperature. The OIT value indicates the amount of antioxidant (not the type) remaining in the test specimen. Howard (1973) showed that OIT is proportional to the antioxidant concentration in the same formulation package. However, for different antioxidant packages, direct comparison between two single OIT values can be misleading.

The test was performed according to ASTM D3895 utilizing a differential scanning calorimeter (DSC). The specimen was heated from room temperature to 200°C at a heating rate of 20°C/min under a nitrogen atmosphere. The gas flow rate was maintained at 50 ml/min. When 200°C was reached, the cell was maintained in an isothermal condition for 5 min. The gas was then changed from nitrogen to oxygen. The pressure and flow rate of oxygen were 35 kPa gauge pressure and 50 ml/min, respectively. The test was terminated after an exothermal peak was detected. Figure 7 shows a schematic diagram of a typical thermal curve with its identified OIT value.

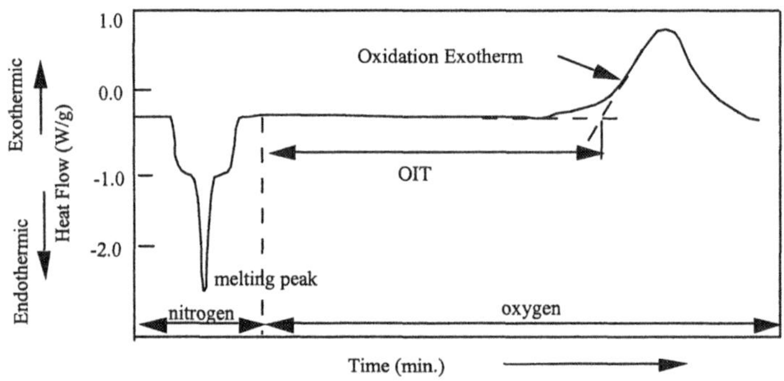

Figure 7: Thermal curve of a standard OIT test

Melt Index (MI) Test

Oxidative degradation of the polymer will induce either a cross-linking or chain scission reaction, resulting in changes in its molecular weight. Cross-linking reactions produce an increase in molecular weight, whereas chain scission reactions produce a decrease in molecular weight. The MI test, ASTM D 1238, is an indirect method to assess molecular weight of the polymer. A high melt index value indicates a low molecular weight, and vice versa. Hence, MI can be considered as an indicator of oxidation degradation.

Mechanical Tests

The mechanical performance of the incubated samples was evaluated using a tensile test according to ASTM D 638 Type V. Four tensile properties are obtained in this: yield stress, yield strain, break stress and break strain. Of the four properties the break strain has a greater sensitivity to molecular changes in the polymer than the other tensile properties.

TEST MATERIALS

A single type of commercially available HDPE geomembrane was used in this study. The antioxidant type is a mixture of hindered phenol (Inganox® 1010) and phosphites (Irgafos® 168), but the exact amount of each type of the antioxidants was not known. The properties of the original material were evaluated according to test methods described in above. The average properties of original non-incubated are as shown in Table 1.

TABLE 1
AVERAGE PROPERTIES OF ORIGINAL HDPE GEOMEMBRANE

OIT	80.5 min	MI	0.23 g/10min
Yield stress	19.2 MPa	Yield strain	13%
Break stress	36 MPa	Break strain	1570%

TEST RESULTS

Figure 8 shows the response of four material properties at 85°C. The other incubation temperatures have similar behavior but the OIT depletion is inverse proportional to the incubation temperature. The melt index and tensile properties do not appear to have any significant changes. Conversely, the OIT exhibit substantial changes with incubation time. The details of the study and data were presented in a prepared by Hsuan and Koerner (1998).

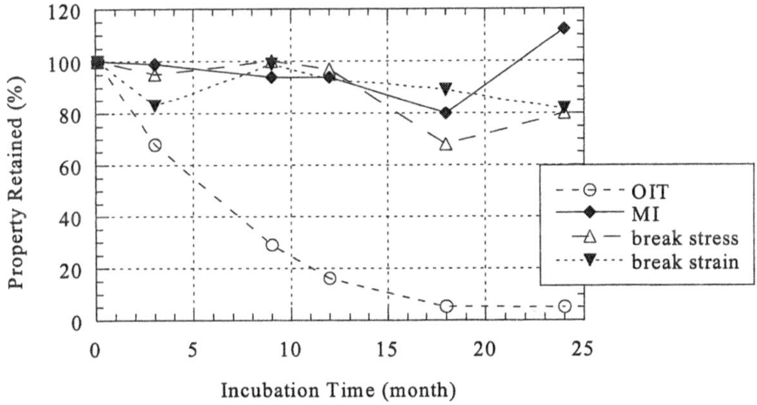

Figure 8 – Changes in properties with incubation time at 85°C

The comparison between OIT and other material properties demonstrates that molecular weight and tensile properties remain unchanged as long as antioxidants exist in the geomembrane. This observation also substantiates the hypothesis of lifetime demonstrating successive stages as shown in Figure 1 earlier. Tests that were performed in forced air oven tests at 115°C (i.e., more aggressive conditions than in this study) have found that the break stress and break strain dropped more than 50% only after OIT values reached negligible level (Hsuan and Guan, 1997). This again demonstrates the key role of antioxidants in that they must be depleted before engineering property degradation is noted.

LIFETIME PREDICTION ANTIOXIDANTS

Since there was no mechanical degradation in the incubated HDPE geomembrane samples after 24 months of incubation, the focus shifted to quantify the antioxidant depletion lifetime of the geomembrane, i.e., duration of Stage A in Figure 1. To provide this quantification, the lifetime of antioxidant is determined based on the depletion rate of the OIT values. Figure 9 shows a graph plotting log of OIT versus incubation time. A set of linear response curves results. The slopes of the lines represent the OIT depletion rate at each particular temperature.

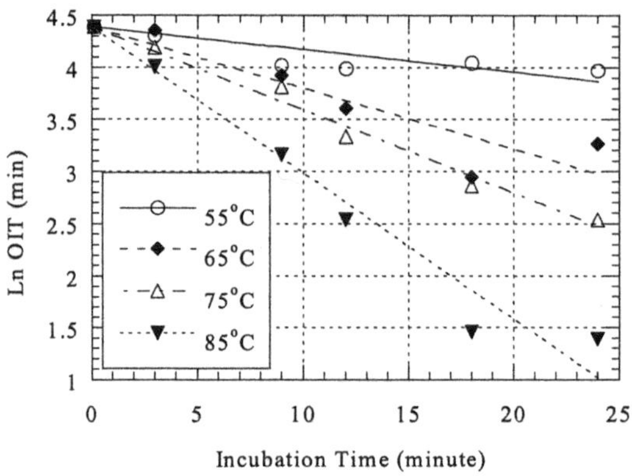

Figure 9 – Ln(OIT) versus incubation time

The generalized equation for each of the straight lines is expressed by Equation (12). The depletion rate at each incubation temperature can be obtained from Figure 9.

$$\ln(OIT) = \ln(P) - (S)(t) \tag{12}$$

Where:

OIT = OIT time (min)
S = OIT depletion rate (min/month)
T = incubation time (month)
P = constant (the original OIT value of the geomembrane)

The next step in the analysis is to extrapolate the OIT depletion rate to a lower temperature, such as the site-specific temperature. As explained by Koerner et al. (1992), Arrhenius equation can be used in such prediction. The Arrhenius equation can be expressed in Equation (13) for this study:

$$\ln(S) = \ln(A) + (-E/R)(1/T) \tag{13}$$

A linear relationship is established between ln(S) and inverse temperature, as shown in Figure 10. The activation energy is taken from the slope of the line, which results in value of 56 kJ/mol. The corresponding Arrhenius equation is expressed in Equation (14). Koerner and Koerner (1995) and Yazadini et al. (1995) found that the temperature at the base of two municipal solid waste landfills in Pennsylvania and California varies between 19 to 22°C. These landfills have been monitored for four and five years, respectively. The average temperature of 20°C is used to demonstrate the extrapolation calculation, as shown in Equation (15).

$$\ln(S) = 17.045 - 6798/T \quad (14)$$
$$S = 0.00212 \text{ at } 20^{\circ}C \quad (15)$$

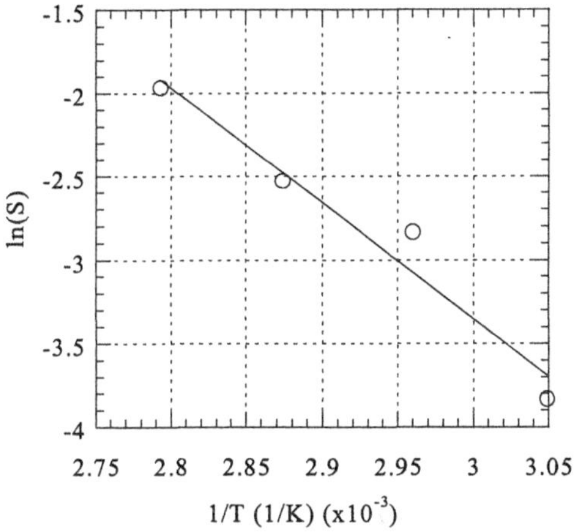

Figure 10 – Arrhenius plot of OIT rate of depletion

To predict the aging time that is required to deplete all the antioxidants in the HDPE geomembrane, a minimum boundary condition must be established. The OIT value for a pure unstabilized HDPE resin was evaluated and found to have OIT value of 0.5 minute. The minimum OIT value, original OIT value and OIT depletion rate at 20°C is now substituted into Equation (12). The lifetime of the antioxidant can then be determined, as shown in Equations (16) and (17):

$$\ln(0.5) = \ln(80.5) - (0.00212)(t) \quad (16)$$
$$t = 2{,}397 \text{ months (or 200 years)} \quad (17)$$

As a result, the predicted antioxidant lifetime at a service temperature of 20°C is approximately 200 years for this particular HDPE geomembrane formulation under the simulated test conditions presented herein.

ON-GOING DURABILITY STUDY

HDPE geomembranes can be considered as isotropic materials. The polymer chains are not stretched during the manufacturing process. On the other hand, the production of the geotextiles and geogrids involves stretching the polymer (i.e. cold drawing). The degree of drawing varies from product to product as well as the draw temperature. The microstructure of oriented polymers is very different from that of non-oriented polymers. The Peterlin Model (1966) describes the polymer orientation mechanism under tensile stress. The drawing results in a much denser amorphous phase and higher crystallinity than that in the comparable isotropic material. This dense structure retards the diffusive mobility of oxygen; thus, the rate of oxidation is retarded.

Using elevated temperatures to accelerate the oxidation degradation of polyolefin geotextiles and geogrids was found non-conclusive (Elias, 1998). They evaluated one PE geogrid and three different PP geotextiles. In the 36 months of incubation period, the tensile properties of the four tested samples exhibited substantial variation; thus, it was impossible to perform appropriate data analysis. In one of the PP geotextiles, micro-cracks that were generated during process were "healed" at elevated temperatures. The high incubation temperature probably altered the microstructure of the polymer, particularly in the amorphous phase.

An alternative method was considered to accelerate the oxidation degradation of oriented geosynthetics. As indicated in Equation (7) previously, the rate of reaction is also governed by concentration of reactants. Increase oxygen concentration of the incubation environment shall enhance the rate of oxidation degradation. On the other hand, the incubation temperature is kept at relatively low (35, 45 and 55°C) to avoid any significant thermal effect on the microstructures. Table 2 shows the current incubation matrix; however, there is insufficient amount of data to be presented at this time.

TABLE 2
TEST MATRIX FOR THE PRESSURE INCUBATION STUDY

Temperature (°C)	Oxygen Pressure (MPa)	Air Pressure (MPa)			
35	1.2	2.1	3.4	4.8	6.2
45	-	-	-	-	6.2
55	-	-	-	-	6.2

Currently, the European standard, under technical committee CEN/TC 189 also evaluates the effect of pressure on oxidation. In their draft proposal, the geotextile is immersed in Na(OH) liquid and pressurized to 5 MPa of pure oxygen at a temperature of 80°C.

CONCLUSION

Utilizing elevated temperatures to accelerate the thermal oxidation of HDPE geomembranes was found to be relevant and appropriate. Elevated incubation temperatures up to 85°C did not cause changes in microstructure that could lead to fluctuation in tensile properties. Thus, the temperature acceleration methodology can be applied to investigate the thermal degradation of other isotropic polymeric materials.

The importance of antioxidants on the longevity of polyolefin geosynthetics is hereby demonstrated by this study using an HDPE geomembrane. For engineering systems that require long service life, appropriate formulated polyolefin product must be required. The OIT test is a simple method to assess importance of the antioxidants. Furthermore OIT can be included in a specification. However, for comparing different types of antioxidants, OIT test must be used together with oven aging to evaluate the depletion rate. This is also possible to specific accordingly

ACKNOWLEDGEMENTS

This ongoing project is supported by the National Science Foundation (Grant No. CMS 9312772); the U.S. Environmental Protection Agency (Contract No. CR 821448); and the consortium of Geosynthetic Research Institute member organizations. Students also participating in the study have been James Fleck, Monica Sculli and Zeqing Quan. Sincere appreciation is extended to all involved.

REFERENCES

Accorsi, J. and Romero, E., (1995). "Special Carbon Blacks for Plastics" Plastics Engineering, April, pp. 29-32.

Elias, V., Salman, A., Juran, I., Pearce, E., and Lu, S. (1999). *Testing Protocols for Oxidation and Hydrolysis of Geosynthetics*, Publication No. FHWA-RD-97-144, 186 pgs.

Fay, J. J. and King R. E., (1994). "Antioxidants for Geosynthetic Resins and Applications", *Geosynthetic Resins, Formulations and Manufacturing*, Edited by Hsuan, Y.G. and Koerner, R.M., GRI Conference Series, Published by IFAI, St Paul, MN., U.S.A., pp. 77-96.

Grassie, N. and Scott, G., (1985). *Polymer Degradation and Stabilization*, Cambridge University Press, New York, U.S.A.

Howard, J.B., (1973). "Data for Control of Stability in Polyolefin Wire and Cable Compounds", Polymer Engineering Science, Vol. 13, No. 6, pp. 429-434.

Hsuan, Y.G., and Guan, Z., (1997). "Evaluation of the Oxidation Behavior of Polyethylene Geomembranes using Oxidation Induction Time Tests", *Oxidative Behavior of Materials by Thermal Analytical Techniques*, ASTM STP 1326, A.T. Riga and G.H. Patterson, eds. ASTM, Philadelphia, PA, pp. 76-90.

Hsuan, Y.G., and Koerner, R.M., (1998) "Antioxidant Depletion Lifetime in High Density Polyethylene Geomembranes", Journal of Geotechnical and Geoenvironmental Engineering, Vol. 124, No. 6, pp. 532-541.

Koerner, G.R. and Koerner, R.M., (1995). "Temperature Behavior of Field Deployed HDPE Geomembranes", Geosynthetics '95 Conference Proceedings, Nashville, TN., pp. 921-938.

Mitchell, D.H. and Spanner, G.E., (1985). "Field Performance Assessment of Synthetic Liners for Uranium Tailings Ponds", Status Report, Battelle PNL, U.S. NRC, NUREG/CR-4023, PNL-5005.

Morrison, R.T., and Boyd, R.N., (1992). *Organic Chemistry*, 3rd ed., Published by Allyn and Bacon Inc, Boston, USA.

Peterlin, A. (1966). "Molecular Mechanism of Plastic Deformation of Polyethylene," *The Meaning of Crystallinity of Polymers*, Ed. by Price, F.P., Journal of Polymer Science, Part C, Polymer Symposia, No. 8, pp. 123-132.

Rapoport, N.Ya and Zaikov, G.E., (1986). "Kinetics and Mechanism of the Oxidation of Stressed Polymer", *Developments in Polymer Stabilization-4*, Chapter 6, edited by Scott, G., Applied Science Publishers Ltd., London, pp. 207-258.

Yazadini, R., Campbell, J.L. and Koerner, G.R., (1995), "Long-Term In-Situ Strain Measurements of a High Density Polyethylene Geomembrane in a Municipal Solid Waste Landfill", Geosynthetics '95 Conference Proceedings, Nashville, TN., pp. 893-906.

Long Term Durability of Structural Materials
P.J.M. Monteiro, K.P. Chong, J. Larsen-Basse, K. Komvopoulos (Eds)

CYCLIC LOADING EFFECTS ON DURABILITY OF POLYMER SYSTEMS

A.M. Vinogradov[1], C.H.M. Jenkins[2] and R.M. Winter[3]

[1] Department of Mechanical and Industrial Engineering, Montana State University
Bozeman, MT 59717 USA
[2] Department of Mechanical Engineering, South Dakota School of Mines and Technology
Rapid City, SD 57701USA
[3] Department of Chemical Engineering, South Dakota School of Mines and Technology
Rapid City, SD 57701USA

ABSTRACT

The paper concerns the response of polymer systems subjected to combined static and cyclic loads. The effects of mean stresses, stress amplitudes, frequencies and temperature on the creep behavior of Nylon 6/6 and polyvinylidene fluoride (PVDF) are investigated. The study is based on a systematic experimental program involving stress-strain tests, creep tests, vibrocreep tests and post-cyclic stress-strain tests. The results of the investigation demonstrate that, under cyclic loading conditions, both polymers exhibit accelerated creep rates. This phenomenon has been observed even in the range of stresses well below the viscoelastic linearity limit. It is shown that the vibrocreep behavior of the polymers is essentially nonlinear, as their response to superimposed sustained and cyclic loads does not represent a simple superposition of the responses to static and fully reversed cyclic loads applied separately.

KEYWORDS

Polymer systems, creep, vibrocreep, viscoelasticity, creep-fatigue interaction, damage, cyclic loading.

INTRODUCTION

Polymeric material systems have been increasingly used in a broad range of applications, including aerospace and automotive industries, electronics, medical products and consumer appliances. The widespread use and rapid growth of these materials result from their superior qualities such as high strength to weight ratio, excellent corrosion resistance, low friction coefficients and tailorability of their properties. Recently, a new generation of "smart" piezoelectric polymer systems has emerged with an ability to actively react to changing stimuli. These materials have been increasingly integrated in structural design as active elements capable of sensing and responding intelligently to external stimuli. A broad range of applications utilizing such functions includes active vibration damping, acoustic suppression, damage detection and self-inspection of structural integrity. However, as the application range of polymers tends to expand, new challenges arise. At present, of critical importance are the issues of durability and long-term integrity of polymeric systems subjected to highly damaging effects of elevated temperatures and cyclic loading regimes.

This paper concerns the creep response of polymeric systems subjected to combined static and cyclic loads under various temperature conditions. Following an overview of major research findings in the field, a systematic parametric study involving two materials, Nylon 6/6 and polyvinylidene fluoride (PVDF), is reported. The obtained results are discussed in the light of related research findings reported in the literature.

AN OVERVIEW OF MAJOR RESEARCH FINDINGS

It is well known that, under cyclic loading conditions, engineering materials undergo progressive deterioration of their qualities, ultimately culminating in the termination of their service life. This phenomenon has been studied for more than a century, primarily, in metals and alloys. In contrast, the state-of-the-art understanding of the cyclic response of polymers and their composites has not been advanced to the same degree, Hertzberg & Manson (1980), Dillard (1991), Suresh (1998).

A systematic review of the studies concerning the response of polymeric solids under cyclic loading conditions indicates that the problem has been treated using two different approaches. The most common approach is to focus on the conditions of material failure due to cyclic loads in the presence of creep. Respectively, the problem is treated in terms of creep-fatigue interaction. The second approach concerns primarily the tendency of materials to alter their time-dependent behavior due to the presence of cyclic loads as compared with static creep. Of prime interest in this regard are time-dependent deformation processes characterized in terms of creep strains or strain rates. As suggested by Courtney (2000), a distinction between these viewpoints can be drawn such as follows: "Whether we should view the problem as creep enhanced by the fatigue environment or vice versa depends on several factors. When the cyclical stress (strain) amplitude is small in comparison to the mean stress, it is proper to view the phenomenon as one of creep perturbed by fatigue. This also holds when the applied frequency is low and/or temperature is high. Under circumstances opposite to these, failure can be considered a fatigue failure accelerated by diffusional processes". It appears that such classification is useful for reference purposes, however, one notes that both approaches deal essentially with the same phenomenon as they concern either the entire process of cyclic creep or the culminating point of that process, i.e. the singular event of material failure.

The concept of creep-fatigue interaction is based on the empirical evidence that failure of polymers due to dynamic fatigue depends on their creep characteristics. Since creep rates tend to increase at elevated temperatures, a number of studies postulate that cyclic life of polymers depends exclusively on hysteretic heating, Riddell, Koo & O'Toole (1966), Tauchert (1967), Higuchi & Imai (1970) and Weaver & Beatty (1978). In particular, Riddell, Koo & O'Toole (1966) stated that fatigue in polymers "does not follow the concepts of crack propagation established for metals and other elastic materials. Therefore the nature of failure cannot be described in terms of mechanical fracture".

However, contrary to this belief, numerous investigations have demonstrated that polymers subjected to cyclic loads undergo progressive damage evolution processes such as crazing and shear band formation, followed by crack nucleation and propagation, Argon & Hannoosh (1977), Hertzberg & Manson (1980), Klaus (1983), Williams (1987), Kramer & Berger (1990), Takemori (1990), Argon & Cohen (1990), Narisawa & Ishikawa (1990), Hristov & Yee (1994), Hristov, Yee & Gidley (1994), Jones & Lesser (1998). At present, it is clear that creep-fatigue interaction involves coupling between several leading factors, i.e., intrinsic viscoelasticity, damage evolution and hysteretic heating. The character of damage evolution appears to have not only an immediate effect on the development of cyclic creep, but also an interactive effect. Thus, by affecting the interrelation between energy dissipation and energy absorption in the material, damage processes tend to stimulate the increase of internal temperature. The latter, in turn, tends to accelerate cyclic creep. The role of an individual factor in the process of material deterioration depends on the loading and temperature conditions as well as the type of the polymer under consideration. In particular, it has been observed that fatigue failure of polymers is dominated by hysteretic heating at higher stress levels and frequencies, whereas damage accumulation processes become critical in the range of lower stresses and frequencies, Constable, Williams & Burns (1970), Lesser (1995).

To date, efforts to establish a correlation between the service conditions and durability of polymer systems remain limited. The problem has attracted attention relatively recently and, unlike metals, no conclusive results have been reported in regard to polymers and polymer based composites, Zhou & Brown (1993), Crocombe & Richardson (1999), Sermage, Lemaitre and Desmorat (2000). Attempts have been made to investigate creep deformations of polymers under various combinations of static and cyclic loads as compared with their respective static creep response, Maksimov & Urzhumtsev (1970) and Kregers & Maksimov (1970). It has been observed that, consistently, under cyclic loading conditions, deformation rates of polymers tended to increase. This phenomenon has been defined as "vibrocreep".

Several elastomers have demonstrated vibrocreep behavior similar to that observed in polymers. Thus, Derham & Thomas (1977) have reported that the natural rubber (SMR 5), characterized as a linearly viscoelastic material under static loading conditions, has exhibited highly nonlinear time-dependent effects under low-frequency load-unload cycles. These affects amounted to considerably accelerated creep rates, up to ten times higher than those measured in static experiments. The cyclic behavior of carbon black filled butyl rubber examined by McKenna & Zapas (1981) has shown a strong dependence on the frequency and the waveform of loading-unloading cycles. In some experiments, inverse effects, i.e. a decrease in cyclic creep rates and increase of the material lifetime have been observed.

Research regarding vibrocreep and creep-fatigue interaction in composite materials has often produced controversial results. Thus, Turtsinsh (1972) has reported that the response of glass fiber reinforced polymer composites subjected to simultaneous torsion and compression combined with axial oscillations at different stress levels, amplitudes, frequencies and temperatures have shown no vibrocreep effects. However, Zhang, Xue & Wang (1994) have reached an opposite conclusion regarding the response of short fiber polymer matrix composites, observing considerable changes in effective material properties and increasingly accelerated creep rates at elevated temperatures. Similarly, a study by Fitzgerald et al. (1992) has demonstrated that samples of poly(dimethylsiloxane) (PDMS) elastomer filled with 35 vol% aluminum oxide (Alcoa T-61) subjected to a static load with superimposed dynamic oscillations developed accelerated degradation involving molecular changes and breakage in the load bearing chains of the polymer. The overall cyclic creep behavior of PDMS has shown to be different and more complex than that predicted by linear viscoelasticity.

EXPERIMENTAL PROGRAM

In this study, a systematic step-by-step experimental methodology has been developed with an objective to determine cyclic loading effects on the time-dependent behavior of polymer systems. Two materials have selected for investigation, Nylon 6/6 and a polyvinylidene fluoride (PVDF). The experimental program has involved the following components: static stress-strain tests, quasi-static creep tests, and cyclic creep tests. In the case of Nylon 6/6, additional static stress-strain tests of samples initially subjected to static and cyclic creep experiments have been performed. Static stress-strain tests have provided the basic mechanical characteristics of the polymers such as the instantaneous Young's modulus, E, yield stress σ_y, yield strain ε_y, ultimate stress σ_u, and ultimate strain ε_u. The results of quasi-static creep tests have been used to characterize the creep response of the materials under sustained loading conditions. Attention has been focused on two viscoelastic characteristics, the creep compliance and viscoelastic linearity limit σ^*. Cyclic creep tests have been designed to determine the degree of vibrocreep effects under various loading and temperature conditions. Throughout the entire program, at least three-five identical experiments have been performed to ensure reproducible results.

The experimental study has focused on the material response under a combined action of static and cyclic loads $\sigma(t) = \sigma_m + \sigma_a \sin \omega t$ depending on the following parameters: mean stress σ_m, stress amplitude σ_a, frequency ω, and temperature T. The results have been characterized in terms of vibrocreep strain $\varepsilon_v(t)$ or normalized vibrocreep creep strain $\varepsilon_v(t)/\sigma_m$. Vibrocreep effects have been assessed in relation to static creep defined in terms of $\varepsilon_m(t)$ or $\varepsilon_m(t)/\sigma_m$, where $\varepsilon_m(t)$ denotes time-dependent strain under sustained stress σ_m. It has been considered that polymers exhibit vibrocreep effects if the cyclic creep curve $\varepsilon_v(t)$ is found to be outside the envelope formed by the creep curves ε_{m+a} (t) and ε_{m-a} (t) corresponding to static stresses (σ_m+σ_a) and (σ_m-σ_a), respectively. In the linear viscoelstic range vibrocreep effects have been assessed through a comparison of normalized vibrocreep strains $\varepsilon_v(t)/\sigma_m$ with the respective creep compliances $\varepsilon_m(t)/\sigma_m$, see Figure 1.

(a)

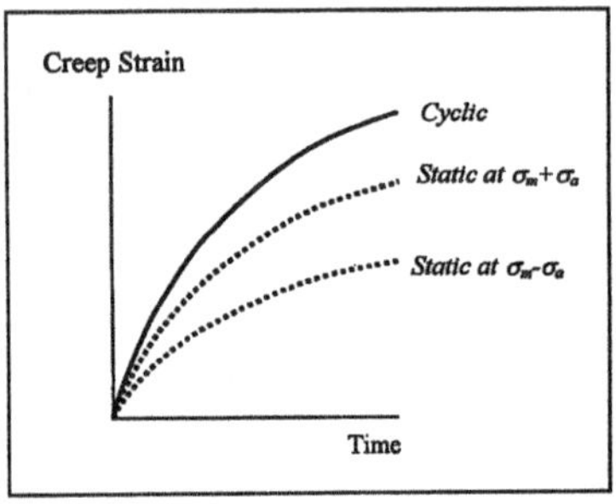

(b)

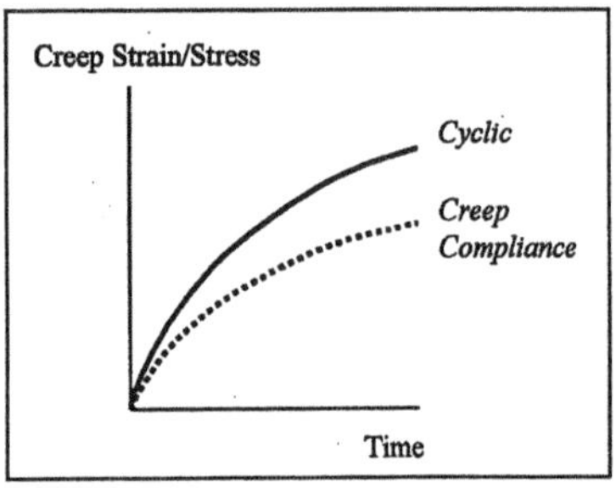

Figure 1: Vibrocreep criteria: (a) general case; (b) linear viscoelstic case.

VIBROCREEP OF NYLON 6/6

The synthetic polymer Nylon 6/6 (DuPont Zytel® 42A NC010 Polyamide) is a high performance thermoplastic with attractive engineering properties. The polymer exhibits excellent wear, abrasion and impact resistance, low friction coefficient, high strength, toughness and stable elastic properties at elevated temperatures. Due to these qualities, Nylon 6/6 has been used in a wide variety of applications including automotive, electrical/electronics and consumer items, Daniels (1989), Strong (2000).

Static creep tests of Nylon 6/6 have been performed using an in-house built testing apparatus, equipped with temperature chamber, load cell, and linear variable differential transformer (LVDT) for displacement measurements. Cyclic creep experiments have been performed on an Instron Model 1350 servo-hydraulic machine equipped with a temperature chamber. In all experiments, environmental conditions have been monitored by thermocouples and a humidity transmitter. Samples of Nylon 6/6 have been prepared in accordance with ASTM D638-96 Type II standards.

The following mechanical properties of Nylon 6/6 at T = 23°C have been determined from tensile stress-strain experiments: the elastic modulus E = 1.42 GPa, yield stress σ_y = 70 MPa; yield strain ε_y = 0.0642, and ultimate stress σ_u = 78 MPa. These results closely agree with published experimental data for Nylon 6/6, Daniels (1989). Stress-strain tests at elevated temperatures T = (35, 41, 50, 59 & 68)°C have shown an almost linear decrease of the elastic modulus and yield stress, such that E = 0.46 GPa and σ_y = 10 MPa at T = 68°C.

Creep tests of Nylon 6/6 under static loading conditions have been performed at six different stress levels, i.e. σ = (0.1, 0.16, 0.2, 0.3, 0.4 & 0.5)σ_y within the same temperature range as the stress-strain tests. It has determined that, at T = 23°C, the material can be treated as linearly viscoelastic at stresses below σ^* = 0.3σ_y. Above this limit Nylon 6/6 tends to exhibit nonlinear stress dependent viscoelastic behavior. The linear viscoelastic limit at elevated temperatures has been found as 0.3 of the yield stress at the respective temperature.

The vibrocreep response of the polymer has been studied within the experimental range of parameters summarized in Tables 1 and 2 at T = 23 °C and T = (35 & 41)°C, respectively. In all experiments, mean stresses have been consistently maintained below the linear viscoelastic limit of $0.3\sigma_Y$ of the respective temperature. This strategy has been adopted in order to exclude the effects of stress induced material nonlinearity.

TABLE 1
CYCLIC TEST CONDITIONS FOR NYLON 6/6 (T = 23°C)

Mean Stress $\sigma_m = n\sigma_y$		Amplitude $\sigma_a = m\sigma_y$				Frequency (Hz)		
n	0.12	m	0.04	0.075	0.1	1	10	20
	0.16		0.04	0.075	0.1	1	10	20
	0.20		0.05	0.075	0.1	1	10	20

TABLE 2
CYCLIC TEST CONDITIONS FOR NYLON 6/6 (T = 35°C & 41°C)

Mean Stress $\sigma_m = n\sigma_y$		Amplitude $\sigma_a = m\sigma_y$			Frequency (Hz)	
n	0.16	m	0.04	0.1	1	10
	0.20		0.05	0.1	1	10

Following creep and vibrocreep experiments, typically performed for 12 hours, Nylon 6/6 samples were subjected to a second-stage tensile stress-strain tests to failure with the objective to determine the material response and any possible changes in material properties as a result of the preceding experimental stage. Second stage tests have been performed at T = (23, 35 & 41)°C.

Some representative results demonstrating the effects of frequency, amplitude and mean stress on the time-dependent response of Nylon 6/6 at various temperatures are shown in Figures 2 - 6. Figure 7 illustrates the tensile stress-strain response of Nylon 6/6 following cyclic and static creep experiments.

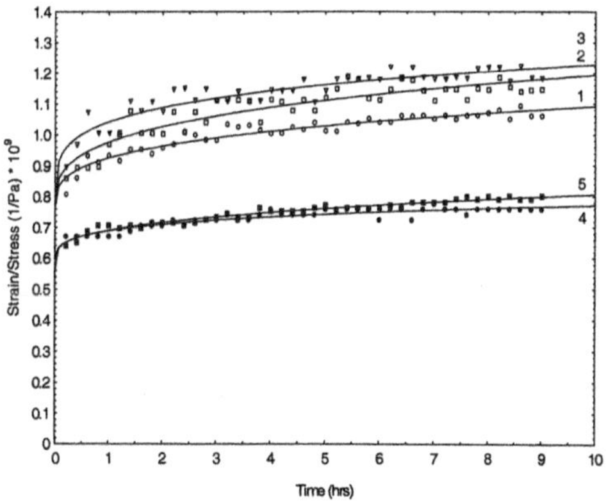

Figure 2: Vibrocreep of Nylon 6/6
Frequency effect (T= 23°C)
Cyclic ($\sigma_m = 0.20\sigma_Y$, $\sigma_a = 0.075\sigma_Y$)
1:ω =1 Hz; **2:**ω =10 Hz; **3:**ω =20 Hz
Static: **4:**$(0.20)\sigma_y$, **5:**$(0.30)\sigma_y$. (σ_y = 70 MPa)

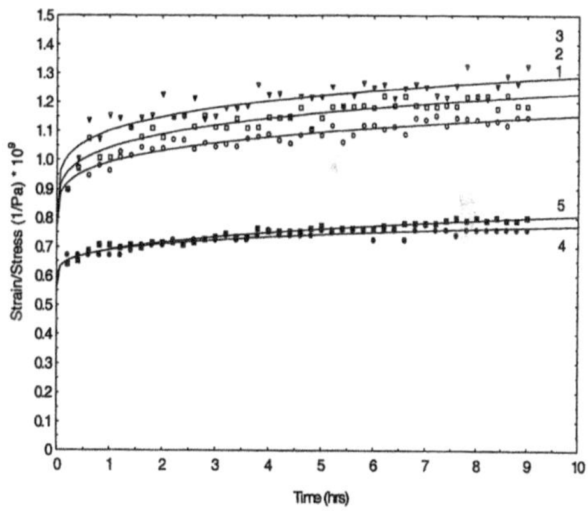

Figure 3: Vibrocreep of Nylon 6/6
Amplitude effect (T= 23°C)
Cyclic ($\sigma_m = 0.20\sigma_y$; ω = 20 Hz)
1:$\sigma_a = 0.05\sigma_y$, **2:**$\sigma_a = 0.075\sigma_y$, **3:**$\sigma_a = 0.1\sigma_y$
Static: **4:**$(0.16)\sigma_y$, **5:**$(0.30)\sigma_y$. (σ_y = 70 MPa)

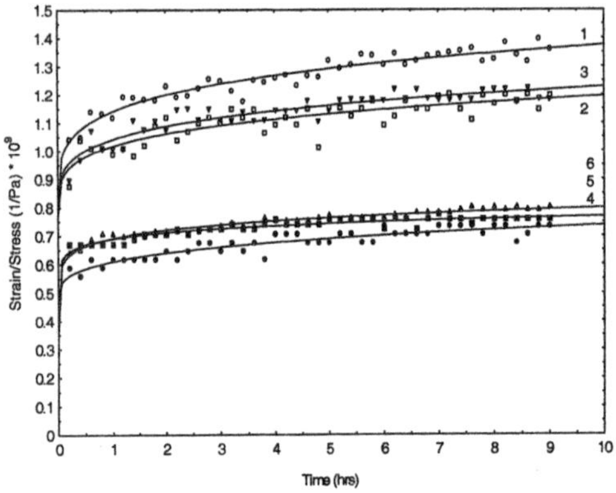

Figure 4: Vibrocreep of Nylon 6/6
Mean stress effect (T = 23°C)
Cyclic ($\sigma_a = 0.075\sigma_y$; ω = 20 Hz)
1:$\sigma_m = 0.12\sigma_y$; **2:**$\sigma_m = 0.16\sigma_y$; **3:**$\sigma_m = 0.20\sigma_y$
Static: **4:**$(0.16)\sigma_y$; **5:**$(0.20)\sigma_y$, **6:**$(0.30)\sigma_y$. (σ_y = 70 MPa)

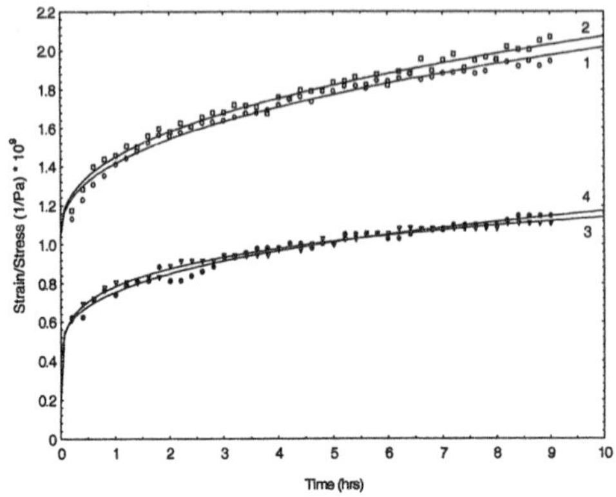

Figure 5: Vibrocreep of Nylon 6/6
Frequency effect (T = 35°C)
Cyclic ($\sigma_m = 0.20\sigma_Y$, $\sigma_a = 0.10\sigma_Y$)
1:ω =1 Hz; **2:**ω =10 Hz
Static: **3:**$(0.20)\sigma_y$, **4:**$(0.30)\sigma_y$. (σ_y = 55 MPa)

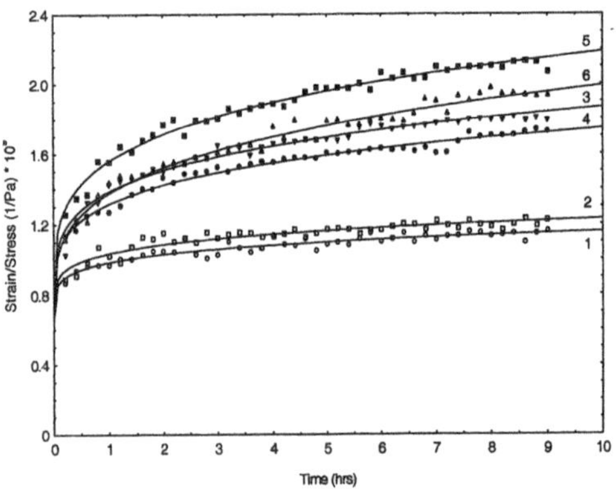

Figure 6: Vibrocreep of Nylon 6/6
Mean stress and temperature effects
Cyclic ($\sigma_a = 0.10\sigma_y$, ω = 10 Hz)
1:$\sigma_m = 0.16\sigma_y$, **2:**$\sigma_m = 0.20\sigma_y$ (T = 23 °C)
3:$\sigma_m = 0.16\sigma_y$, **4:**$\sigma_m = 0.20\sigma_y$ (T = 35°C)
5:$\sigma_m = 0.16\sigma_y$, **6:**$\sigma_m = 0.20\sigma_y$ (T = 41°C)
σ_y = 70 MPa at 23 °C

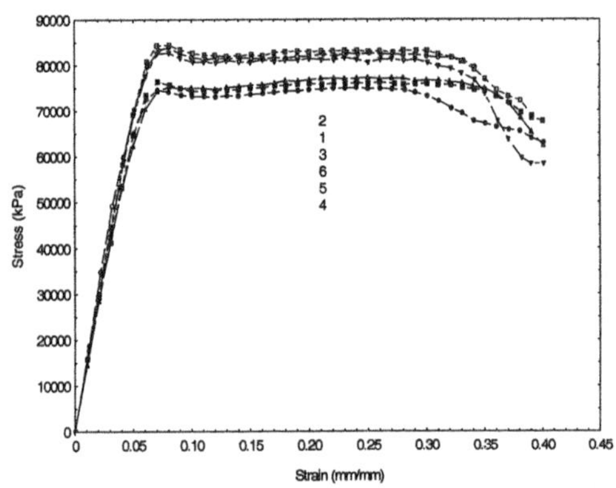

Figure 7: Tensile stress-strain response
of Nylon 6/6 following cyclic and static creep
experiments (T = 23 °C)
Cyclic ($\sigma_m = 0.16\sigma_y$, $\sigma_a = 0.075\sigma_y$)
1:1 Hz, **2:**10 Hz, **3**- 20 Hz; Static Creep **4:**$0.16\sigma_y$,
5:$0.20\sigma_y$, **6:**Virgin Material; σ_y = 70 MPa at 23 °C

VIBROCREEP OF PVDF

Polyvinylidene fluoride (PVDF) is a piezoelectric semi-crystalline polymer with typical crystallinity of, approximately, 50%. The amorphous phase of the polymer has the properties of a supercooled liquid with the glass transition temperature of about -50°C. The molecular structure of PVDF consists of a repeated monomer unit $-CF_2-CH_2-$. The atoms are covalently bonded together, forming long molecular chains. Since the hydrogen atoms are positively charged and the fluoride atoms are negatively charged with respect to the carbon atoms, PVDF is inherently polar. However, the net polar moment of the material is zero due to the random orientation of the crystallites. Permanent dipole polarization of PVDF is obtained through a technological process that involves stretching and polling of extruded thin sheets of the polymer. An applied electric field of up to 100 kV/mm at an elevated, typically, 103°C temperature causes permanent polarization maintained after the material is cooled to room temperature. Typically, PVDF is produced in the form of thin films of thickness ranging from 9 to 800 μm (10^{-6} m). A thin layer of nickel, silver or copper is deposited on both material surfaces in order to provide electrical conductivity when an electric field is applied, or to allow measurements of the charge induced by mechanical deformations. Thus, the product represents a composite polymer system with effective properties determined by the characteristics of its constituents.

In this study, thin film 7.62 mm x 76.2 mm PVDF samples of 28 μm thickness including 10 μm silver layers deposited on both surfaces have been tested in compliance with the ASTM D882-95a standards. To implement the experimental program, a special apparatus has been designed involving a load pre-tension device, MB Electronics Model PM-50 electro-dynamic vibration exciter, power amplifier, signal generator, and temperature chamber. Displacements have been measured by LVDT. The instrument has been equipped with software supported data acquisition system. Cold room facilities have been used to test the material at below freezing temperatures.

Since PVDF is a semicrystalline polymer with unidirectional alignment of molecular chains, the material has shown strong orthotropic in-plane properties in the direction coinciding with (direction 1), and perpendicular to (direction 2) the orientation of molecular chains. A summary of the results obtained from stress-strain tests of PVDF is provided in Table 3.

TABLE 3
MECHANICAL PROPERTIES OF PVDF

Properties	Direction 1	Direction 2
Elastic Modulus, E	1.96 GPa	1.69 GPa
Yield Stress, σ_y	30.4 MPa	23.5 MPa
Strength, σ_u	245.6 MPa	31.1 MPa

Based on creep tests, the viscoelastic linearity limits for PVDF have been identified such that $\sigma^*_1 = 0.6\sigma_{y1}$ and $\sigma^*_2 = 0.75\sigma_{y2}$ in the directions 1 and 2, respectively. Linear viscoelastic properties of PVDF have been characterized in Holloway (1997), Vinogradov and Holloway (1997, 1999) and Vinogradov (1999). In this study, cyclic tests of PVDF have been performed in the material direction 1 at two temperatures, T = 23°C and T = -25°C over an experimental range of loading parameters summarized in Tables 4 and 5.

TABLE 4
CYCLIC TEST CONDITIONS FOR PVDF (T = 23°C)

Mean Stress $\sigma_m = n\sigma_y$		Amplitude $\sigma_a = m\sigma_y$					Frequency (Hz)		
n	0.30	m	0.10	0.20			5	10	20
	0.45		0.10	0.20	0.35	0.40	5	10	20
	0.60		0.10	0.20	0.40	0.50	5	10	20

TABLE 5
CYCLIC TEST CONDITIONS FOR PVDF (T = -25°C)

Mean Stress $\sigma_m = n\sigma_y$		Amplitude $\sigma_a = m\sigma_y$				Frequency (Hz)	
n	0.30	m	0.20			5	10
	0.45		0.20	0.35	0.40	5	10
	0.60		0.20	0.40	0.50	5	10

Some representative results demonstrating the effects of frequency, amplitude and mean stress on the time-dependent response of PVDF in material direction 1 are shown in Figures 8 - 11. From these diagrams, cyclic loading effects can be observed through a comparison of normalized cyclic creep curves with the respective linear viscoelastic creep compliances.

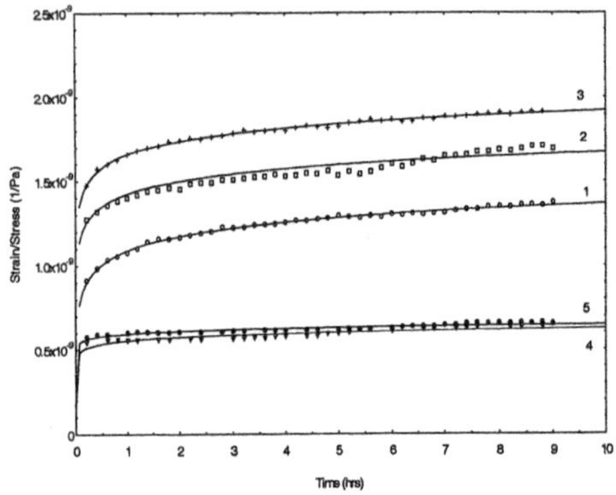

Figure 8: Vibrocreep of PVDF
Frequency effect (T= 23°C)
Cyclic ($\sigma_m = 0.45\sigma_{y1}$, $\sigma_a = 0.10\sigma_{y1}$)
1: $\omega = 5$ Hz; **2:** $\omega = 10$ Hz; **3:** $\omega = 20$ Hz
Static: **4:**$(0.45)\sigma_{y1}$, **5:** $(0.60)\sigma_{y1}$
($\sigma_{y1} = 30.4$ MPa)

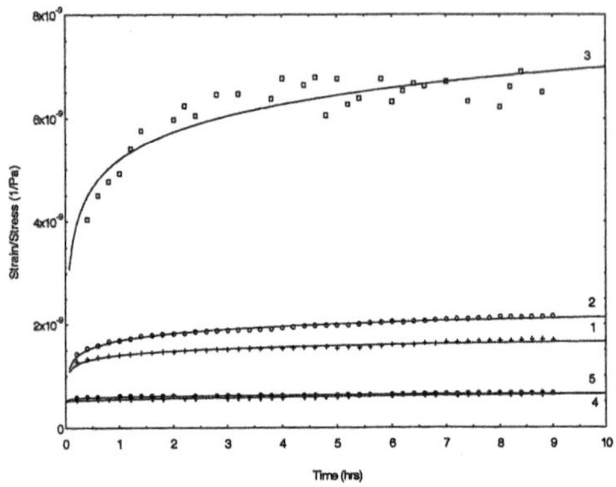

Figure 9: Vibrocreep of PVDF
Amplitude effect (T= 23°C)
Cyclic ($\sigma_m = 0.45\sigma_{y1}$, $\omega = 10$ Hz)
1: $\sigma_a = 0.10\sigma_{y1}$; **2:** $\sigma_a = 0.20\sigma_{y1}$; **3:** $\sigma_a = 0.40\sigma_{y1}$
Static: **4:**$(0.45)\sigma_{y1}$, **5:**$(0.60)\sigma_{y1}$
($\sigma_{y1} = 30.4$ MPa)

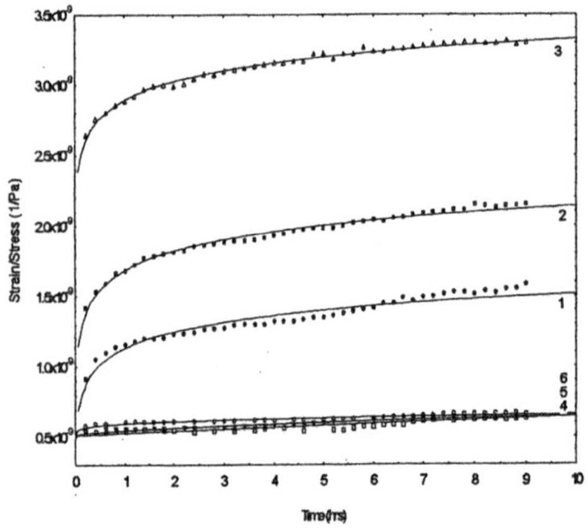

Figure 10: Vibrocreep of PVDF
Mean stress effect (T= 23°C)
Cyclic ($\sigma_a = 0.20\sigma_{y1}$, $\omega = 10$ Hz)
1: $\sigma_m = 0.30\sigma_{y1}$; **2:** $\sigma_m = 0.45\sigma_{y1}$; **3:** $\sigma_m = 0.60\sigma_y$
Static **4:** $(0.30)\sigma_{y1}$, **5:** $(0.45)\sigma_{y1}$,**6:** $(0.60)\sigma_{y1}$
($\sigma_{y1} = 30.4$ MPa)

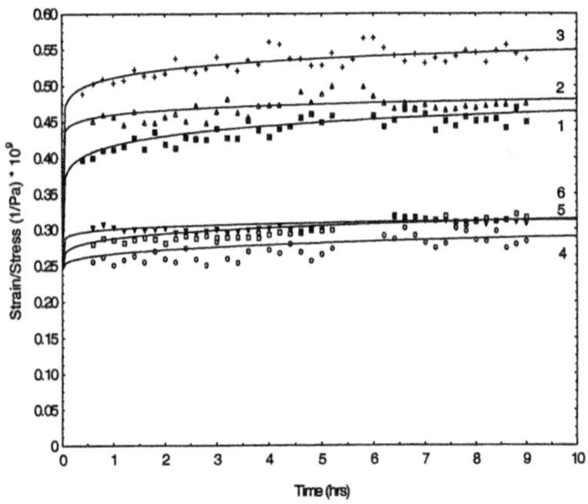

Figure 11: Vibrocreep of PVDF
Mean stress effect (T= -25°C)
Cyclic ($\sigma_a = 0.20\sigma_{y1}$, $\omega = 10$ Hz)
1: $\sigma_m = 0.30\sigma_{y1}$; **2:** $\sigma_m = 0.45\sigma_{y1}$; **3:** $\sigma_m = 0.60\sigma_{y1}$
Static **4:** $(0.30)\sigma_{y1}$, **5:** $(0.45)\sigma_{y1}$,**6:** $(0.60)\sigma_{y1}$
($\sigma_{y1} = 30.4$ MPa)

DISCUSSION OF RESULTS AND CONCLUSIONS

The results of this investigation demonstrate that Nylon 6/6 and PVDF exhibit accelerated creep under the conditions of superimposed static and cyclic loads. Creep acceleration due to cyclic loading effects has been observed even in the range of stresses well below the viscoelastic linearity limit. It is clear that the cyclic response of the polymers is essentially nonlinear, since it does not represent a simple superposition of the responses to static and fully reversed cyclic loads applied separately.

In the temperature range from 23°C to 41°C, Nylon 6/6 has consistently demonstrated an increase of creep rates with an increase of cyclic amplitudes and frequencies. The same tendency has been observed in PVDF. Similar results have been reported in the literature for other polymers, Maksimov and Urthumtsev (1970). In regard to the effects of mean stresses on the cyclic creep behavior of both polymers no conclusive results have been obtained. Thus, as shown in Figures 4 and 6, Nylon 6/6 has demonstrated either an increase in vibrocreep rates with an increase of the mean stress or an inverse effect, depending on the stress magnitude. The inverse effect appears to be stronger at higher stresses and temperatures. In contrast, according to the diagrams in Figures 10 and 11, vibrocreep rates of PVDF appear to increase with an increase of mean stresses at both temperatures, T = 23°C and T = -25°C. The latter result, however, must be confirmed by further experiments since, the observed vibrocreep response of PVDF at higher mean stresses may also reflect an increase of creep rates due to stress induced material nonlinearity.

The diagrams shown in Figure 7 characterize the stress-strain properties of Nylon 6/6 following static creep tests (curves 4 and 5), vibrocreep tests (curves 1-3). The stress-strain diagram for virgin material is represented by curve 6. It can be observed from these diagrams that, within the experimental range of loading conditions, static creep has not produced significant changes in material properties, whereas samples subjected previously to superimposed static and cyclic loads have demonstrated a tendency towards material hardening observed through an increase of the yield and ultimate stresses.

Under cyclic loading conditions, Nylon 6/6 samples have demonstrated no hysteretic heating. Respectively, the observed vibrocreep phenomenon has been attributed to damage development in the material. A microstructural investigation of Nylon 6/6 using a JEOL 840A scanning electron microscopy (SEM) operating at 15 eV and 4 E-10 amp has demonstrated that, although no damage in the material has been detected after 12 hours of cycling when mean stresses remained below the viscoelastic linearity limit, a regularly spaced array of parallel crazes aligned in the direction perpendicular to the axis of the applied load has been observed in samples tested at the mean stress $\sigma_m = 0.8\sigma_y$, amplitude $\sigma_a = 0.2\sigma_y$, and cyclic frequencies 10 and 20 Hz, see Figure 12. Jones & Lesser (1998) have reported similar observations in isotactic polypropylene cycled at 35% of the yield stress for the duration of 10^5 cycles.

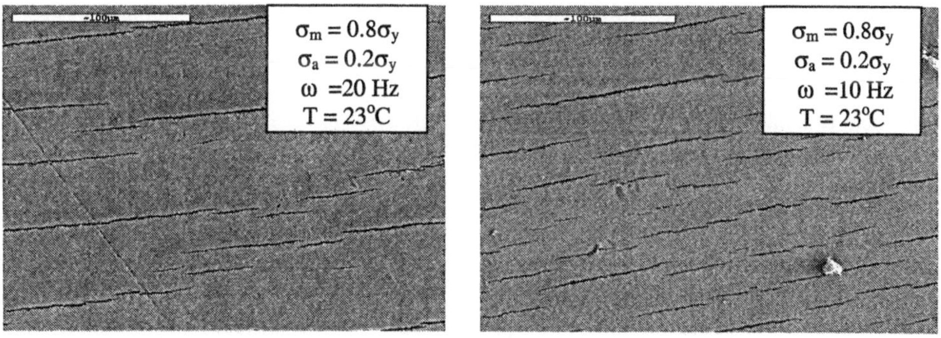

Figure 12: Damage in Nylon 6/6 caused by superimposed static and cyclic loads

Using a Digital Instruments MiltiMode™ atomic force microscopy (AFM), considerable uplifting of the material has been observed in the vicinity of crazes, as demonstrated in Figure 13.

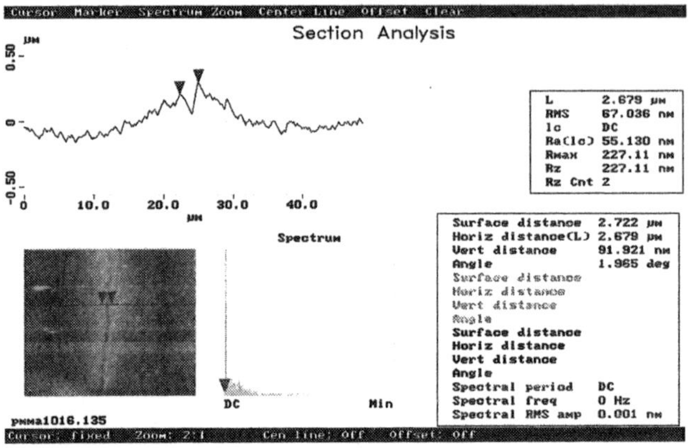

Figure 13: An AFM section analysis across a single craze.

It is of interest to evaluate the vibrocreep responses of Nylon 6/6 and PVDF in the light of other related research findings reported in the literature. Such findings concern primarily the effects of cyclic loading conditions on the fatigue life of polymers. In particular, the effect of cyclic frequency on the fatigue life of polymers has been examined by Hertzberg, Manson & Skibo (1975). In the latter study, the authors have observed that many polymers have shown a surprising enhancement of their fatigue resistance in terms of fatigue crack propagation (FCP) rates as a result of increased cyclic frequencies. Although, according to Hertzberg, Manson & Skibo (1975), both, Nylon 6/6 and PVDF have demonstrated an opposite trend, the study by Arad, Radon & Culver (1973) has shown that, similarly to other polymers, an increase of cyclic frequency has resulted in a decrease of FCP rates in Nylon 6/6.

The latter observations must be assessed in the light of the fact that FCP in polymers represents only one of various mechanisms that determine their fatigue life. Thus, Rabinovitz, Krause & Beardmore (1973) have demonstrated that fatigue damage processes in polymers tend to evolve through several consecutive stages involving craze formation, craze growth, crack nucleation and crack propagation. As shown by Bouda, Zilvar & Staverman (1976), even before the appearance of crazes, glassy polymers subjected to cyclic loading tend to undergo measurable changes of their properties such as internal damping, shear modulus, and density. In this regard, experiments demonstrate that an increase in cyclic frequency tends to accelerate material deterioration at the initial stages of damage development. The study by Hertzberg, Manson & Skibo (1975) has provided convincing evidence of direct correlation between the sensitivity of polymers to cyclic frequency effects and the processes of craze nucleation and propagation.

Concerning the effects of mean stresses and cyclic amplitudes it has been observed that the fatigue life of polymers depends on the relative proportion between these parameters. Specifically, it has been observed that, at higher mean stresses and frequencies (thermally dominated region), the cyclic response of polymers is governed by creep processes, whereas at low mean stresses (mechanically dominated region), progressive damage evolution in polymers becomes a major cause of failure, Hertzberg & Manson (1980), Crawford (1987). Hertzberg and Manson (1980) indicate that, as a result of the synergistic interaction between intrinsic creep and damage evolution, "surprisingly beneficial influence of mean stress on fatigue crack propagation behavior is not without precedent".

The phenomena of material hardening or softening due to cyclic loading conditions have been discussed by Bouda et al. (1976) and Lesser (1995), indicating that the degree of these effects tends to vary depending on the polymer type and the cyclic loading regime. In particular, Lesser (1995) has reported that in the thermally dominated region Nylon 6/6 has shown material softening in terms of a decrease in the storage modulus and corresponding increases in loss modulus and loss tangent. Conversely, in the mechanically dominated region the polymer has demonstrated a certain degree of material hardening.

ACKNOWLEDGEMENT

Funding of this research by the National Science Foundation (NSF Award #9872352) is gratefully acknowledged.

The authors whish to acknowledge the contributions of graduate students Shane Schumacher, Zhiyu Liu, Isamu Kitahara and Carl Thrasher, and a Visiting Professor, Dr. I.M. Klebanov.

REFERENCES

Arad S., Radon J.C. and Culver L.E. (1973). Strain Rate Dependence of Failure Processes in Polycarbonate and Nylon. *J. Appl. Polymer Sci.* **17**, 1467-1478.

Argon A.S. and Hannoosh J.G. (1977). Initiation of Crazes in Polystyrene. *Philosophical Magazine* **36:5**, 1195-1216.

Argon A.S. and Cohen R.E. (1990). Crazing and Toughness of Block Copolymers and Blends. *Crazing in Polymers* **2**, Springer-Verlag, 301-351.

Bouda V., Zilvar V. and Staverman A.J. (1976). The Effects of Cyclic Loading on Polymers in a Glassy State. *J. Polym. Sci.* **14**, 2313-2323.

Constable I., Williams J.G., and Burns D.J. (1970). Fatigue and Cyclic Thermal Softening of Thermoplastics. *J. Mech. Engng. Sci.* **12:1**, 20-29.

Courtney T.H. (2000). *Mechanical Behavior of Materials*, 2-nd ed., McGraw-Hill.

Crawford R.J. (1987). *Plastics Engineering*, 2-nd ed., Pergamon Press.

Crocombe A.D. and Richardson G. (1999). Assessing Stress State and Mean Load Effects on the Fatigue Response of Adhesively Bonded Joints. *Int. J. Adhesion and Adhesives* **19**, 19-27.

Daniels C.A. (1989). *Polymers: Structure and Properties*, Technomic.

Derham C.J. and Thomas A.G. (1977). Creep of Rubber Under Repeated Stressing. *Rubber Chemistry and Technology* **30**, 397-402.

Diaz-Calleja R., Riande E. and Guzman J. (1986). Influence of Static Strain on Dynamic mechanical Behavior of Amorphous Networks Prepared from Aromatic Polyesters. *J.Polymer Sci.: Polymer Physics Ed.* **24**, 337-344.

Dillard D.A. (1991). Viscoelastic Behavior of Laminated Composite Materials. *Fatigue of Composite Materials*, (K.L. Reifsnider, ed.), Elsevier, 339-384.

Fitzgerald J.J., Martellock A.C., Nielsen P. and Schillace R.V. (1992). The Effect of Cyclic Stress on the Physical Properties of a Poly(Dimethylsiloxane) Elastomer. *Polym. Engng. Sci.* **32:18**, 1350-1357.

Hertzberg R.W. and Manson J.A. (1980). *Fatigue of Engineering Plastics*, Academic Press.

Hertzberg R.W., Manson J.A. and Skibo M. (1975). Frequency Sensitivity of Fatigue Processes in Polymeric Solids. *Polymer Engng. Sci.* **15:4**, 252-260.

Higuchi M. & Imai Y. (1970). Rheological Interpretation of Heat Generation Associated with Fatigue of Polycarbonate. *J. Appl. Polymer Sci.* **14**, 2377-2383.

Holloway F.C. (1997). *Material Characterization of Poly(vinylidene Fluoride): A Thin Film Piezoelectric Polymer*, M.S. Dissertation, Montana State University, Bozeman,

Hristov H.A. and Yee A.F. (1994). Fatigue Craze Initiation in Polycarbonate: Study by Small-Angle X-Ray Scattering. *Polymer* **35:20**, 4287-4292.

Hristov H.A., Yee A.F. and Gidley D.W. (1994). Fatigue Craze Initiation in Polycarbonate: Study by Transmission Electron Microscopy. *Polymer* **35:17**, 3604-3611.

Jones N.A. and Lesser A.J. (1998). Morphological Study of Fatigue Induced Damage in Isotactic Polypropylene. *J. Polymer Science: Part B: Polymer Physics* **36**, 2751-2760.

Klaus F. (1983). Crazes and Shear Bands in Semi-Crystalline Thermoplastics. *Crazing in Polymers, Advances in Polymer Science* **52/53**, Springer-Verlag, 225-274.

Kramer E.J. and Berger L.L. (1990). Fundamental Processes of Craze Growth and Structure. *Crazing in Polymers* **2**, Springer-Verlag, 1-68.

Kregers A.F. and Maksimov R.D. (1970). Vibrocreep of Polymeric Materials, Polyurethane, Recovery. *Mech Composite Materials.* **6:5**, 708-713.

Lesser A.J. (1995). Changes in Mechanical Behavior During Fatigue of Semicrystalline Thermoplastics. *J. Appl. Polymer Sci.* **58:5,** 869-879.

Maksimov R.D. and Urzhumtsev Yu.S. (1970). Vibrocreep of Polymeric Materials, Polyurethane. Vibrocreep Coefficient. *Mech. Composite Materials.* **6:4**, 561-564.

Maksimov R.D. and Urzhumtsev Yu.S. (1970).Vibrocreep of Polymeric Materials, Polyurethane. Combined Loading. *Mech. Composite Materials* **6:6**, 862-870.

McKenna G.B. and Zapas L.J. (1981). Response of Carbon Black Filled Butil Rubber to Cyclic Loading. *Rubber Chemistry and Technology* **54**, 718-733.

Narisawa I. and Ishikawa M. (1990). Crazing in Semicrystalline Thermoplastics. *Crazing in Polymers* **2**, Springer-Verlag, 353-391.

Rabinowitz S., Krause A.R. and Beardmore P (1973). Failure of Polystyrene in Tensile and Cyclic Deformations. *J. Mater.Sci.* **8**, 11-22.

Riddell M.N., Koo G.P. and O'Toole J.L. (1966). Fatigue Mechanisms in Thermoplastics. *Polymer Engng. and Sci.* **6**, 363-368.

Sermage J.P., Lemaitre J. and Desmorat R. (2000). Multiaxial Creep-Fatigue Under Anisothermal Conditions. *Fatigue Fract. Engng. Mater.Struct.* **23**, 241-252.

Schumacher S.C. (2000). *Parametric Study of Cyclic Loading Effects on the Creep Behavior of Polymers and Polymer Based Composites*, MS Dissertation, Montana State University, Bozeman.

Schumacher S.C., Vinogradov A.M., Liu Z., Jenkins C. H. M., Kitahara, I. and Winter R.M. (In press). Time-Dependent Response of Polymer Systems Under Cyclic Loading Conditions. *Proc.*

Strong A.B. (2000). *Plastics. Materials and Processing*, 2-nd ed., Prentice Hall.

Suresh S. (1998). *Fatigue of Materials*, 2-nd ed., Cambridge Univ. Press.

Takemori M.T. (1990). Competition Between Crazing and Shear Flow During Fatigue. *Crazing in Polymers* **2**, Springer-Verlag, 263-300.

Tauchert T.R. (1967). The Temperature Generated During Torsional Oscillations of Polyethylene Rods. *Int. J. Engng. Sci.* **5**, 353-365.

Turtsinsh R.P. (1972). Influence of Vibration on Creep in Glass-Fiber-Reinforced Plastic in a Complex State of Stress. *Mech. Composite Mater.* **9:2**, 317-319,

Vinogradov A. M. (1999). Constitutive Model of Piezoelectric Polymer PVDF. *Mathematics and Control in Smart Structures, Proc. SPIE* ***3667***, 711-718.

Vinogradov A.M. and Holloway F. (1999). Electro-Mechanical Properties of the Piezoelectric Polymer PVDF. *Ferroelectrics* **226**, 169-181.

Vinogradov A.M. and Holloway F. (1997). Mechanical Testing and Characterization of PVDF, a Thin Film Piezoelectric Polymer. *J. Adv. Mater.* **29:1**, 11-17.

Vinogradov A.M. and Schumacher S. (In press). Cyclic Creep of Polymers and Polymer-Based Composites. *Mech. Composite Materials.*

Weaver J.L. & Beatty C.L. (1978). The Effect of Temperature on Compressive Fatigue of Polystyrene. *Polymer Engng. And Sci.* **18:14**, 1117-1126.

Williams J.G. (1987). Fracture Mechanics of Polymers, Ellis Horwood.

Zhang S.Y., Xue, Q., and Wang D. (1994).Creep Effect on Dynamic Viscoelastic Properties of Polymetric Matrix Composite. *Appl. Composite Mater.* **1**, 125-133.

Zhou Y. and Brown N. (1993). Evaluating the Fatigue Resistance of Notched Specimens of Polyethylene. *Polymer Engng. Sci.* **33:21**, 1421-1425.

Long Term Durability of Structural Materials
P.J.M. Monteiro, K.P. Chong, J. Larsen-Basse, K. Komvopoulos (Eds)

Analysis of Physical and Chemical Deterioration of Polymeric Coatings for Structural Steel

[1]Y.C. Jean, [1]R. Zhang, [1]H.M. Chen, [1]C.M. Huang, [1]P. Mallon, [1]Y. Li, [1]Y. Huang, [1]T. C. Sandreczki, [2]J. R. Richardson, [2]Q. Peng

[1]Department of Chemistry, University of Missouri--Kansas City, Kansas City, MO, 64110
[2]Department of Civil Engineering, University of Missouri-Columbia, Kansas City, MO, 64110

ABSTRACT

The photo-degradation of several commercial bridge coatings is being investigated by exposing samples to accelerated and natural UV irradiation. Primary microscopic techniques include positron annihilation spectroscopy (PAS), which detects and characterizes nanometer-scale physical holes/defects, and electron spin resonance (ESR) spectroscopy, which detects broken chemical bonds. For the PAS tests, significant decreases of sub-nanometer defect parameters are observed as a function of exposure time and of depth from the surface to the bulk. This is interpreted as being a result of a loss of free-volume and hole fraction and an increase in cross-link density of the polymers during the degradation process. For the ESR tests, direct free radical signals are observed as a function of irradiation time and chemical environments. A high sensitivity of PAS and ESR tests to the early stage of degradation is observed. The obtained results are used to interpret the degradation mechanism of polymeric coatings. Investigations of chemical degradation and environmental durabilities are in progress.

KEYWORDS

Positron Annihilation, Free Volume, Polymer, Coatings, Free Radicals, Durability, Photo-Degradation, Accelerated Testing, Florida Weathering

INTRODUCTION

Service life is relatively short for bridge coatings because bridges are subjected to ice, wind, de-icing salts, sunlight, water, industrial plant exhaust, and automotive chemicals[1]. Furthermore, bridges are subjected to higher frequency dynamic vibration due to traffic loads than non-bridge structures. This repeated dynamic stressing of the steel elements and surface coatings is likely an additional factor contributing to coating degradation. Protective coatings for steel and structural systems typically used for transportation often consist of a multilayered structure with three or four elements: a topcoat (finish), an intermediate coat, a primer, and in some cases a surface sealer for the substrate, which is either metal or concrete [2-3]. Coatings are multifunctional materials, which contain organic or inorganic binder, metals, oxides, and other minerals.

The purpose of this research is to provide fundamental knowledge, which will lead to long-term durability of protective coating systems for structural materials. This research aims directly at the areas of (1) identification of degradation mechanisms of polymers in coatings and subsequent development of appropriate models for performance predictions, and (2) investigation of synthesis /structure /property /performance relationships for coating systems.

Durability is a primary concern for protective coating systems [4]. There is an incomplete understanding of the origins of poor durability in most coating systems. Existing methods of assessment of durability and degradation are chiefly macroscopic approaches, measuring mechanical properties: adhesion, hardness, pulling strength, etc. Most knowledge of coating degradation and failure is based on these evaluations of performance. However, the microscopic origins, mechanisms, and progress of degradation are not yet ascertained for coating systems. Only recently have spectroscopic and physical methodologies begun to be used to investigate the underlying cause of coating degradation [5].

There are two fundamental processes involved in coating degradation, chemical and physical changes, although they are interrelated in most cases [6]. For example, photo-degradation is a chemical process well known to degrade paints [6]. Polymers in paint absorb UV radiation, which leads to bond breakage yielding free radicals. These radicals initiate chain reactions in polymers, which eventually lead to the degradation and failure of the polymer. This chemical process is a complicated one, which depends on the composition/formulation of the coating. The process becomes more complicated in the presence of other environmental factors, such as oxygen, moisture, pollutants, temperature changes, and long-term dynamic stress loadings.

A series of microscopic physical defects occurs simultaneously with, and is caused by, chemical reactions. The defects involved with chemical degradation initiate at a very small scale, on the order of 0.1 nm, as bonds are broken and atoms are displaced from polymer structures. Essential information about these atomic- and molecular-level defect properties is currently unavailable because of the extremely small scale and very brief duration of the phenomena involved.

Defect sizes and distributions as a function of distance from the top-most surface layer down to the bulk of the coating system are being investigated by using an innovative physical method, positron annihilation spectroscopy (PAS). Radical formation leading to degradation of topcoat/finish coat systems is being investigated using electron spin resonance (ESR) spectroscopy. In this way, the origins of degradation are investigated in terms of chemical defects at the atomic level.

In this paper, we report the recent results from a series of PAS, ESR, crosslink density and glossiness studies on commercial coatings which are subjected to the accelerated UV irradiation, i.e. xenon-lamp light and QUV lamp, and to the Florida natural weathering.

EXPERIMENTAL

Eighteen coating samples as shown in Table 1 were prepared, of which seventeen are topcoats and one (sample 3) is midcoat. Those coatings are classified into five categories according to chemical composition: polyurethane, acrylic, epoxy, metallic-aluminum-containing paint, and calcium sulfonate. The detailed description of the eighteen commercial coating materials can be found in the product data sheets available at companies' websites. The model polyurethane (PU) sample was prepared by mixing commercially supplied raw materials of polisocyanate and polester-polyol (Bayer Chemical Allentown, PA). The PU sample was prepared by the solvent-casting method onto Al plates. The rest of the seventeen commercial coatings were thoroughly mixed and then applied to aluminum sheets using a pressurized spray gun (Binks 95, Binks Manufacturing Company, Franklin Park, IL) connected with a nitrogen gas tank regulated at 60 psi. The thicknesses of the coatings were determined to be ~ 20 μm using profilometry.

Table 1. Coating samples studied

No	Polymer base	Coating Name	Vender	Description
1	Polyurethane	MC-Ferrox A	Wasser, Seattle, WA	Moisture-cured aliphatic polyurethane
2	Polyurethane	MC-Luster	Wasser	Moisture-cured aliphatic polyurethane
3	Polyurethane	541-D-101	Valspar, Baltimore, MD	Moisture-cured urethane intermediate coat
4	Polyurethane	Acrolon 218 HS	Sherwin-Williams, Cleveland, OH	Polyester modified acrylic polyurethane
5	Polyurethane	Poly-lon 1900	Sherwin-Williams	Polyester-aliphatic polyurethane
6	Polyurethane	Carboline 133 HB	Carboline, St. Louis, MO	Aliphatic polyurethane
7	Polyurethane	Carbothane 134 HG	Carboline	Acrylic aliphatic polyurethane
8	Polyurethane	Devthane 359	ICI Devoe, Louisville, KY	Acrylic aliphatic polyurethane
9	Polyurethane	PU	Bayer, Allentown, PA	N-100 polyisocyanate +631A-75 polyol
10	Acrylic	Carboline 3359	Carboline	Waterborne acrylic
11	Acrylic	Devflex 4218	ICI Devoe	Waterborne acrylic
12	Acrylic	Devflex 4206	ICI Devoe	Waterborne acrylic
13	Epoxy	Epolon II	Sherwin-Williams	Catalyzed polyamide epoxy
14	Epoxy	Macropoxy 646	Sherwin-Williams	High solid polyamide epoxy
15	Epoxy	Devran 224 HS	ICI Devoe	Catalyzed polyamide epoxy
16	Alkyd	Silver-Brite	Sherwin-Williams	Metallic aluminum in petroleum resin
17	Alkyd	GridGard 2600	Carboline	Alkyd/Calcium sulfonate
18	Alkyd	GridGard 2901	Carboline	Alkyd/Calcium sulfonate modified with epoxy polyester

Three types of artificial light sources were applied in the accelerated UV irradiation of the coating samples: a QUV chamber with UVA-340 or UVB-313 fluorescent lamps, and a xenon arc-lamp light source (SLM Instrument, Inc., Urbana, IL; with a luminescence 32,000 W/m^2 over the wavelength range 250 nm to 1,200 nm, corresponding to 1,900 W/m^2 from 250 to 350 nm). The natural exposure was performed in Florida. The wavelength spectra of these sources are shown in Figure 1

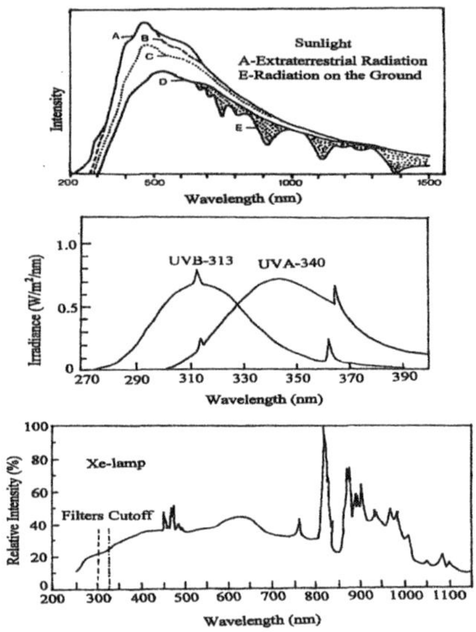

Fig. 1. Wavelength distributions of sunlight, UVA-340, UVB-313, and Xe-lamp light sources.

The technique of Doppler broadened energy spectra (DBES) of positron annihilation coupled with a slow positron beam was employed to measure the first three coating samples. The DBES experiments were performed at the University of Missouri-Kansas City[7]. The energy resolution was 1.5 keV at 0.511 MeV (corresponding to the positron 2γ annihilation peak). The total counts for each DBES spectrum was 0.5 million with the counting rate of 4000 cps. The obtained DBES data are characterized by the S-parameter, a measure of the momentum broadening. The S-parameter is defined as the ratio of the central area to the total counts after the background is properly subtracted. It provides a qualitative measurement of sub-nanometer defects, such as free volumes and holes, of polymers in coatings. The change of S-parameter, $-\Delta S = S_t - S_0$, where S_t and S_0 are the value after and before irradiation, respectively, gives information about the change in physical defects due to weathering.

The positron annihilation lifetime (PAL) experiments were performed at the intense slow-positron facility in the Electrotechnical Laboratory (ETL) in Japan [8]. The PAL data were fitted into four lifetimes: τ_1 (~0.125 ns, p-Ps); τ_2(~0.4 ns, positron), τ_3 (1-3 ns, o-Ps in coatings) and τ_4 (>10 ns, o-Ps in vacuum). Detailed description of DBES and PAL can be found in our previous paper s [9-11].

ESR experiments were performed on the topcoats to detect the free radicals involved in the photochemical reactions. The spectra were recorded using a Bruker ER-200-D X-band ESR spectrometer. Spectral scans were 15 mT, and were averaged from 3 to 10 scans at a scan rate of 0.30 to 0.75 mT/s. Modulation amplitude was typically 0.5 mT.

A statistical glossmeter (Novo-Gloss, Rhopoint Instr. East Sussex, England) was used to measure the glossiness of the identical samples used in the PAS experiments. The maximum sensitivity of glossiness was obtained at a 60° angle reflection of incident green light (515 nm). The results of glossiness were expressed on a relative scale based on the reading originally calibrated by the manufacturer.

Crosslink density determination: For the swelling determination of crosslink density, a 1×1 mm^2 piece was cut from the free film. The sample was sandwiched between two microscope slides and then exposed to a drop of methylene chloride. Swelling was observed on a microscope with a TV monitor.

RESULTS AND DISCUSSION

PAS as an Accelerated Test of Durability:

Bridge coatings are multi-layered systems and each layer is a complex mixture of polymers, solvents, pigments, stabilizers, binding agents, and other additives used to achieve desired properties. In our previous paper [9], a three-layer depth profile of the S-parameter in aircraft topcoat MIL-C-85285B was observed: a surface skin polymer layer, an intermediate layer and the bulk. Figures 1-3 show the results of S defect parameter in three different coatings both un-irradiated (virgin) and one hour xenon lamp light irradiated as a function of positron incident energy.

The variation of S-parameter vs. positron incident energy is: a sharp increase near the surface (<20 nm), then a decrease. This variation is a general feature for polymeric coating systems. This has been interpreted in terms of a multi-composition system[9]. This multi-composition feature of the topcoat can be interpreted as being due to a concentration gradient of the pigment from the surface to the bulk. This feature is also observed in the slow positron lifetime results discussed later. Below 20 nm from the surface, there exists a polymer skin layer. The small value of S at the surface is due to the back-diffusion of slow positrons implanted and not detected by the solid state detector. This is a general phenomenon for neat polymers. The decrease of the S parameter inside 20 nm is a result of positron annihilation with paint, which contains pigments. For example, the 3d electrons from Ti in TiO_2 has a higher electron momentum which will lead to a smaller value of the S parameter than that in polymers, which contain lower momentum O, H, and N atoms.

It is interesting to observe a smaller value of the S-parameter in Xe-irradiated samples than in virgin samples. A decrease of the S parameter due to UV irradiation is observed.

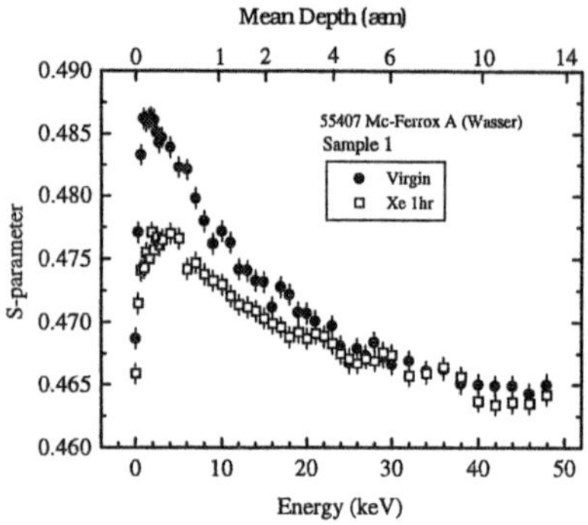

Fig. 2. S-defect parameter vs positron incident energy in un-irradiated (virgin) and irradiated coating sample 1.

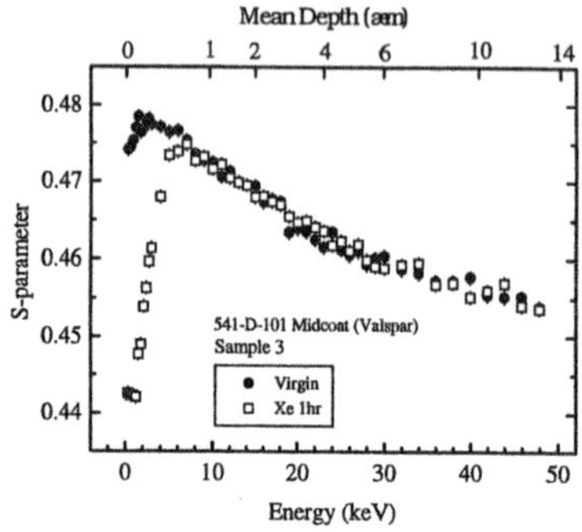

Fig. 3 S-defect parameter vs positron incident energy in un-irradiated (virgin) and irradiated coating sample 3.

This observation is also consistent with our existing findings for aircraft coatings under different UV wavelengths [9-11]. The magnitude of the S-parameter reduction, $-\Delta S = S_t - S_0$, is calculated and $-\Delta S$ vs. the depth from the surface is plotted in 4 for different samples. The decrease of $-\Delta S$ with depth (d) indicates an attenuation of UV intensity as the light enters the samples.

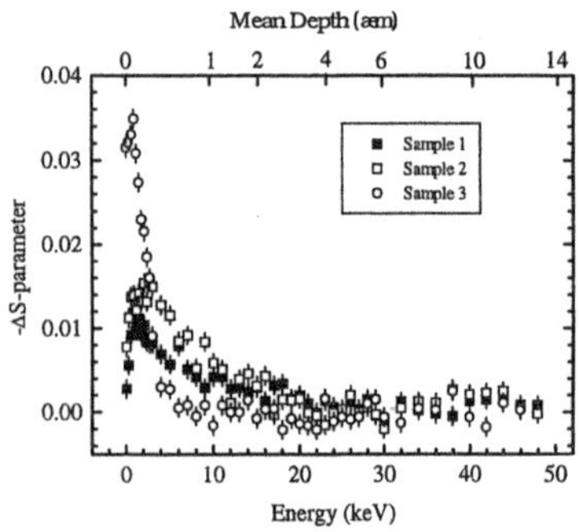

Figure 4. The reduction of S parameter ($-\Delta S$) due to 1 hr of Xe light irradiation vs. positron incident energy in three coating samples 1,2, and 3.

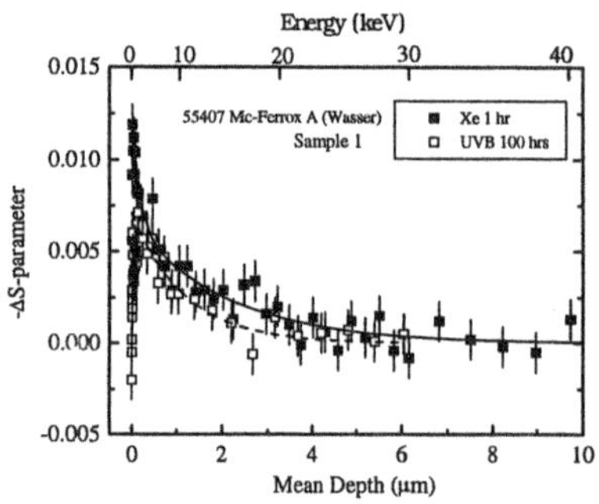

Figure 5. The reduction of S parameter ($-\Delta S$) due to 1 hr of Xe light and 100 hrs of UVB irradiation vs. mean depth from the surface in sample 1 coating. The solid line is fitted with two exponential functions for Xe-irradiated data. The dashed line is fitted with single exponential for UVB-irradiated data

The absorption of UV light can follow an exponential function according to Beer's law: $-\Delta S_t = -\Delta S_{max}(1-e^{-\varepsilon d})$, where $-\Delta S_{max}$ is maximum change inside the thin polymer layer (>20 nm). However we found that the variation of $-\Delta S$ vs d is more rapid than single exponential can describe for coating systems. For a good fit to the data, two exponential functions are required for Xe-irradiated data. Figure 5 shows such a two-exponential fit with $-\Delta S_{max}$=0.053, and 0.065 with ε=5.7 and 0.46 μm^{-1}, respectively.

The need of two exponentials is due to the spread of UV wavelengths from a Xe-lamp light source. In general for a monochromic light source, $-\Delta S$ vs d can be fitted by a simple exponential function according to Beer's law. For example, in Figure 5, we also present $-\Delta S$ vs the mean depth for the sample subjected to a QUVB source (which peaks at 313 nm with a spread of about 40 nm). In the case of 313 nm UV light, we found $-\Delta S_{max}$=0.071 and ε=0.85 μm^{-1}. The ε value agrees with the extinction coefficient measured by using UV spectroscopy. The current results support our original idea that the S parameter is a measure of defects induced by UV irradiation. From the magnitude of $-\Delta S$ shown in Figure 4 for these three different paints, we rank the quality of durability against UV irradiation as sample 1 > sample 2> sample 3.

From these observations, we obtain the following results for the PAS method: (1) S is a new physical parameter to test the durability of coatings in terms of sub-nanometer defects induced by UV irradiation. (2) The positron technique provides a defect profile from the surface to the bulk at a depth precision about 10% of the mean depth. (3) The sensitivity for detecting deterioration is a few minutes for the accelerated Xe-light source, 10 hrs for the 313 nm UVB source, and 50 hrs for the 340 nm UVA source and natural weathering. (4) The detection can be applied to both transient and permanent defects by performing *in situ* or *ex situ* experiments, respectively.

Slow positron annihilation lifetime (PAL) experiments were also performed using the same series of samples to obtain the free-volume and hole sizes. In this method, the measured ortho-Ps (o-Ps) lifetime, τ_3 has a direct correlation with the free-volume and hole size which is typically a few tenths of a nm [12]. The o-Ps lifetime (τ_3) and its intensity (I_3) were resolved from the data using a multi-exponential analysis. From these results, we then calculated the hole radius R of free volumes according to a correlation equation[10] as shown in Figures 6. In virgin coatings, o-Ps lifetime (τ_3 ~ 2 ns) shows different variations as a function of depth: it can be larger or smaller near the surface. However Xe-light irradiation significantly reduces the defect size as seen from the consistently smaller values compared with the virgin data at the same energy. This is interpreted as smaller holes being formed after UV irradiation. Similarly, a large change in o-Ps intensity as a function of irradiation time is observed. I_3 systematically decreases with Xe light irradiation. Since o-Ps intensity has been suggested to be correlated with free-volume fraction in polymers [12], the decrease of o-Ps intensity can be interpreted as a reduction of free-volume and hole fraction induced by photodegradation of the coating material. The decrease of I_3, which is a measure of o-Ps formation, is consistent with the decrease of the S-parameter, which is a measure of p-Ps formation. A possible explanation for the decrease of the free-volume fraction is that there is an increase in crosslink density during the degradation process.

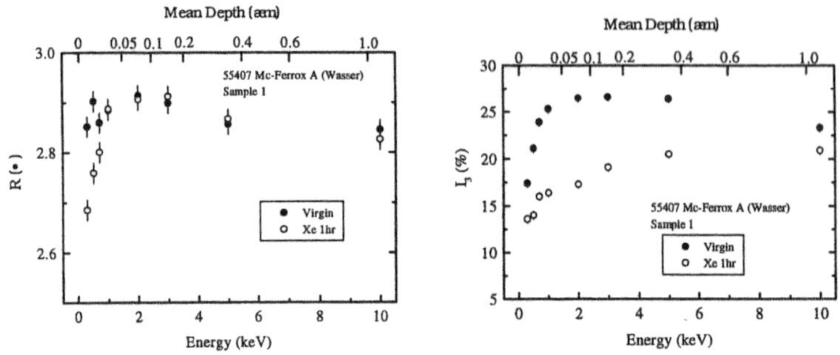

Fig. 6. The free-volume hole radius (left) and its intensity (right) vs positron incident energy in coating sample 1.

For testing the durability of paints, we employ the defect parameter S as a measure of UV degradation. The decrease of free volumes, -ΔS, for 8 commercial samples under UVB irradiation, are calculated from the obtained S vs. depth. Figure 7 shows the result of -ΔS at 100 hrs UVB irradiation for those coatings studied at different depth. A large value of -ΔS indicates a low durability against UV irradiation. Among these samples, polyurethane based coatings show a better durability over alkyd-base paints. Epoxy based paints are among the poorest durability against UV irradiation. These observations are consistent with the general descriptions of coating performance in commercial paints. PAS emerges as a novel method for accelerated test for coating durability.

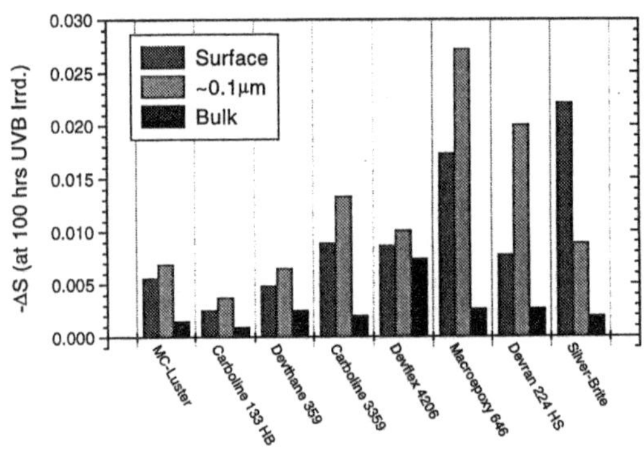

Fig. 7. The decrease of defect parameter -ΔS in various coatings.

Correlation between Loss of Gloss and Loss of Free Volumes:

One important physical property regarded by the coating industry is gloss, which is a measure of surface roughness. Since PAS has a capability to measure atomic scale defects from the surface to the bulk, we conducted parallel experiments of glossiness and PAS measurements in a model polyurethane coating (sample #9 of Table1). Figure 8 shows S defect parameter vs. positron incident energy for polyurethane under different durations of QUVB irradiation.

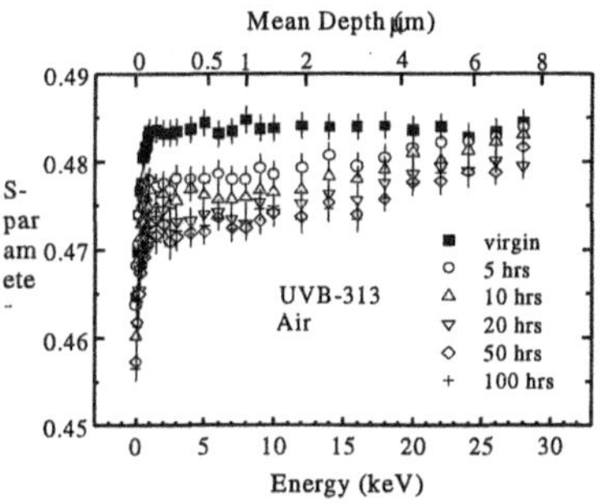

Figure 8. S defect parameter vs positron incident energy in polyurethane under UVB irradiation of polyurethane (sample #9).

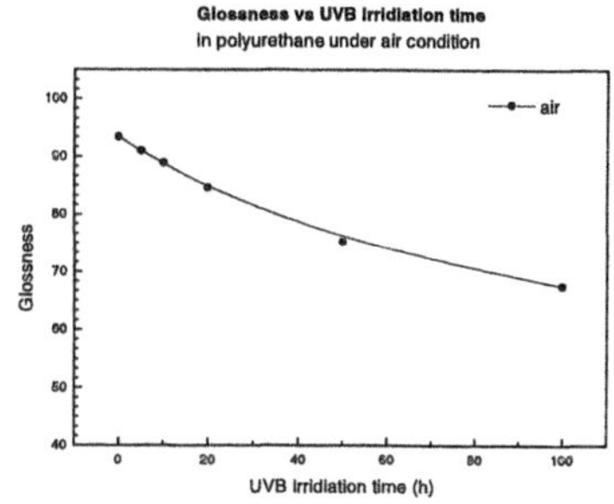

Figure 9. Glossiness vs hours of UVB irradiation of polyurethane(sample #9).

As shown in Figure 8, a decrease of the S parameter due to UV irradiation is observed. Figure 9 shows the resulting glossiness of these irradiated samples at different periods of irradiation. As shown, irradiation reduces glossiness as a function of time. The reflection of light from the polymer surface is reduced due to degradation induced by UV irradiation. The loss of glossiness, -ΔG =glossiness(irradiated)-glossiness(virgin), is then plotted against the reduction of S defect parameter in Figure 10. In Figure 10, a direct correlation between the -ΔS and -ΔG is observed. A better sensitivity of -ΔS to detect UV degradation is expected near the surface (20 nm) than in the bulk. This correlation further supports the usefulness of S parameter to detect UV degradation in polymers.

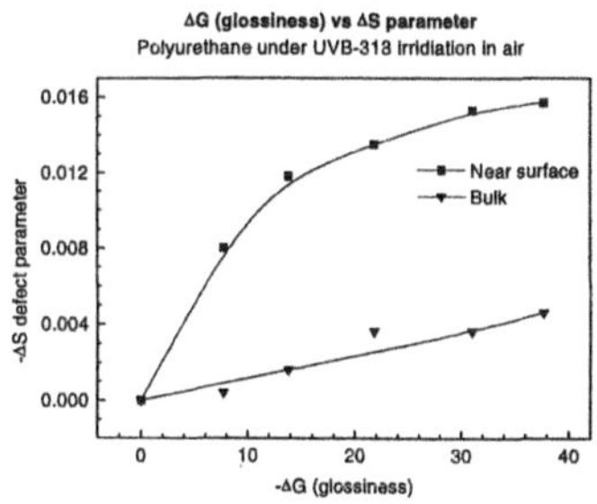

Fig. 10. Correlation between loss of glossiness and loss of free-volume under UVB irradiation in a polyurethane coating (#9 in Table 1).

Cross-link Density and Free-volume from PAS:

The S parameter in polymers depends on the fraction of Ps formed in holes and the hole size. More Ps will produce a larger S parameter. A decrease of free volume may reduce formation of Ps, and hence reduce the S parameter. Figure 11 shows the change in the S parameter before and after Xe irradiation of the polyurethane coating (sample #9 of Table 1). It indicates the loss of free volume under UV exposure. In the slow positron lifetime experiment, no large change of o-Ps lifetime (τ_3) was observed. There is, however, a significant decrease in the o-Ps intensity (I_3), as shown in Figure 12. The similarity of S parameter and I_3 in these two figures reveals their physical coherence.

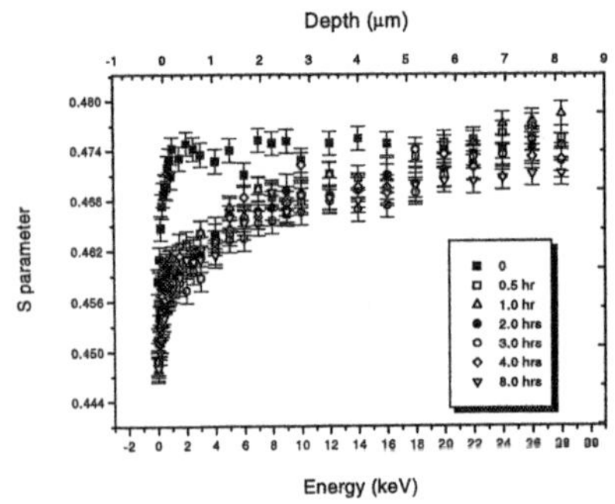

Fig. 11. S parameter vs positron energy

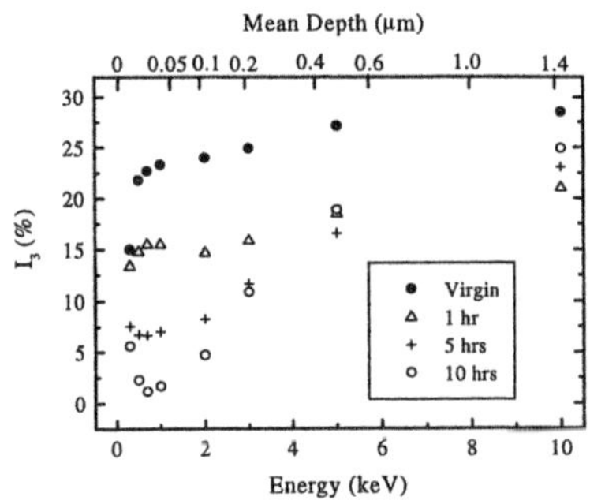

Fig. 12. Intensity of o-Ps vs positron energy

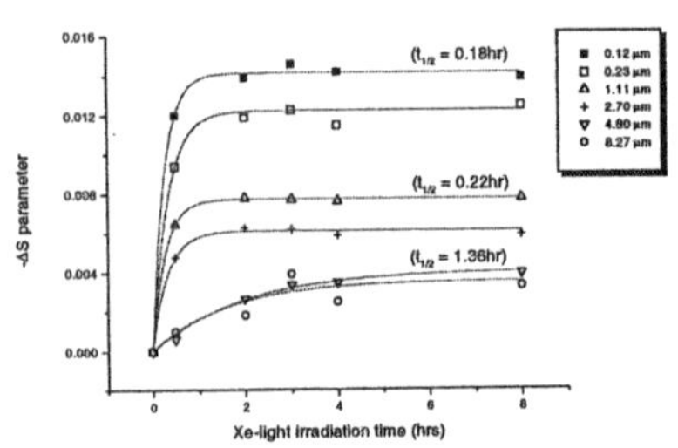

Fig. 13. ΔS vs irradiation time at different depth

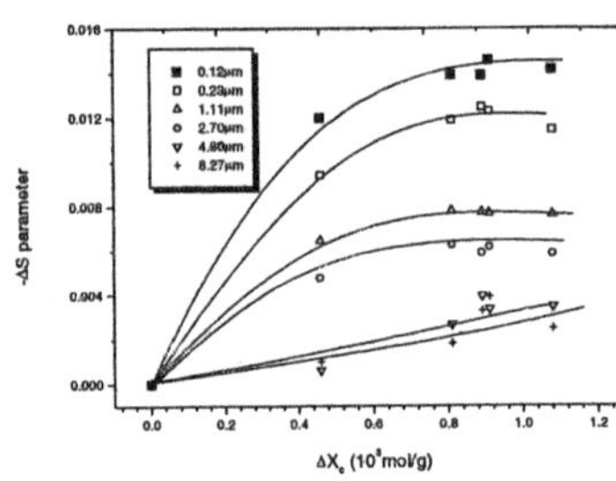

Fig.14. Correlation between ΔS and ΔXc

The change of crosslink density indicates some new bonds are formed between the polymer chains. The formation of these new bonds and the increase in crosslink density will decrease the free volume in the polymer. The ΔS parameter describes the change of free volume in polymeric material. Figure14 expresses the correlation of ΔS and ΔXc at different depths of polyurethane. ΔXc is the overall change of crosslink density for the whole sample. ΔS on the surface and near the surface increases significantly while ΔS in the bulk shows little change. This means that crosslink formation and the resultant collapse in free volume at the surface contributes the most to the crosslink density changes in the early stages of exposure. Crosslink density (Xc) of the coating material--polyurethane--increases after irradiation under a Xe-light source. This indicates the collapse of free volume during degradation. The S parameter from DBES and I_3 from PAL describes the depth profile of free volume from surface to bulk. There is a correlation between ΔXc and ΔS at different depths. The interpretation is in accordance with the observation we obtained previously.

ESR:

Samples for use in the ESR studies were applied to aluminum sheets. They were then exposed to xenon-lamp irradiation under two different sets of conditions. In the first case, fourteen commercial topcoat materials were irradiated for 30 minutes under vacuum at 77K. They were then examined by ESR to detect the presence of free radicals, indicating photo-induced bond cleavage. Currently, we attempt to correlate the change of ESR spectral amplitudes, Δ, with susceptibility-to-bond cleavage, which is an important process in photodegradation. However, the data presented here have not yet been correlated with changes in coating performance (e.g. loss of gloss). These experiments are currently in progress. The results are summarized in the bar chart shown in Figure 15. The seven different polyurethane samples identified in the chart had significantly different numbers of free radicals following irradiation,

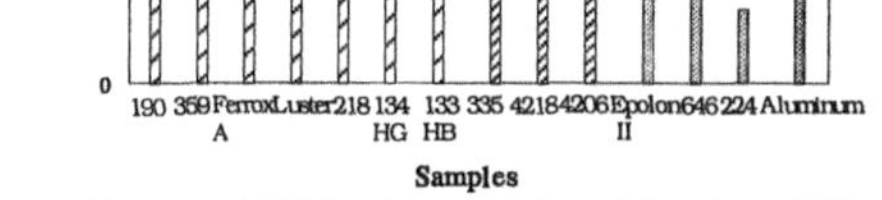

Fig. 15. Amplitude of ESR signals after 30 min of Xe-light irradiation on 14 commercial samples (Table 1) in vacuum at 77 K.

with spectral amplitudes varying by as much as a factor of three. The Poly-lon 1900 sample appeared to be the most susceptible to bond cleavage, in that it had the highest number of free radicals. The Carboline 133 HB had the lowest number of radicals. The three epoxy samples displayed an even wider range of susceptibilities to bond cleavage, with spectral amplitudes varying by as much as a factor of four. The Epolon II sample had the highest number of free radicals, and the Devran 224 HS had the least. The three different acrylics displayed a narrower range of susceptibilities, with the spectral amplitudes varying by less than ten percent among the samples. All the acrylics had a low-to-moderate susceptibility to bond cleavage compared to the polyurethane and epoxy samples. Finally, the metallic aluminum-containing sample, Silver Brite, showed the highest susceptibility to free radical formation of all the samples, with a spectral amplitude approximately five times greater than that of the most stable sample of the fourteen, Devran 224 HS. The above data was acquired in order to quickly determine the relative stabilities of commercial paint formulations to short-wavelength light. The results in Figure 15 indicate that knowledge of a topcoat's chemical class (e.g., polyurethane or epoxy) is not enough to predict its photo-stability, and that is possible to formulate highly stable polyurethanes and epoxies. The performance of the acrylics was nearly as good at those of the best epoxy and polyurethane.

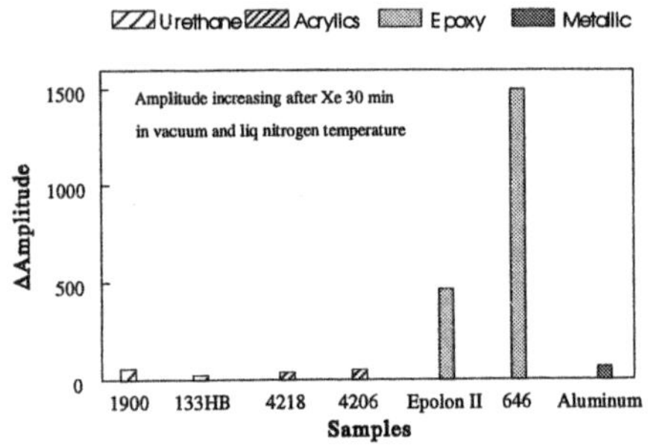

Fig. 16. Amplitude of ESR signals after 30 min of Xe-light irradiation on 7 commercial samples (Table 1) at room temperature and under 1 atm oxygen.

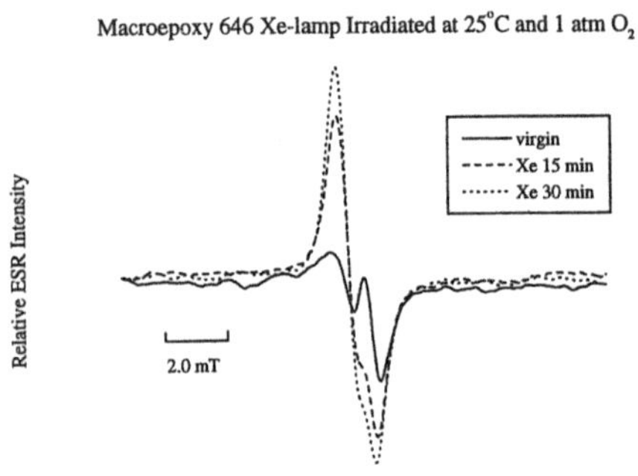

Fig. 17. ESR spectra from sample 14 coating following different times of Xe-irradiation.

Seven of the above topcoats were down-selected for further investigation. These include two each of the polyurethanes (Poly-lon 1900 and Carboline 133HB), epoxies (Epolon II and Macropoxy 646), and acrylics (Devflex 4218 and Devflex 4206), plus the one aluminum-containing topcoat (Silver Brite). This down-selected set of samples was exposed to xenon-lamp irradiation under a second set of conditions, viz., up to 30 minutes at room temperature under 1 atmosphere of O_2. From a set of related experiments on polyurethane clearcoats, it is known that the temperature alone (up to 150 °C) does not result in measurable radical production. Under these conditions, photo-generated free radicals exist in a mobile environment, and therefore freely react with their surroundings. The result is that the radicals observed are only those net radicals that remain after a sequence of reactions with neighboring chains. These are often secondary (or later) radicals formed from the reaction of the initial photo-generated primary radicals. A reason for performing this experiment is that the added molecular mobility and the presence of oxygen enable the process of photo-oxidative degradation to occur. This process can include branching chain reactions that result in extensive bond cleavage per photon absorbed. The results of this investigation are shown in Figure 16. There are at least two main features to be noted. First, most of the samples have a very similar net number of radicals following exposure to the photo-oxidative environment. This is in marked contrast to the widely varying numbers of radicals observed under vacuum at 77K (Figure 17). Second, one of the epoxy samples (Macropoxy 646) produced an unusually high number of observable radicals under these exposure conditions, with the net number of radicals for this sample being nearly three times higher than the average for this group of samples. ESR spectra from the Macropoxy 646 sample are shown in Figure 13, where the large increase in signal intensity (due to broken chemical bonds) with exposure time is evident. Without examining a second set of these same samples following xenon irradiation at room temperature in vacuum it is not possible to comment on the role of O_2 in determining the lifetimes of the radicals in the coatings. For example, two alkyl-peroxy radicals can combine to form a stable non-radical product, plus molecular oxygen. This decay pathway is only available for radicals produced in a photo-oxidative environment, and not in an oxygen-free environment.

As a second phase to this project, the sample matrix was modified to include three kinds of primers (inorganic-zinc filled, simple epoxy, and calcium sulfonate) and four kinds of topcoats (acrylic, epoxy, urethane, and sulfonate). (See Table 1.) The enlarged sample set includes both alkyd sulfonates and long-oil epoxy sulfonates, which are commonly used as over-coatings for previously painted structures. Each of the eight topcoats was applied directly to cleaned aluminum.

Aluminum specimens were exposed to QUVA, Prohesion/QUVA, and Florida weathering. In addition, UVB and xenon-lamp irradiation are to be included in the near future. Exposures involving QUVA and Prohesion/QUVA included a one-week exposure to 12 hr cycles made up of a dry exposure to UVA for 8 hr followed by a condensing exposure in the dark for 4 hr. For the Prohesion/QUVA samples, this week was followed by one week under Prohesion conditions. For the simple QUVA samples, the second week was full time exposure to dry UVA.

Samples were analyzed using ESR. Virgin samples and samples having 800 and 1600 hour's exposure to QUVA or Prohesion-QUVA, and 800 hours of real-time Florida weathering have been analyzed. Contrary to some of the first-phase ESR studies involving xenon-lamp irradiation, ESR signals from the QUVA and Florida-exposed samples were very weak, when they were observable at all. For the samples that did have observable signals, the signals were typically strongest for QUVA exposure, strong or intermediate for Florida exposure, and weakest for Prohesion-QUVA exposure. This is reasonable because the longer wet cycle of the Prohesion-QUVA exposure would be expected to enhance destruction of free-radicals. Typical results are shown in Figure 18 below. These data indicate that exposures for 800 hours, including several hundred hours exposure to wavelengths longer than ~300 nm, produces only small accumulations of broken bonds in this series of topcoats. However, it is likely that more severe irradiation conditions (experiments in preparation) are capable of producing radicals. These include: prolonged exposure under the above conditions, exposures to QUVA (and QUVB) without a high humidity condensing cycle, and finally exposure within the ESR cavity to high intensity Xe-lamp irradiation in vacuum or oxygen at 77K and ambient temperature. The latter conditions do not simulate natural weathering, but do help determine the most photolysible sites in the samples.

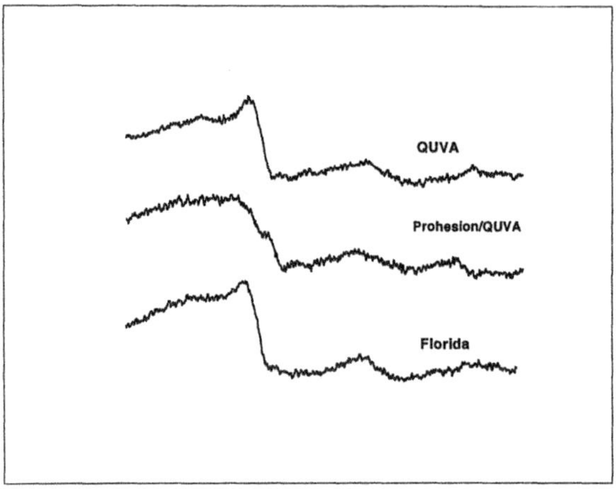

Figure 18. ESR spectra from E1 (epoxy) topcoat following exposure to 800-hr QUVA, Prohesion/QUVA and Florida sunlight. Note the presence of a weak, narrow line in the Florida and QUVA spectra.

A major aspect of this project involves the correlation of photo-sensitivity of the different coatings with their molecular structures. To accomplish this, topcoat resins are being analyzed using high performance liquid chromatography (HPLC) along with nuclear magnetic resonance (NMR) and Fourier-transform infra-red (FTIR) spectroscopies. Chromatography allows separation of commercial formulations into their separate chemical components, and the spectroscopic techniques identify the molecular structures of the components. In a related study, we have determined that the polyester polyol component of a particular polyurethane clearcoat is the component susceptible to photo-induced bond cleavage, and that an aliphatic radical is a major component of the ESR spectrum.

CONCLUSION

In this paper, we have presented a series of results on the photodegradation of polymer-based coatings induced by UVA, UVB, Xe-light and natural light irradiation as studied by slow positron annihilation and electron spin resonance methods. The S-parameter and o-Ps intensity from the positron method are found to decrease with exposure time. Current results show that the slow positron technique is a sensitive tool for detecting coating degradation in a time much earlier than any existing testing methods. The electron spin resonance results show promise for identifying the specific chemical bonds or locations which are responsible for the deterioration of coatings. Combining these two atomic and molecular testing methods will be pursued next in a systematic way for both commercial products and model compounds. The next major research activity is to perform engineering tests and to correlate change in macroscopic engineering properties with observed nano-scale changes detected by PAS and ESR.

ACKNOWLEDGEMENT

This research is supported by the National Science Foundation (CMS-9812717). We also appreciate collaborations with Drs. R. Susuki, T. Ohdaira, and B. Nielsen. Fruitful discussions with Profs. K.L. Cheng and E.W. Hellmuth are acknowledged.

References

1. Federal Highway Administration, "Recording and Coding Guide for the Structural Inventory and Appraisal of the Nation's Bridges," U.S. Department of Transportation, Federal Highway Administration, Report No. FHWA-PD-96-001, 1995.
2. Weismantel, G.E. *Paint Handbook*, Ed. Charlesworth, G.B. and Weismantel, G.E., Chapter 18, New York: McGraw-Hill, 1981.
3. D. Satus, *Coating Technology Handbook*, New York: M. Dekker,1991.
4. *Polymer Durability: Degradation, Stabilization, and Lifetime Prediction*; Clough, R.E.; Billingham, N.C.;Gillen, K.T. Eds.; *Adv.Chem.Ser. #249*, Washington, D.C.: Amer.Chem.Soc., 1996.
5. For example, see: *Polymer Spectroscopy*, Ed. Fawcett, A.H. New York: Wiley & Sons Sci., 1996.
6. Rabek, J.R. *Polymer Degradation and Stabilization: Photodegradation of Polymer, Physical Characterization and Applications*, New York: Springer Pub., 1996.
7. Zhang, R.; Cao, H.; Chen, H.M.; Mallon, P.; Sandreczki, T.C.; Richardson, J. R.; Jean, Y.C.; Nielsen, B.; Suzuki, R.; Ohdaira,T. Rad. Chem. Phys. **2000**, 58, 639-644.
8. Suzuki R.; Kobayashi Y.; Mikado T.; Ohgaki H.; Chiwaki M.; Yamazaki T.; Tomimasu T., Mater. Sci. Forum, **1992**, 105-110, 1993-1996.
9. Cao H.; Zhang R.; Sundar C. S.; Yuan J.-P.; He Y.; Sandreczki T.C.; Jean Y.C.; Nielsen B. Macromolecules, **1998**, 31, 6627-6635.
10. Cao H.; He Y.; Zhang R.; Yuan J.-P.; Sandreczki T. C.; Jean Y.C.; Nielsen B. J. Polym. Sci. B: Polym. Phys. Ed., **1999**, 37, 1289-1305.
11. Cao H.; Yuan J.-P.; Zhang R; Huang C.-M.; He Y.; Sandreczki T. C.; Jean Y. C.; Nielsen B.; Suzuki R.; Ohdaira T. Macromolecules, **1999**, 32, 5925-5933.
12. Jean Y.C., In Positron Spectroscopy of Solids, pp.563-580, Dupasquier, A., Mills, Jr., A.P. Eds. IOS Press, Amsterdam, 1995.

PIEZOELECTRIC ACTUATION OF FATIGUE CRACK GROWTH ALONG POLYMER/METAL INTERFACE

T. Du[1], M. Liu[2], S. Seghi[1], K. J. Hsia[2], J. Economy[1] and J. K. Shang[1]

[1]Department of Materials Science and Engineering
[2]Department of Theoretical and Applied Mechanics
University of Illinois at Urbana-Champaign, Urbana, IL 61801, U.S.A.

ABSTRACT

A new experimental technique for determining durability of the polymer-metal interface under cyclic loading was developed by piezoelectric actuation of fatigue crack growth in an adhesive bond. The piezoelectric actuation was provided by a ferroelectric ceramic plate actuator bonded to a polymer-metal bi-layer structure. Under alternating electric fields, fatigue crack growth was observed along epoxy/aluminum interface. The crack growth rate was found to depend on the magnitude of the applied electric field and decrease with testing frequency.

KEYWORDS:

Piezoelectric actuator, crack, fatigue, adhesive, durability.

INTRODUCTION

Polymer-metal interface is found in a wide range of engineering materials and structures. In adhesive bonding, a polymeric adhesive is used to provide the connection between two different materials and/or parts of an engineering structure. At the material interface, potential defects may develop because of mismatches in chemical and physical properties between the polymer adhesive and the metal adherend. Under external stresses, the defect may grow gradually to cause slow degradation of the adhesive bond. Therefore, understanding growth of these defects, such as a crack, along the polymer-metal interface is essential to predicting the durability of an adhesive bond.

In the past, crack growth in adhesive bonds under cyclic loads has been studied by conducting fatigue experiments on various specimens such as lap shear, double-cantilever beam, cracked lap shear and flexural peel (Mostovoy et al, 1971; Kinloch, 1979; Johnson, 1987; Lai et al., 1996; Zhang and Shang, 1996a). These tests are typically performed on a mechanical testing machine where the fatigue load is provided by a mechanical actuator. For most testing machines, the response of such mechanically actuated systems is limited to relatively low loading frequencies, below about 100 Hz.

In this study, we developed a new technique to conduct fatigue experiments on polymer-metal interface in adhesive bonds. The technique was based on piezoelectric actuation of the fatigue crack growth in an adhesive bond. Because of the fast response of the piezoelectric actuator, experiments could be performed at high frequencies. The feasibility of the piezoelectric actuation was demonstrated for an epoxy/aluminum system at frequencies up to 5 kHz. The fatigue crack growth rate was found to decrease with cyclic frequency.

EXPERIMENTAL PROCEDURE

The piezoelectric testing system is shown in Fig. 1, where a tri-layer specimen was controlled by external power supplies. The specimen was made of a lead zirconate titanate (PZT) actuator, a thin polymer interlayer, and aluminum substrate. The PZT samples were obtained from a commercial source in 0.375 mm thick plates. The PZT plates were poled along the thickness direction, metallized with a nickel coating and cut down to 5.5 cm × 1.5 cm actuators.

The adhesive bond was formed between 6061 aluminum alloy and a toughened epoxy adhesive. The properties of the adhesive have been reported elsewhere (Zhang and Shang, 1996b). Aluminum adherend was machined into 90 mm x 15 mm x 25 mm blocks, polished to a surface finish of 1 μm using polishing powder, and cleaned in an ultrasonic bath and rinsed with acetone. A precrack, 3 mm long through the thickness, was introduced to the one end of the adhesive bond by masking the rest of the surface with tape and spraying PTFE release agent onto the exposed precrack area. The bond thickness was controlled to be 330 μm.

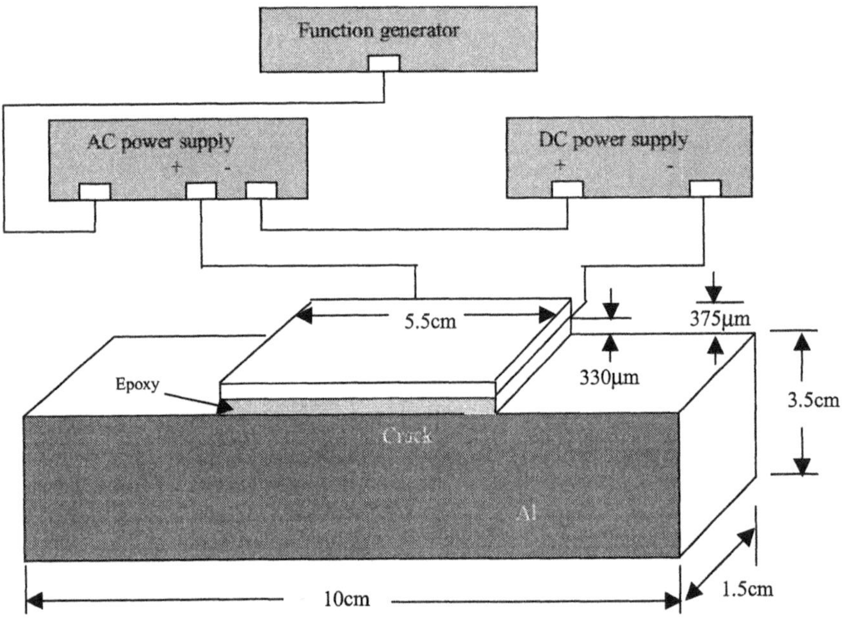

Figure 1. Schematic of the piezoelectric testing system.

Fatigue testing of the adhesive bond was conducted by applying an electric field to the PZT actuator. The field was sinusoidal and the minimum field (and the load-ratio, R) was zero. The testing frequency ranged from 5 Hz to 5000 Hz. At high frequencies (above 1000 Hz), the specimen was water-cooled through cooling channels in the aluminum block to minimize excessive heating of the polymer. During fatigue testing, crack development in the adhesive bond was monitored by a video microscopy system and the crack length was recorded as a function of number of cycles. To assist crack detection, the edge of the sample was polished and coated with the correctional fluid. After fatigue testing, the specimen was examined using a scanning electron microscope (SEM) and X-ray photoelectron spectroscopy (XPS).

RESULTS AND DISCUSSION

As the applied field was increased gradually, cracking of the adhesive bond was observed at field levels above a critical value. The crack started from the precrack and grew steadily along the adhesive bond. The crack path developed under piezoelectric loading is shown Fig. 2 where the crack was seen to propagate between epoxy and aluminum. The variation of the crack length with number of fatigue cycles is shown in Fig. 3 at a constant peak electric field of 8 kV/cm. The crack growth curve may be divided into three segments. The initial segment corresponded to the opening of the precrack area, with the crack length jumping 3-4 mm almost immediately after the application of the electric field. In the last segment, as the crack length approached the total length of the adhesive bond, the crack growth decelerated and eventually stopped. In the second segment, the crack growth was table and the crack length increased linearly with number of cycles. From the slope of the line, the crack growth rate was determined for a given electric field.

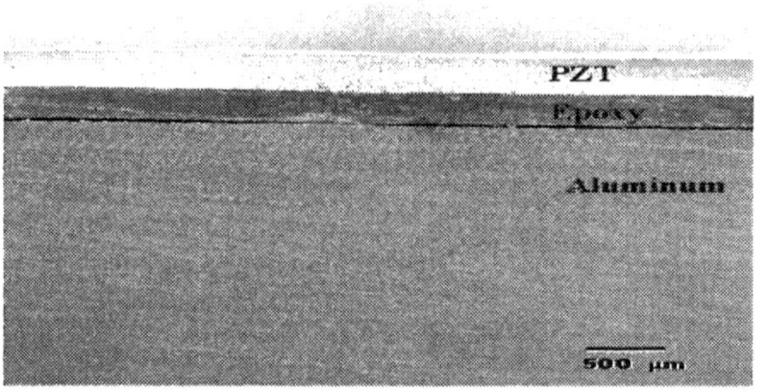

Figure 2. Crack growth along epoxy/aluminum interface.

The dependence of the crack growth rate on the applied field was examined by increasing the applied field in a step-wise manner, with each increment less than 10 pct of the previous field. The results were analyzed by converting the applied field to an equivalent mechanical driving force, the strain energy release rate, using the finite element method (FEM). For the specimen configuration in Fig. 1, FEM results could be fitted to a parabolic function between the total energy release rate, G, and the applied field, E:

$$G = 1.185\text{x}10^{-2} E^2 \qquad (1)$$

where E is in V/m and G in J/m^2. Crack growth rate was then correlated with the range of the strain energy release rates calculated from the field range for an electric cycle.

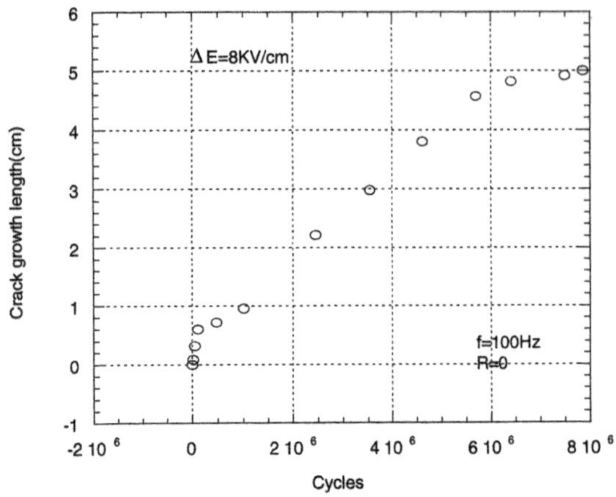

Figure 3. Variation of crack length with number of cycles.

The results for the fatigue crack growth experiments conducted at different frequencies are shown in Fig. 4. Similar to fatigue crack behavior in metals, the fatigue curve may be divided into the near-threshold, intermediate and high growth-rate regimes. In the intermediate growth rate regime, the crack growth rate was fitted into the Paris law:

$$\frac{da}{dN} = C(\Delta G)^n \qquad (2)$$

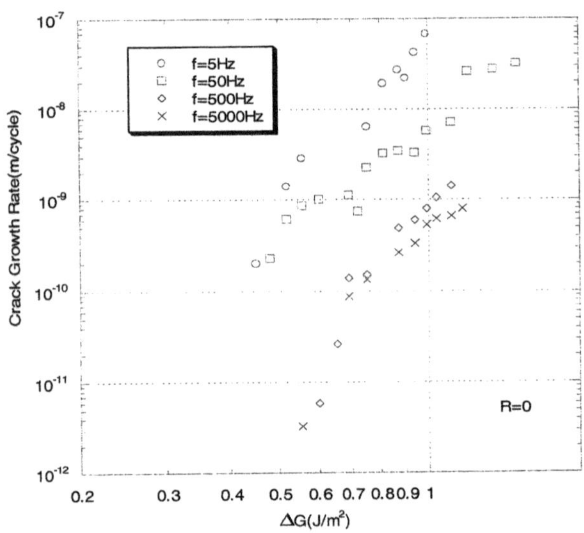

Figure 4. Fatigue crack growth behavior at different testing frequency.

where da/dN is the crack growth rate, ΔG is the difference of maximum and minimum energy release rates, C and n are material constants. From the data in Fig. 4, we obtained n = 6-7, which is higher than 2-4 for epoxy/aluminum interface and polymers obtained by the conventional mechanical fatigue (Kinloch, 1987; Shang, 1996; Hertzberg and Manson, 1980; Hertzberg and Mason, 1986). As the frequency increased from 5 Hz to 5000 Hz, the crack growth rate decreased. At low strain-energy release rates, the crack growth curves for the 5 Hz and 50 Hz converged to about the same nominal fatigue threshold of 0.4 J/m^2. At high strain energy release rates, the crack growth rate decreased from 5 Hz to 500 Hz but remained about the same from 500 Hz to 5000 Hz. At all strain energy release rates, the difference in the crack growth behavior between 500 Hz and 5000 Hz was small, suggesting that fatigue tests may be accelerated by testing at higher frequencies as long as proper cooling is used.

The fatigue crack growth mechanism was studied by examining the fatigue crack surfaces after the test. The aluminum side of the crack was marked by continuous straight polishing lines. A replica of those polishing lines was also evident on the polymer side of the crack, indicating that the fatigue failure was interfacial. The morphology of the polymer fracture surface is shown in Fig. 5, where deformation bands were evident. These deformation bands may come from either creep or shear-induced slip during fatigue process. Since creep deformation is time-dependent rather than cycle-dependent, a decrease in the test frequency prolongs the creep time for a fixed number of cycles, resulting in faster crack growth, in agreement with the experimental results. XPS studies of the crack surfaces indicated that the polymer chemistry was modified during the fatigue experiment. From the intensity of the C_{1s} peak obtained before and after fatigue experiment, it was concluded that an oxidation reaction occurred during the fatigue experiment. The oxidation could cause embrittlement of the polymer, reducing the resistance of the interface to fatigue crack growth.

Figure 5. SEM micrograph of the fracture surface on the polymer side tested at 50 Hz.

CONCLUDING REMARKS

A new testing technique based on the piezoelectric property of PZT was developed to measure fatigue properties of polymer-metal interface in an adhesive bond. Our experiments have demonstrated that it was possible to produce sufficient driving force for interfacial crack growth using piezoelectric actuators made from PZT ceramic. The driving force for the crack growth was calculated from the electric field and crack length, and correlated with the crack growth rate. Using the piezoelectric testing system, the fatigue crack growth behavior of an epoxy/aluminum interface was found to depend strongly on testing frequency.

ACKNOWLEDGEMENT

Support for this work was provided by the National Science Foundation under the grant CMS-9872306.

REFERENCES

Hertzberg R.W. and Manson J.A. (1980). *Fatigue of Engineering Plastics*, Academic Press, New York.

Hertzberg R.W. and Manson J.A. (1986). Fatigue and Fracture, in *Encyclo. of Polymer Science Engineering*, **6**, 2nd ed., Wiley, New York.

Johnson W.S. (1987). *J. Testing and Evaluation*, **15**, 303.

Kinloch A.J. (1979). *J. Adhesion*, **10**, 193.

Kinloch A.J. (1987). *Adhesion and Adhesives: Science and Technology*, Chapman and Hall, London.

Lai Y-H., Rakestraw M.D. and Dillard D.A. (1996). *Int. J. Solids Struct.*, **33**, 1725.

Mostovoy S., Ripling E.J. and C. F. Bersch C.F. (1971). *J. Adhesion*, **3**, 125.

Shang J.K. (1996). in "*Fatigue'96*", Eds., G. Lutjering and H. Nowack, Pergamon Press, Oxford, UK, 43.

Zhang Z. and Shang J.K. (1996a). *Metall. Mater. Trans.*, **27A**, 205.

Zhang Z and Shang J.K. (1996b). *Metall. Mater. Trans.*, **27A**, 221.

Test Methods

Long Term Durability of Structural Materials
P.J.M. Monteiro, K.P. Chong, J. Larsen-Basse, K. Komvopoulos (Eds)

ACCELERATED LIFE PREDICTION AND TESTING OF STRUCTURAL POLYMERS UNDER CYCLIC LOADING

Hongbing Lu[1], Bo Wang[1], Guixiang Tan[1] and Weinong Chen[2]

[1]School of Mechanical & Aerospace Engineering
Oklahoma State University
Stillwater, OK 74078
[2]Department of Aerospace and Mechanical Engineering
The University of Arizona
Tucson, AZ 85721

ABSTRACT

Based on the framework of Lemaitre's continuum damage model, a new damage variable is defined in terms of the remaining fracture strain, and a new evolution relation is derived to characterize the extent of fatigue damage after a certain number of loading cycles on the polymer specimen. A new fatigue damage model is proposed to establish a predictive formula for the fatigue service life of polycarbonate under certain amplitude of cyclic stress. This formula indicates that in the double logarithmic coordinate system the fatigue life N_f has a linear relation with the stress amplitude $\Delta\sigma$. Fatigue tests with a stress ratio $r = 0.1$ were conducted at room temperature to construct the stress amplitude versus fatigue life curve. After different numbers of cycles of fatigue, the values of the new damage variable for polycarbonate were measured by pulling damaged specimens to fracture under monotonic loading. The non-contact deformation measurement technique—digital image correlation method was used to measure full-field surface deformation on the specimen. Experimental results on the damage evolution and the fatigue life have a good agreement with those predicted by the proposed model. Fatigue tests at different frequencies were also performed and the results show that fatigue life increases with frequency for polycarbonate under both stress and strain controlled conditions within the test frequency range.

KEYWORDS

Damage Variable, Damage Model, Fatigue, Polymer, Lifetime, Frequency

INTRODUCTION

Polymer components, like metal components, may fail at a stress level that is much lower than their static strength when a cyclic stress is applied (Hertzberg and Manson, 1980; Schutz, 1996). Such behavior, known as fatigue, is usually associated with the accumulation of permanent fatigue damage in the materials. Polymer fatigue has distinct features. Because all polymeric materials exhibit significant time-dependent behavior, or equivalently, viscoelastic behavior, polymer fatigue is inevitably accompanied by creep or relaxation when a cyclic stress or strain around a mean value is applied to polymer components. A simply increasing in frequency of loading cannot provide reliable acceleration of testing to predict realistic failure behavior over long periods of time because viscoelastic effects in polymers cannot be scaled with frequency. Polymers are also very sensitive to environmental conditions such as temperature and moisture content, physical aging and degradation (chemical aging), and their previous loading and environment histories. Because of these distinct features, the accelerated life testing of polymer fatigue is fundamentally different from that of metal fatigue and must be addressed differently. So far, little research has been done on polymer fatigue involving time-dependent feature and environmental conditions, and the available research work has been limited to experimental investigation using empirical relations (Li, et al., 1995).

In this study we focus on the investigation of polymer fatigue by considering damage evolution based on some available experimental results obtained for polycarbonate. From the micro-mechanics point of view, the fatigue failure of structures and materials results from the nucleation and growth of initial defects such as microvoids, microcracks and second phase inclusions. With the development of Continuum Damage Mechanics (CDM) (Lemaitre, 1984, 1985, 1990), it has become possible to understand the fatigue damage evolution behavior. Based on CDM, the failure behavior of materials can be described by means of a damage evolution model with an internal variable representing the degradation state of the material (Wang, et al., 2000). Under cyclic loading, the damage process and final failure of materials are caused by the evolution of initial damage. Recently, damage mechanics theory has been introduced to study fatigue failure for ductile metals, and the fatigue damage model for strain-hardening materials was proposed under cyclic fatigue loading (Jiang, 1995).

In this study, in the context of CDM theory, a new damage variable based on the remaining fracture strain is proposed for the damage of polymers under fatigue loading. The formula is provided for the prediction of fatigue life and damage evolution equation is derived for fatigue damage process. Tests for polymer materials are conducted to measure material characteristic parameters and damage variable. Fatigue and damage experiments are performed to examine the theoretical model for fatigue damage. The effect of frequency of cyclic loading on fatigue life is investigated for polycarbonate materials to understand the time- or frequency-dependent fatigue behavior.

FATIGUE DAMAGE EVOLUTION MODEL

In damage mechanics (Lemaitre, 1984, 1985, 1990), damage variables play an important role in damage evolution model. Since any state variable could be potentially used to measure fatigue damage and the concept of ductility exhaustion has been introduced in fatigue damage model, a new damage variable is defined as

$$D = \frac{\varepsilon_f - \tilde{\varepsilon}_f}{\varepsilon_f} = 1 - \frac{\tilde{\varepsilon}_f}{\varepsilon_f}, \tag{2-1}$$

where ε_f is the fracture strain for a virgin sample, and $\tilde{\varepsilon}_f$ is the remaining fracture strain after a certain number of loading cycles. Based on Eqn. (2-1), for a virgin sample $\tilde{\varepsilon}_f = \varepsilon_f$, $D = 0$, which represents the undamaged state of material. When the material is in fully damaged state, in which $\tilde{\varepsilon}_f =$ 0, then $D = 1$. Therefore, the definition of the new damage variable D is viable in the framework of damage mechanics (Lemaitre, 1984).

Based on Lemaitre's damage model, together with the hypothesis of isotropy of damage and isotropy of plasticity, the constitutive equation for damage evolution is given by the normality property of the potential (Lemaitre, 1985),

$$\dot{D} = -\frac{\partial \varphi^*}{\partial y}, \tag{2-2}$$

where $\dot{D}$ is the damage growth rate, φ^* the potential of dissipation and y the damage strain energy release rate. For damage evolution under fatigue loading, the potential of dissipation can be assumed as (Lemaitre, 1985)

$$\varphi^* = \frac{S_0}{s_0+1}\left(\frac{-y}{S_0}\right)^{s_0+1} f(D), \tag{2-3}$$

where s_0 and S_0 are the material characteristic constants. For a polymeric material, fracture strain ε_f can describe the ductility of material and is related to strain rate. Therefore, considering the effect of inherent ductility of material on damage evolution, $f(D)$ is introduced as

$$f(D) = [1-(1-D)^{1+2s_0}]^{\varepsilon_f}. \tag{2-4}$$

The damage strain energy release rate y can be expressed as

$$-y = \frac{\Delta\sigma_{eq}^2}{2E(1-D)^2}\left[\frac{2}{3}(1+\nu)+3(1-2\nu)\left(\frac{\Delta\sigma_H}{\Delta\sigma_{eq}}\right)^2\right], \tag{2-5}$$

where E and ν are Young's modulus and Poisson's ratio of the material, respectively, $\Delta\sigma_{eq}$ is the amplitude of the Von-Mises stress and $\Delta\sigma_H$ is the amplitude of isotropic stress. In uniaxial stress state, $\Delta\sigma_H = \frac{1}{3}\Delta\sigma$ and $\Delta\sigma_{eq} = \Delta\sigma$, where $\Delta\sigma$ is the stress amplitude. Therefore, from Eqn. (2-2) to Eqn. (2-5), we have

$$\dot{D} = -\frac{\partial\varphi^*}{\partial y} = \left(\frac{-y}{S_0}\right)^{s_0} f(D) = \left[\frac{\Delta\sigma^2}{2ES_0(1-D)^2}\right]^{s_0} f(D). \tag{2-6}$$

Substituting Eqn. (2-4) into Eqn (2-6), the damage evolution per cycle can be expressed as

$$\frac{dD}{dN} = \left[\frac{\Delta\sigma^2}{2ES_0(1-D)^2}\right]^{s_0}[1-(1-D)^{1+2s_0}]^{\varepsilon_f}. \tag{2-7}$$

Assuming that the damage variable D is zero at the beginning of the cyclic loading, that is, when N = 0, $D = 0$. Then the damage value at any cycle can be determined by integrating Eqn. (2-7).

$$\int_0^D \frac{d[1-(1-D)^{1+2s_0}]}{[1-(1-D)^{1+2s_0}]^{\varepsilon_f}} = \int_0^N (1+2s_0)\left(\frac{\Delta\sigma^2}{2ES_0}\right)^{s_0} dN. \tag{2-8}$$

Thus, the relation between the damage variable D and the number of cycles N is

$$[1-(1-D)^{1+2s_0}]^{1-\varepsilon_f}=(1-\varepsilon_f)(1+2s_0)(\frac{\Delta\sigma^2}{2ES_0})^{s_0}N. \tag{2-9}$$

When fatigue rupture occurs, $D=D_c=1$. At this moment, the number of cycles $N=N_f$ and the fatigue life N_f can be obtained as

$$N_f=\left[(1-\varepsilon_f)(1+2s_0)(\frac{\Delta\sigma^2}{2ES_0})^{s_0}\right]^{-1}=\frac{(2ES_0)^{s_0}}{(1-\varepsilon_f)(1+2s_0)}\Delta\sigma^{-2s_0}. \tag{2-10}$$

The above equation can be readily used to predict the fatigue life. Eqn. (2-10) can also be expressed in the following form:

$$\log N_f=\log\frac{(2ES_0)^{s_0}}{(1-\varepsilon_f)(1+2s_0)}-2s_0\log\Delta\sigma=B+A\log\Delta\sigma, \tag{2-11}$$

where

$$B=\log\frac{(2ES_0)^{s_0}}{(1-\varepsilon_f)(1+2s_0)}\text{ and }A=-2s_0. \tag{2-12}$$

From Eqn.(2-11), it can be concluded that in the double logarithmic coordinate system, fatigue life N_f is linear to stress amplitude $\Delta\sigma$. This formula has the same form as empirical ones for metals. The coefficients A and B in Eqn. (2-11) are related to E, ε_f, s_0 and S_0 through Eqn. (2-12).

Young's modulus E and fracture strain ε_f can be determined from material tests. The stress amplitude $\Delta\sigma$ versus fatigue life N_f curve can be obtained by fatigue tests for polymeric materials. Thus coefficients A and B can be determined through this curve. Finally s_0 and S_0 can be calculated by Eqn. (2-12).

After determining all constants in the model, the damage evolution behavior can be studied for fatigue damage of materials. From Eqn. (2-9) and (2-10), the damage evolution equation can be represented as

$$D=1-\left[1-(\frac{N}{N_f})^{\frac{1}{1-\varepsilon_f}}\right]^{\frac{1}{1+2s_0}}. \tag{2-13}$$

Eqn. (2-13) can be used to evaluate the material damage under fatigue loading. This damage evolution equation will be examined by fatigue tests of polymers.

MECHANICAL PROPERTY AND FATIGUE TESTS

Material and Specimen

The material used in this study is amorphous polycarbonate (PC), with a glass transition temperature of 140*°C*, manufactured by GE Plastics, named commercially as Lexan® 121 (clear). Test specimens were made according to ASTM Standard D638-97; their geometry and dimensions are shown in Figure 1. Laxan® 121 sheets of 6.3 *mm* thick were cut into strips and annealed at 120*°C* for 10 hours to remove residual stresses resulting from both manufacturing and cutting procedures. Standard flat specimens with a gauge length of 50 *mm* were machined from the strips. The surface roughness of the machining side surface is 1.9 *μm*. After machining, the PC specimens were annealed again at 120*°C* for 10 hours to eliminate residual stress.

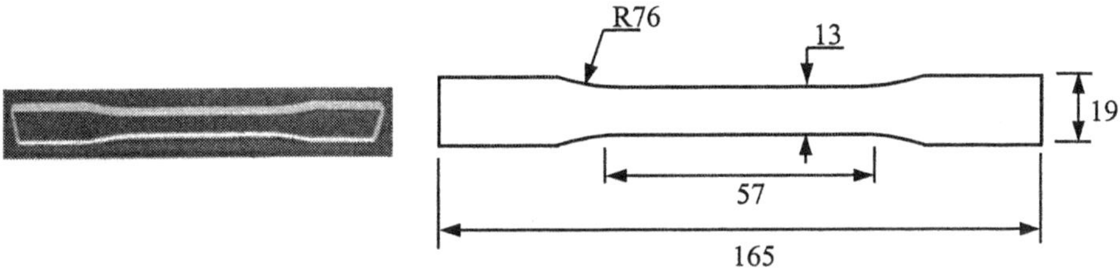

Figure 1: A Photo of a Specimen and the Specimen Geometry (in *mm*)

Experimental Setup

Both quasi-static uniaxial tension and fatigue tests were carried out on an Instron 4202 material test system. A National Instrument's data acquisition system was used to record the load and displacement and a digital image acquisition system with the use of a Kodak Megaplus ES 1.0 digital camera was employed to acquire images of the specimen surface for non-contact strain measurements.

Experimental Results

Uniaxial tensile tests were performed at room temperature at a constant crosshead speed of 1.27 *mm/min*. The load versus crosshead displacement curve is plotted in Figure 2. The tensile stress is 65 *MPa* at point *A*. After point *A*, the load decreases significantly to point *B* due to necking in the specimen. Necking continues until point *C* while the load is constant and the crosshead displacement keeps increasing from *B* to *C*. The final rupture occurs at point *D*.

During simple tension tests, images of the specimen at different loads were acquired and analyzed using the digital image correlation code (Lu and Cary, 2000) to determine engineering strains. For example, at point *F*, the measured strain is 0.014 at a load 2.3 *kN* (the stress is 29.5 *MPa*). The fracture strain ε_f and Young's modulus of the PC material are 0.12 and 2.1 *GPa*, respectively.

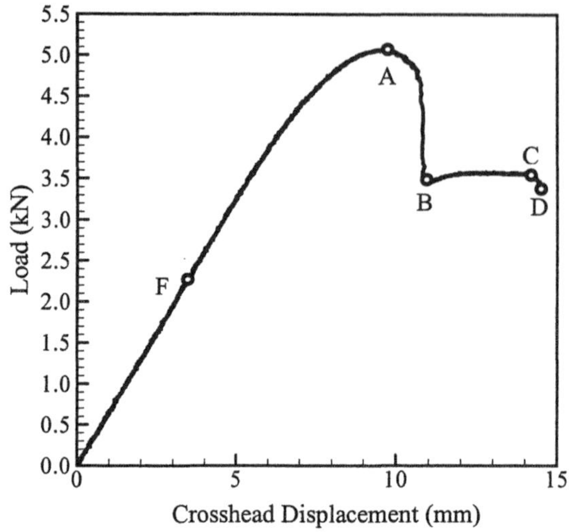

Figure 2: Load vs. Crosshead Displacement under Uniaxial Tension

Tension-tension fatigue tests under load control were conducted at room temperature at a constant crosshead speed of 12.7 *mm/min*. The ratio of the minimum stress σ_{min} to the maximum stress σ_{max}, r, is 0.1 for all fatigue tests. Figure 3 shows a typical fatigue fracture surface of PC. There are three distinct regions on the fracture surface.

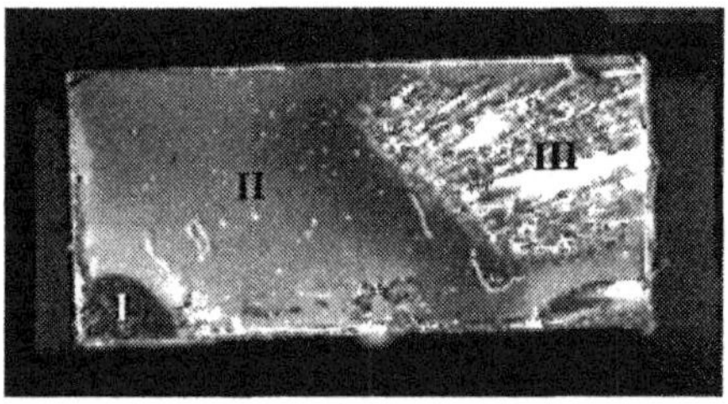

Figure 3: A Typical Fatigue Fracture Surface of PC

The images of the speckled surface of the specimen acquired by the digital image acquisition system under different loads were analyzed by the digital image correlation code to determine the surface deformations. Figure 4 shows the contour of the normal strain in the axial direction of the specimen at a maximum stress of 15.9 *MPa* under fatigue. It is clear that the strain is not uniformly distributed on the surface. Crazes could most likely form at high strain sites.

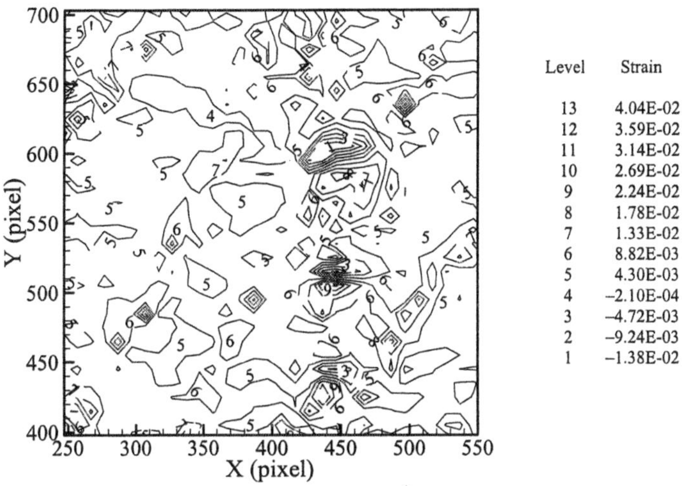

Figure 4: Contour of the Axial Strain on the Specimen Surface

In Figure 5, the symbols represent the experimental data, and the solid line is the fitted curve. Experimental results indicate that in the double logarithmic coordinate system the fatigue life N_f has a linear relation with the stress amplitude $\Delta\sigma$; this behavior is consistent with the prediction by Eqn. (2-11). The correlation coefficient between the fitted straight line in logarithmic scale and experimental results is 0.95, indicating that the proposed predictive formula for the fatigue life is suitable for polycarbonate materials. The relation can be expressed as

$$\log N_f = 7.62 - 2.71 \log \Delta\sigma . \tag{3-1}$$

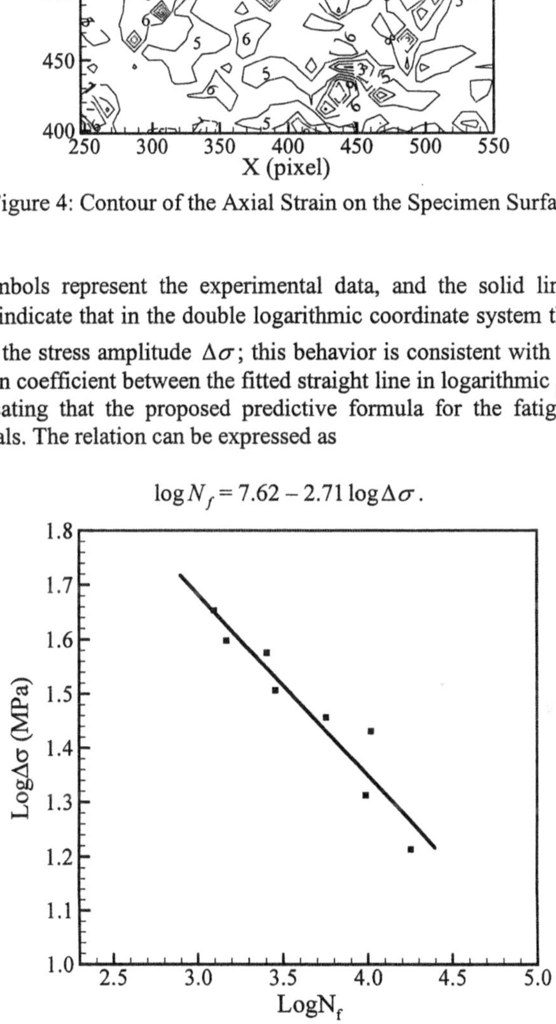

Figure 5: $\text{Log}\,\Delta\sigma$ -$\text{Log}\,N_f$ Curve of PC

Based on Eqn (3-1), the two parameters *A* and *B* introduced in Eqn. (2-11) can be determined as –2.71 and 7.62, respectively. Substituting *A* and B into Eqn. (2-12), the material parameters s_0 and S_0 can be calculated as 1.36 and 196 *MPa*, respectively. Therefore, the damage evolution equation can be determined as

$$D = 1 - \left[1 - \left(\frac{N}{N_f}\right)^{1.13}\right]^{0.27}. \tag{3-2}$$

Six sets of tests were conducted to determine the damage variable defined in Eqn. (2-1). At first, the specimens were subjected to a certain number of cycles of fatigue loading, *N*, under stress amplitude of 22 *MPa*. These specimens were damaged after fatigue loading. The damaged specimens were then pulled until failure under quasi-static loading condition at a crosshead speed 1.27 *mm/min* to obtain the remaining fracture strain $\tilde{\varepsilon}_f$. Damage variable *D* can be calculated based on its definition in Eqn. (2-1). The stress-strain curves of the damaged samples at different N/N_f ratios are shown in Figure 6, where N_f can be determined for the given stress amplitude in Figure 5.

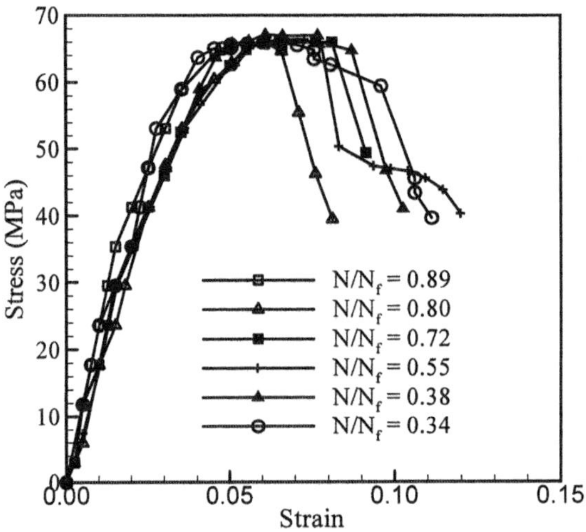

Figure 6: Stress-Strain Curves for Damaged Samples

Figure 6 shows that Young's modulus is independent of the extent of fatigue damage. This may be because that the maximum stress in fatigue is less than yield stress and does not cause global plastic deformation leading to massive damage formation. In fact, there are only a limited number of damaged sites. The majority part of the material is still in elastic state such that the global elastic response represented by Young's modulus is not significantly affected. However, the small amount of damage can indeed cause localized damage evolution leading to rupture.

The theoretical values of damage variable *D* calculated in Eqn. (3-2) are compared with experimental ones under different cycle fractions N/N_f in Figure 7. These two results have a reasonably good agreement.

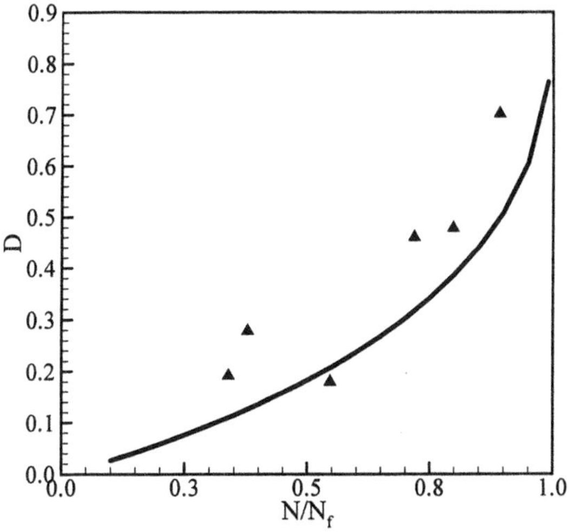

Figure 7: Comparison of Theoretical and Experimental Damage Variables at Different N/N_f

Effects of Frequency on the Fatigue Life

We also conducted experiments to look at the effects of frequency on the fatigue life. Fatigue tests, under either stress or strain control, were preformed at different frequencies on a MTS 809 material test system. Figure 8 plots the fatigue life versus frequency curves under either stress or strain controlled loading conditions. The curves show that the fatigue life generally increases with frequency for the range of frequency investigated, which is between 0.25 and 2 *Hz* under a cyclic stress loading and between 1 and 10 *Hz* under a cyclic strain loading. Under the stress controlled loading condition, the fatigue life increases by 13.7% from 0.25 *Hz* to 2 *Hz*; under the strain control condition the fatigue life increases by 28.4% within the frequency range between 1 *Hz* to 10 *Hz*.

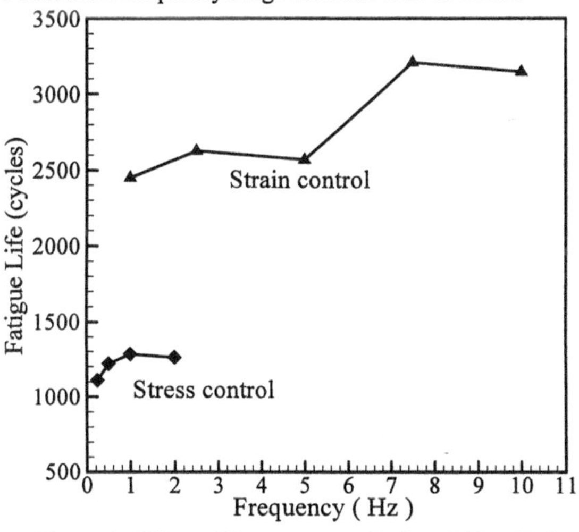

Figure 8: Effect of Frequency on Fatigue Life of PC

Observation of the specimen surface after fatigue failure indicates that there were more crazes formed at a higher frequency than at a lower one. The crazes on the surface of the specimen at frequency 7.5 *Hz* are shown in Figure 9. The initiation and propagation of the crazes is one of the major sources for energy absorption, more craze formation tends to suppress the initiation and propagation of the major crack leading to the eventual fatigue failure of the material, thus extends the fatigue life. The dependence of fatigue life on frequency indicates that the time- or frequency-dependent behavior in polymers must contribute to the fatigue behavior. A simply increasing in frequency in polymer fatigue tests tends to give a fatigue life longer than the actual fatigue life when the polymer is used at a lower frequency. Further research needs to be carried out for the development of a method to extract actual service life based on accelerated life testing results.

Figure 9: Crazes on the Specimen Surface at Frequency 7.5 *Hz*

During these fatigue tests, the temperature on the specimen surface was measured by a thermocouple. Figure 10 shows the temperature as a function of number of loading cycles. As the frequency increases, the temperature on the specimen surface increases as the result of higher dissipated energy during cyclic loading. The surface temperature suddenly goes up when the test is near the final failure in the last a few cycles.

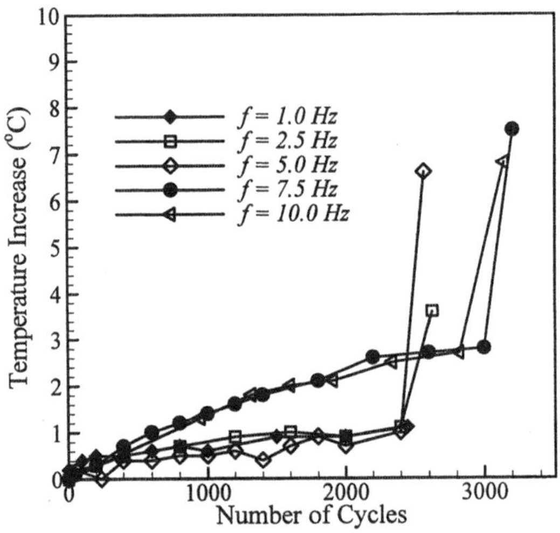

Figure 10: Effect of Frequency on Local Temperature at the Specimen Surface

CONCLUSIONS

In this study, following the Lemaitre's damage model, a new damage variable is defined, a new damage evolution model is proposed, and a formula is derived to predict the fatigue life under the given stress amplitude for structural polymers. The predictive formula indicates that the fatigue life N_f has a power law relation with the stress amplitude $\Delta\sigma$.

Experiments for polycarbonate materials were carried out to determine material parameters and the damage variable. Fatigue tests were performed at a stress ratio of 0.1 at room temperature to construct the stress amplitude versus fatigue life curve. Theoretical and experimental results are in good agreement. It is thus verified the effectiveness of the proposed fatigue damage model. The model is also very useful in the evaluation of the fatigue damage evolution for polymers. The results of fatigue tests at different frequencies indicate that temperature increases with loading frequency, even when the frequency is very low, and that the fatigue life of polycarbonate increases with loading frequency. A simply increase in loading frequency in accelerated life testing tends to give a longer fatigue life than the actual fatigue life for polymeric structures actually used at a lower frequency.

ACKNOWLEDGEMENTS

The authors acknowledge the support of the National Science Foundation under Grant CMS-9872350.

REFERENCES

Hertzberg R. W. and Manson J.A. (1980). Fatigue of Engineering Plastics, *Academic Press, Inc.*

Jiang M. (1995). A damaged evolution model for strain fatigue of ductile metals, *Engineering Fracture Mechanics* **52:6**, 971-975.

Lemaitre J. (1984). How to use damage mechanics, *Nuclear Engineering Design* **80**, 233-245.

Lematre J. (1985). A continuous damage mechanics model for ductile fracture, *Journal of Engineering Material Technology* **107:1**, 83-89.

Lematre J. (1990). Micro-mechanics of crack initiation, *International Journal of Fracture* **42**, 87-99.

Li X., Hristov H.A. and Yee A.F. (1995). Influence of cyclic fatigue on the mechanical properties of amorphous polycarbonate, *Polymer* **36:4**, 759-765.

Lu H. and Cary P.D. (2000). Deformation measurements by digital image correlation: implementation of second order displacement gradient, to appear in *Experimental Mechanics* **40:4**.

Schutz W. (1996). A history of fatigue, *Engineering Fracture Mechanics* **54:2**, 263-300.

Wang B., Hu N., Kurobane Y., Makino Y. and Lie S.T. (2000). Damage criterion and safety assessment approach to tubular joints, *Engineering Structures* **22**, 424-434.

Long Term Durability of Structural Materials
P.J.M. Monteiro, K.P. Chong, J. Larsen-Basse, K. Komvopoulos (Eds)

ACCELERATED DURABILITY TESTING OF GAS TURBINE COATINGS EMPHASIZING OXIDE-METAL INTERFACES

M. J. Stiger[1], R. Handoko[2], J. L. Beuth[2], F. S. Pettit[1], and G. H. Meier[1]

[1] Department of Materials Science and Engineering, University of Pittsburgh, Pittsburgh, PA 15261, USA
[2] Department of Mechanical Engineering, Carnegie Mellon University, Pittsburgh, PA 15213, USA

ABSTRACT

This paper describes the basis for a mechanism-based approach for establishing a protocol for accelerated durability testing of oxidation resistant coatings and thermal barrier coatings. Tests to evaluate the individual factors controlling spallation are described and preliminary short-term acoustic emission experiments, which seem to correlate with long term cyclic oxidation are described.

KEYWORDS

Alumina scales, thermal barrier coatings, adherence, durability.

INTRODUCTION

The adherence and durability of oxide scales, particularly alumina scales, is critical to the long-term performance of most high temperature alloys. For example, the adherence of oxide scales to hardware in the hot section of gas turbines is essential for their long-term corrosion resistance. Oxides also play a key role in other types of coating systems. In particular, the resistance of ceramic thermal barrier coatings (TBCs) to spallation from nickel-based substrates is dependent on the adherence of the alumina scale which forms between the TBC and the substrate. The loss of adherence along the oxide/substrate interface is a fundamental barrier to more widespread use of TBCs on rotating components such as turbine blades and the significant increases in turbine efficiency that would result. The problem of in-service loss of adhesion in TBC systems is thus a major one in the aircraft and power generation gas turbine industries.

Despite the importance of issues associated with oxide adherence and durability, state-of the-art oxidation testing in industry relies on time-consuming and expensive tests (e.g. burner-rig tests), exposing alloys to highly simulative cyclic thermal environments. Such tests are required because of a lack of understanding of fundamental chemical and mechanical processes controlling oxide degradation. Existing limits in the understanding of oxide degradation serve as a barrier to lower-cost accelerated testing methods and rapid development of new alloy and coating systems.

Recent developments in two fields have the potential to alter this situation. First, in the study of oxidation, an increased understanding of fundamental oxidation mechanisms has evolved. Meanwhile, fracture mechanics theory and modeling methods for the failure of thin films and coatings have advanced tremendously. The work described in this paper is an attempt to exploit these advances to gain an increased fundamental understanding of the degradation of oxide scales and thermal barrier coating systems. Figure 1 shows schematic cross-sections of the two systems of interest in this paper. State-of-the-art coatings for oxidation resistance (Fig. 1a.) form a protective alumina film during exposure. The length of time that this film can resist breakdown by chemical or mechanical means determines the life of the coating. Thermal barrier coatings (Fig. 1b.) pose similar issues in that an alumina layer (often called a "thermally grown oxide", TGO) forms between the metallic bond coat and ceramic topcoat. This situation has the additional feature that strain energy accumulated in the ceramic layer can contribute to the mechanical degradation of the TGO.

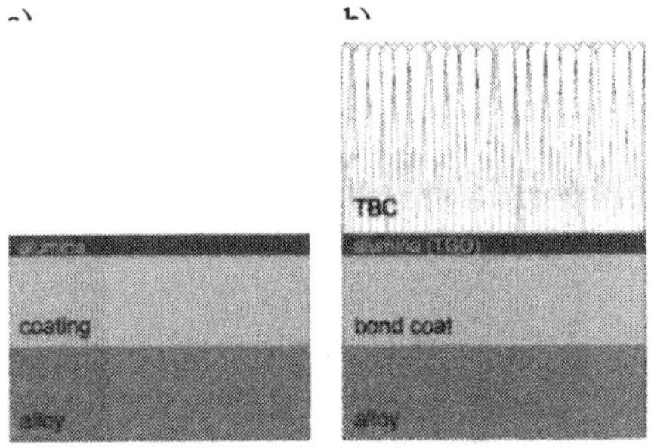

Figure 1. Schematic diagram of cross-sections of the coatings considered in this paper. (a.) An alumina-forming coating for oxidation protection, deposited on a nickel-base superalloy. (b.) A thermal barrier coating on a nickel-base superalloy.

The lifetime of thermal barrier coatings illustrates some of the important issues associated with accelerated testing. Figure 2 presents a macroscopic photograph of an EB-PVD TBC on a Pt-aluminide bond coat, which has failed after 1287 cycles (45 min. at high temperature and 15 min. cooling) at 1100°C in dry air. The failure initiated at a specimen edge and propagated as an elongated buckle. Figure 3 indicates the effect of temperature on the time to failure. Here the inverse of the time at elevated temperature (0.75 times the number of cycles) is plotted versus reciprocal temperature. The data follow a reasonable straight line with failure times decreasing from about 1000 h at 1100°C to 50 h at 1200°C. Importantly, if these data are extrapolated to the temperatures currently experienced by bond coats ($\approx$ 1000°C) the failure times are predicted to be on the order of 50,000 hours. Experiments of this duration are generally not feasible and the effectiveness of changes made to bond coats must be evaluated on a much shorter time scale. This example clearly shows the need for a reliable accelerated method for determining failure times.

Philosophy of Accelerated Testing

A reliable accelerated testing protocol requires, as a prerequisite, a knowledge of the failure mechanism and the important parameters which control failure. The accelerated test must then modify the important parameters to produce failure in a reasonable time. However, it must be clear that the changes do not move the system into a regime where a different failure mechanism becomes operative. For example, increasing the exposure temperature is one common way of accelerating a test. However, there are a number of well-known cases where changing the temperature also changes the mechanism. One such case is the oxidation of the intermetallic compound $MoSi_2$, Grabke and Meier (1995). This compound oxidizes very rapidly at 500°C by growing a surface layer of MoO_3. Increasing the temperature to 600°C actually decreases the oxidation rate because it results in the formation of a slowly-growing surface film of SiO_2. The following is a description of work directed toward the accelerated testing of alumina-forming coatings and thermal barrier coatings.

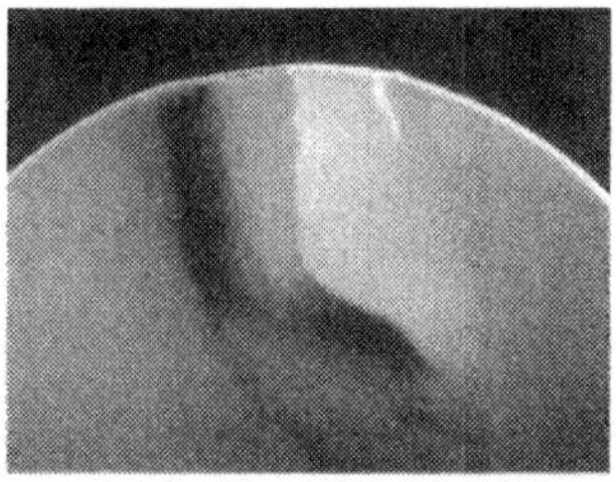

Figure 2. Macroscopic photograph of the failure of a thermal barrier coating consisting of an EB-PVD topcoat with a Pt modified aluminide bond coat on a disk of a nickel-base superalloy.

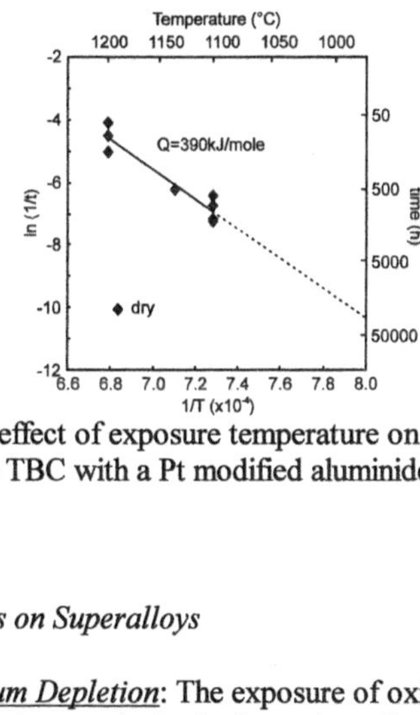

Figure 3. The effect of exposure temperature on the time to failure of an EB-PVD TBC with a Pt modified aluminide bond coat

Important Parameters

Alumina-Forming Coatings on Superalloys

Oxide Growth and Aluminum Depletion: The exposure of oxidation resistant coatings, such as that depicted in Fig. 1a., results in the formation of an alumina film, which depletes aluminum from the coating. The factor which controls life is the loss of Al in forming alumina on the surface and through spallation during thermal cycling, as well as interdiffusion with the substrate. This is illustrated for a NiCoCrAlY coating in Figure 4. The Al-depletion is manifested by dissolution of the Al-rich β phase and, eventually the appearance of spinels in the oxide scale. After selecting the failure criterion the rate of Al loss is needed. Nesbitt and coworkers, Smialek et al. (1995) have analytically modeled this type of a process and predicted the lifetimes associated with Al losses from alumina-forming systems. This will not be addressed further in this paper, rather the factors associated with mechanical breakdown of alumina scales will be addressed i. e. loss of adherence. The adherence of protective oxide scales to substrates is governed by the stored elastic energy in the scale which drives delamination and the fracture toughness of the alloy/oxide interface which quantifies the resistance to fracture.

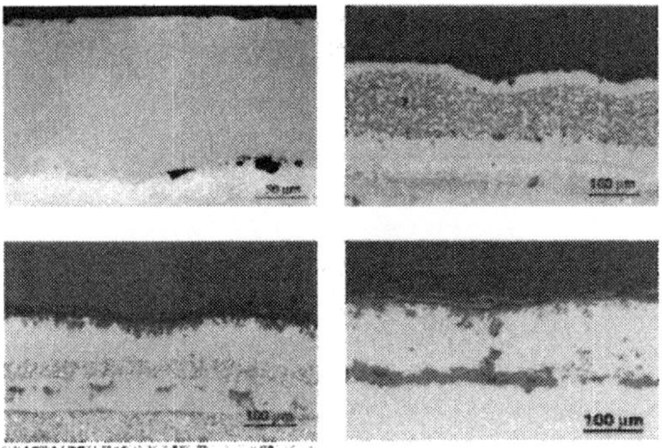

Figure 4. Cross-sections of a NiCoCrAlY coating on a nickel-base superalloy showing the Al-depletion (indicated by the dissolution of the β phase), which occurs during cyclic oxidation at 1100°C.

Elastic Energy Stored in the Scale: The stored elastic energy is determined primarily by the scale thickness h and the stress level σ in the scale such that ($E_{elastic} \alpha h\sigma^2$). The scale thickness is determined by the rates of diffusion of metal and/or oxygen through the scale and is generally represented by the parabolic rate constant k_P. The growth rates of continuous alumina scales are strongly dependent on oxidation temperature but are only influenced slightly by alloy composition. However, for systems which require extremely long lifetimes, a moderate change in k_P can result in measurable changes in oxide thickness and, therefore, metal consumption and elastic strain energy in the scale.

The stress state in the scale is determined by stresses which arise during the oxide formation (growth stresses), stresses produced during temperature changes as the result of thermal expansion mismatch between the oxide and the alloy (thermal stresses), and any stress relaxation which may occur. Growth stresses can, in principal, be determined by the growth mechanism of the alumina and, therefore, by doping with elements such as yttrium. Thermal stresses are primarily a function of the temperature is change (ΔT) but can be influenced by the rate of temperature change if stress relaxation processes occur. Stress relaxation processes arise primarily from creep of the oxide and/or substrate and are influenced by the relative thicknesses and creep strengths of the oxide and substrate.

Fracture Toughness of the Alloy/Oxide Interface: The fracture energy of the interface is a function of the composition at the interface, the microstructure in the interfacial region, and the composition of the exposure environment. It is now well established that small additions of reactive elements, such as yttrium, hafnium, and cerium, substantially improve the adherence of alumina films to alloy substrates, Sarioglu et al. (2000). While the effects produced by the reactive elements are widely known, the mechanisms whereby they improve adherence are not completely understood. Over the last fifty years a number of mechanisms have been proposed.

The precious metals, such as Pt, have also been known for many years to have beneficial effects on the cyclic oxidation of alumina-forming alloys and this effect is the basis for the platinum-aluminide coatings which are widely used to protect Ni-base superalloys. However, there is still much which is not understood with respect to the mechanisms by which Pt affects alumina adhesion, Schaeffer, et al. (1989).

A critical step in understanding mechanisms leading to oxide adhesion loss and in matching adhesion loss mechanisms between industry standard and accelerated oxidation tests is periodic measurement of interfacial fracture toughness. A reliable interfacial toughness test is needed, suitable for application to relatively small-sized specimens.

Response to Stresses: Generally, alumina scales will be loaded in compression when cooled to room temperature. Possible responses to these stresses are shown in Figure 5. The mechanisms shown if Fig. 5(a), buckling, and Fig. 5(b), wedge cracking, are the most damaging in that they expose fresh metal to the oxidizing gas.

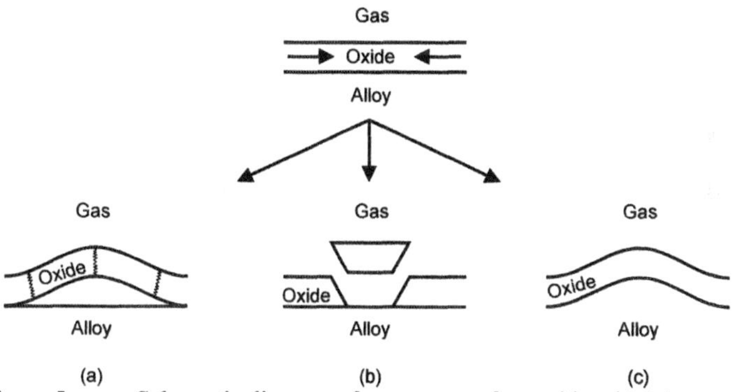

Figure 5. Schematic diagram of responses of an oxide which is loaded in compression. (a) buckling of the oxide, (b) shear cracking of the oxide, and (c) plastic deformation of the oxide and alloy.

Thermal Barrier Coatings

There are a number of degradation modes which can limit the life of a TBC and these must be understood in order to make lifetime predictions for existing systems and to provide the basis for the development of improved TBC systems, Stiger et al. (1999). The modes most important for the thermal cycling degradation of EB-PVD systems are:

(i) Cracking along the interface between the TBC and the bond coat for EBPVD coatings which results in spalling of the entire TBC

(ii) Sintering of the TBC at the outer surface where the temperature is highest which can affect the strain energy stored in the TBC.

It is now generally accepted that oxidation of the bond coat is a critical factor controlling the lives of EBPVD TBCs. It is now well established that the ability of a bond coat to form an α-alumina layer with negligible transient oxidation and the adherence of the alumina to the bond coat are critical factors in controlling the durability of TBCs. Experience with aircraft engines has shown that bond coat oxidation and the ability to resist spalling of the TBC from the bond coat are critical factors determining coating life. The bond coats develop a thermally grown oxide (TGO) layer during fabrication. The TGO grows thicker during exposure of the TBC. Therefore, this problem is related to that for the alumina-forming coatings with the added effect of strain energy stored in the TBC also influencing the spallation behavior.

EVALUATION OF IMPORTANT VARIABLES

Long Term Cyclic Oxidation Testing

Figure 6 shows a comparison of the cyclic oxidation behavior of the various NiCrAl alloys at 1100°C. All of the reactive element doped alloys show substantially better cyclic oxidation resistance than the undoped alloy. There are, nevertheless, quantitative differences among the various alloys.

The concentrations and distributions of the reactive elements play significant roles in the cyclic oxidation resistance of coatings alloys. When concentrations are too high, preferential oxidation of the reactive elements can occur which adversely affects protective alumina formation. Low concentrations of these elements may not provide the desired effects on cyclic oxidation resistance. Moreover, control of reactive element concentrations is often difficult because of their high reactivity. It may be possible to predict cyclic oxidation resistance by definition of reactive element concentrations and distributions on alloys.

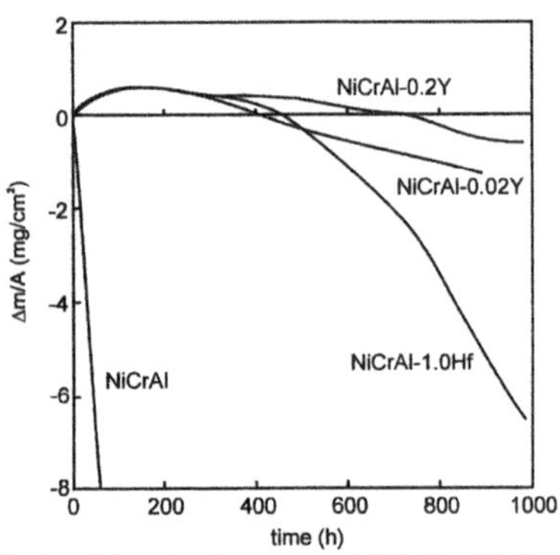

Figure 6. Cyclic oxidation kinetics for several Ni-Cr-Al alloys exposed at 1100°C.

XRD Stress Measurements

XRD stress measurements are based on the determination of the strain by measuring the d-spacing of specific lattice planes. The following relationship, which is subject to several experimentally verifiable assumptions, including a biaxial stress state, Noyan and Cohen (1987), exists between strain and stress:

$$\varepsilon_{\Psi} = \frac{d_{\Psi} - d_0}{d_0} = \tfrac{1}{2} S_2(hkl)\sigma \sin^2 \Psi + 2S_1(hkl)\sigma \qquad (1)$$

where ε_ψ is the strain measured for a particular set of (hkl) planes in a laboratory coordinate system, d_ψ is the lattice spacing of (hkl) planes when the specimen is tilted by the angle ψ, d_0 is the strain-free lattice spacing and σ is the stress. S_1(hkl) and S_2(hkl) are the X-ray elastic constants. In this $\sin^2\psi$ technique the stress is calculated from the slope of the d_ψ vs. $\sin^2\psi$ plot.

The $\sin^2\psi$ method allows the calculation of the stress without measuring d_0, which could yield great errors when using one point methods. In order to calculate the stress from the slope of d vs. $\sin^2 \psi$ plots the knowledge of S_2(hkl) is sufficient. For randomly oriented grains S_2 can be calculated from single crystal data using the Reuss,Voigt and other approximations. In the following S_2 is calculated within the Reuss approximation for hexagonal crystals

$$S_2(hkl) = (2s_{11} - s_{12} - s_{13}) - (5s_{11} + s_{33} - 3s_{44} - s_{12} - 5s_{13})a_{33}^2$$
$$+(s_{11} + s_{33} - s_{44} - 2s_{13})a_{33}^4 + 6s_{14}a_{22}a_{33}(3a_{11}^2 - a_{22}^2) \qquad (2)$$

with

$$a_{33}^2 = \frac{l^2}{\tfrac{4}{3}(c/a)(h^2 + hk + k^2) + l^2}.$$

The s_{ij} are the single crystal elastic compliances and a_{33} is the direction cosine between the c-axis and the normal of the selected (hkl). It should be noted that for (hk0) planes $a_{33} = 0$ and equation (2) is reduced to its first term.

Standard techniques based on equation (1) are the Rocking and Tilting methods, illustrated schematically in Figures 7a. and 7b., respectively.. In the Tilting technique, the tilt axis is the intersection of the sample surface with the diffraction plane. The diffraction plane is fixed and by definition contains the incident and reflected X-ray beams. The angle ψ is zero when the normal to the sample surface is within the diffraction plane. ψ increases as the sample is tilted and its surface normal is rotated out of the diffraction plane. In the Rocking technique the sample normal is tilted within the diffraction plane and ψ is the angle between the surface normal and the normal [hkl] of the measured set of (hkl) planes. In both techniques only one selected set of (hkl) planes is used for the measurement. To achieve high accuracy during stress measurements, only reflections with high 2 θ_B (Bragg angle) can be used, since errors in the measurement of d decrease with increasing θ_B.

<u>High Temperature Stress Measurements by XRD</u>: Tilting out of the diffraction plane is generally not possible with a hot stage for high-temperature measurements, so the Tilting technique cannot be used. There are also concerns that during rocking of the hot-stage at high temperature the sample might shift or fall off. Therefore, a new method, the Fixed

Incidence Multiplane technique (FIM) was developed, Sarioglu et al. (1997). Instead of tilting the sample, the angle Ψ is varied by measuring different (hkl) reflections (Fig . 7c). The angle Ψ is defined as $\Psi = \theta_B - \alpha$ where α is the angle between the incident beam and the specimen surface. Since this method uses different reflections, the cell dimension a calculated from the d-spacings is plotted vs. $\sin^2\Psi$. Furthermore, in order to calculate a stress from the slope of this line all (hkl) planes must have the same S_2(hkl) value. For hexagonal crystals this is fulfilled for planes of the type (hk0), for which a_{33} is zero, as can be seen in equation 2. One consequence of this method is that only a few (hk0) planes are available for hexagonal substances and high Bragg angles.

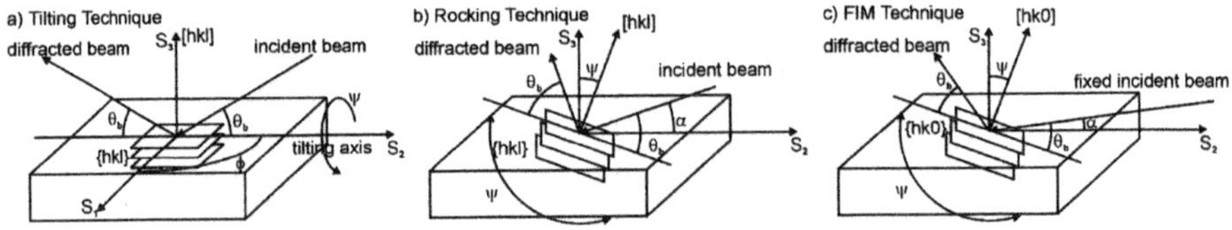

Figure 7. Geometry of stress measurement by XRD (a) using the Tilting Technique, (b) using the Rocking Technique, (c) using the FIM Technique.

In this study XRD measurements of the stresses formed in the alumina on a NiCrAlY alloy were performed in a Philips X'pert diffractometer with a CuK_α line source. Parallel beam optics with a flat graphite monochromator and a proportional counter were used. The room temperature residual stresses were measured using the Tilting technique.

The high temperature stress measurements where conducted with a resistively heated Pt hot-stage (Anton Paar KG) in ambient air using the rocking technique. The specimens were fixed to the Pt-strip using a special bonding agent (Zapon-Lacquer). Temperature control was achieved with a thermocouple welded to the underside of the Pt stripe. This thermocouple was correlated with the desired sample surface temperature by use of a Laser-pyrometer.

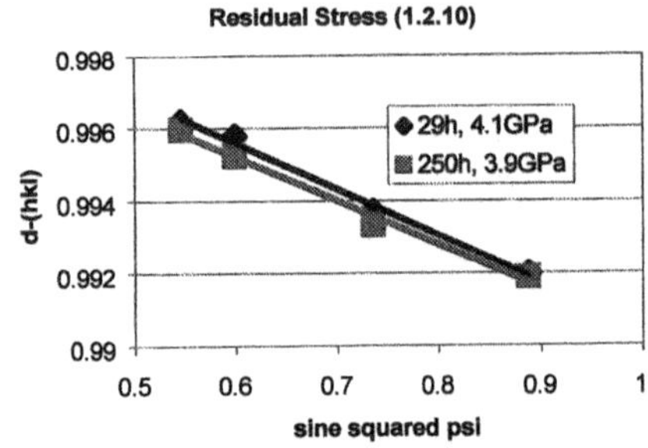

Figure 8. Residual stress measurement ($\sin^2\psi$ plot) for alumina scale that formed on NiCrAlY exposed to 1000°C.

The residual stress in the alumina layer that grew on NiCrAlY during an isothermal exposure to 1000°C for 250h was measured by the tilting technique on the (1.2.10) plane. A compressive stress of 3.90GPa was determined from the $\sin^2\psi$ plot in Figure 8. This value agrees with the calculated thermal stress value of 4.10GPa form the CTE mismatch between the oxide and the substrate. In support of the above results, high temperature work on the same material exposed to1000°C revealed that there is no growth stress.

Acoustic Emission Measurements

Acoustic emission measurements can be used to detect cracking and spalling of oxide scales in-situ. The apparatus used in this study has been described in detail previously, see Ashary et al. (1983). A 1mm diameter Pt rod was spot welded to the oxidation specimens. This rod served both to support the specimen in a vertical oxidation furnace and to provide a wave guide to transmit the acoustic signal from the specimen to a stainless steel cone, outside the furnace, which was connected to a transducer. The acoustic emission monitoring system used was the Dunegan/Endevco, 3000 series. The parameters measured in the present study were mainly acoustic emission counts, which scale with the amount of energy released by the event(s) producing the emission, and the peak amplitude of the acoustic emission event. The specimens had the same dimensions as those used for the oxidation studies

Figure 9 illustrates an important aspect of acoustic emission measurements. Here cumulative A.E counts were measured after a 24-hour exposure at 1100°C for various alloys. Comparison of these data with long term cyclic oxidation data in Figure 6 show there is a correlation between short term acoustic emission measurements and long term cyclic oxidation behavior i.e. less emission indicates better resistance. It is believed that acoustic emission measurements can provide an important component to an accelerated testing protocol.

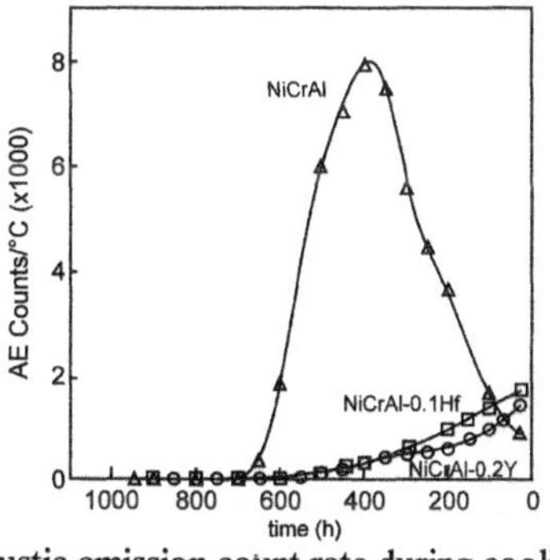

Figure 9. Acoustic emission count rate during cooling of three NiCrAl based specimens as a function of temperature after oxidation at 1100°C in air for 24 hours.

Fracture Toughness Measurements

A critical step in understanding mechanisms leading to oxide adhesion loss and in matching adhesion loss mechanisms between industry standard and accelerated oxidation tests is periodic measurement of interfacial fracture toughness. A reliable interfacial toughness test is needed, suitable for application to relatively small-sized specimens. Despite its importance, no tests have been developed for quantifying toughness loss in oxide scales or TBC/oxide systems. As a result, existing observations of oxide adherence loss have not been linked to measured changes in interfacial toughness.

Indentation Test for Fracture Toughness: The authors have developed and applied an indentation test for quantifying degradation of interfacial toughness in TBC systems (Vasinonta and Beuth, 2001, Handoko et al., 2000). The principal advantages of this type of test are that it is easy to perform and that it can be performed on relatively small specimens. The dimensions of specimens used thus far by the authors are approximately 2.5 cm in diameter and 0.3 cm thick, and smaller specimens can be used.

The indent test is diagrammed in Figure 10. In the current form of the test, a specimen is placed in a Rockwell hardness tester using a brale C indenter. The coating (in this case the TBC and the alumina scale beneath it) is penetrated by the indenter and the plastic deformation of the underlying substrate induces compressive radial stresses in the substrate, away from the indent crater. This compressive radial stress is transferred to the coating and acts to drive the extension of an axisymmetric interface crack (shown in cross section in Fig. 10). Figure 11 shows optical micrographs of a typical debond produced in an Electron Beam Physical Vapor Deposited (EBPVD) TBC system, as viewed from the side (Fig. 11a) and from the top (Fig. 11b). The toughness of the interface where debonding occurs (in this case the interface between the alumina scale and the PtAl bond coat) can be determined from a mechanics analysis of the indentation problem and a measurement of the delamination radius.

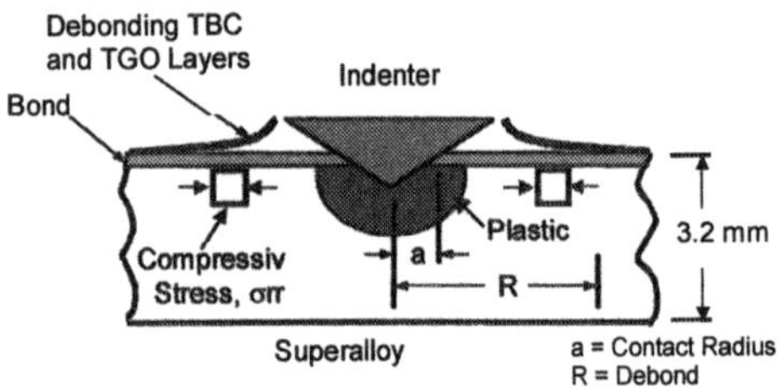

Figure 10. Schematic diagram of the indentation test for measuring interfacial fracture toughnesses.

The indent test has also been used to study oxide-only systems (with no TBC on top). For such systems, indentation is not accompanied by a single axisymmetric delamination, but instead induces either localized delaminations or, in the case of adherent scales, compressive failures of the oxide. By measuring the radial extent of such failures, interfacial toughnesses or compressive strengths of the oxide can be estimated. Regardless of the failure mechanism, the goal of this type of test is to induce an oxide scale failure via compressive applied stresses. In this way, it mimics in-service failure mechanisms caused by compressive stresses in the oxide, whatever they may be.

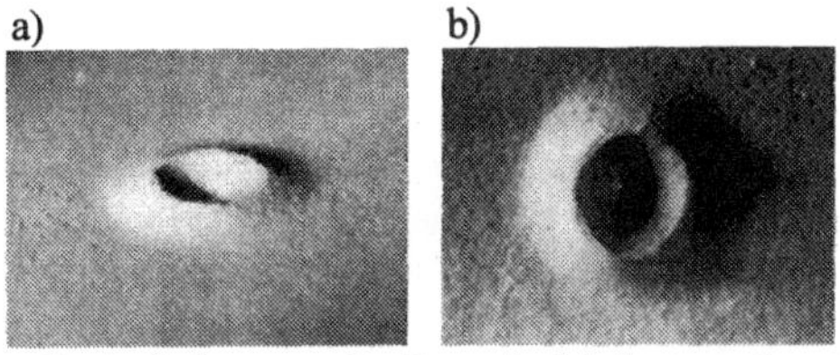

Figure 11. Optical photograph of a typical indent produced in a TBC after high temperature exposure.

<u>*Exposure-Induced Toughness Loss in TBC Systems*</u>: Figure 12 shows a plot of "apparent" losses of interfacial toughness for EB-PVD TBC systems subjected to 1100°C, 1135°C, and 1200°C isothermal exposures in dry air, as determined by the indent test. They are designated as apparent toughness losses because calculations used to obtain toughness values from measured delamination radii do not include the effects of known changes in the TBC system that could affect adherence. For instance, oxide growth or increases in stress magnitudes in the debonded coatings could cause an apparent decrease in toughness even if the interface itself were not weakened or embrittled. Toughnesses (in the form of a critical stress intensity factor, K_C) in the as-processed state are in the range of 3.1 – 3.7 MPa√m or higher. The applied K, resulting from residual stresses alone (with no indentation) is approximately 1.0 MPa√m, so that times to failure are designated as occurring when K_C reaches this value.

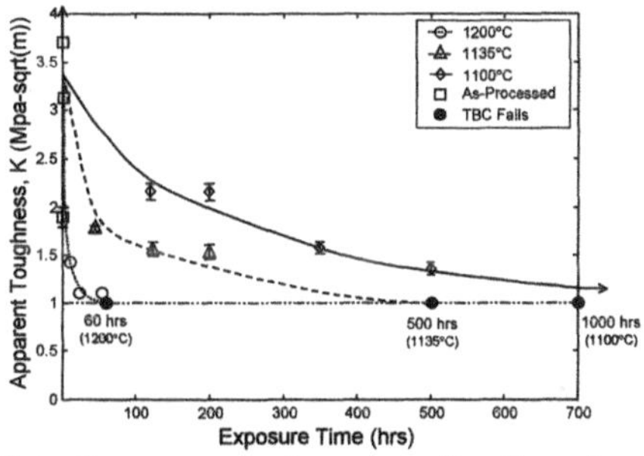

Figure 12. Plot of apparent toughness as a function of exposure time for TBC systems at various temperatures.

The results plotted in Figure 12 are the first available data quantifying toughness degradation vs. time for TBC systems and they show that much of the loss in toughness occurs at times that are a fraction of the time needed for spontaneous failure (Fig. 3). This loss in toughness could be due to one or more of a number of mechanisms, including chemical or mechanical damage at the interface, oxide growth and sintering of the TBC at high temperature (which changes the overall stiffness and residual stress in the TBC).

Each of these mechanisms is thermally activated (where mechanical damage is likely linked to thermally activated creep deformation). Of these mechanisms only chemical or mechanical damage at the interface results in a "true" loss of toughness at the interface.

Insights into Accelerated Testing Techniques: Insight into accelerated testing methods for TBC systems can be obtained by plotting the times to reach a given interfacial toughness (taken from Figure 12) on an Arrhenius plot similar to that in Figure 3. Such a plot is shown in Figure 13. In Figure 13, the line at the bottom of the plot reproduces average failure times plotted in Figure 3, except these have now been related to an apparent interfacial toughness of 1.0 MPa√m. Times to reach higher values of apparent interfacial toughness are plotted as three additional lines, with all data in the temperature range of 1100°C to 1200°C.

Two types of accelerated testing methods for TBC systems have been considered in this research. They are high-temperature testing to failure and mechanical testing for interfacial toughness loss. The first method allows shorter testing times by shortening the time to failure. The second method allows shorter testing times by probing toughness loss at early times, before failure occurs. The plot of Figure 13 and on-going research by the authors gives insight into the validity of these two methods. First, because the slopes of all four lines are similar, it is suggested that the mechanisms leading to apparent toughness loss are the same as those that lead to TBC system failure. As a result, measurement of interfacial toughness losses in TBC systems at early exposure times appears to be a valid accelerated testing method for understanding TBC failure.

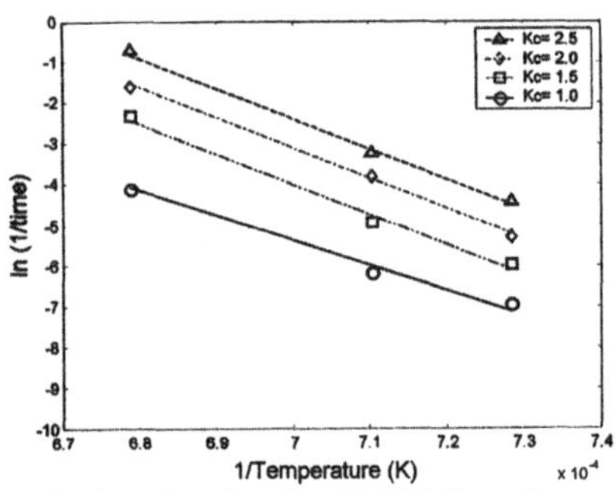

Figure 13. Arrhenius plot showing TBC failure times as a function of exposure temperature (similar to Fig. 3) and the time to reach a given level of fracture toughness as a function of temperature.

In order for high temperature testing to be valid, the failure times at operating temperatures (near 1000°C) need to fall on the bottom line of Figure 13, extrapolated to these lower temperatures. As previously noted, determining whether this occurs (requiring tests on the order of 50,000 hours) is not feasible. However, exposing a TBC sample at 1000°C to reduce the apparent interfacial toughness to 2.5 - 2.0 MPa√m could be accomplished in a much shorter time. If such data points were to lie on lines extrapolated from the top two lines of Figure 13, it could be presumed that the bottom line could also be accurately extrapolated to operating temperatures. The authors are currently carrying out low temperature toughness tests of this type.

In this way, the plot of Figure 13 can serve a roadmap for understanding these two types of accelerated testing methods for TBC systems. Although it gives feedback on whether mechanisms leading to failure are independent of the test method, it does not give insight into what mechanisms may or may not be dominant for individual TBC systems. Identifying the relative contribution of failure mechanisms is the goal of current research by the authors, which includes optical and scanning electron microscopy, stress measurements, acoustic emission monitoring and other techniques.

REFERENCES

Ashary, A., Meier, G. H., and Pettit, F. S. (1983) Acoustic Emission Study of Oxide Cracking During Alloy Oxidation in *High Temperature Protective Coatings*, S. C. Singhal, ed., AIME, p. 105.

Grabke, H. J. and Meier, G. H. (1995) Accelerated Oxidation, Internal Oxidation, Intergranular Oxidation, and Pesting of Intermetallic Compounds. *Oxid. Metals*, **44**, 147.

Handoko, R. A., Beuth, J.L., Meier, G.H., Pettit, F.S. and Stiger, M.J., "Mechanisms for Interfacial Toughness Loss in Thermal Barrier Coating Systems," accepted for the Proceedings of the Materials Division Symposium on *Durable Surfaces*, 2000 ASME International Mechanical Engineering Congress and Exposition, Orlando, November, 2000.

Noyan, I. C., and Cohen, J. B., *Residual Stresses*, Springer-Verlag, 1987.

Sarioglu, C., Blachere, J. R., Pettit, F. S., and Meier, G. H. (1997) "Room Temperature and In-Situ High Temperature Strain (or Stress) Measurements by XRD Techniques, *Microscopy of Oxidation 3*, S. B. Newcomb and J. A. Little, eds., The Institute of Materials. London, p. 41.

Sarioglu, C., Stiger, M. J., Blachere, J. R., Janakiraman, R., Schumann, E., Ashary, A., Pettit, F. S. Pettit, and Meier, G. H. (2000) The Adhesion of Alumina Films to metallic Alloys and Coatings, *Materials and Corrosion*, **51**, 1-15.

Schaeffer, J. S., G. M. Kim, G. M., G. H. Meier, G. H.,and F. S. Pettit, F. S. (1989) The Effects of Precious Metals on the Oxidation and Hot Corrosion of Coatings, in *The Role of Active Elements in the Oxidation Behavior of High Temperature Metals and Alloys*, E. Lang ed., Elsevier, p. 231.

Smialek, J. L., Nesbitt, J. A., Brindley, W. J., Brady, M. P., Doychak, J., Dickerson, R. M.,and Hull, D.R. (1995) Service Limitations for Oxidation Resistant Intermetallic Compounds", Mat. Res. Soc. Symp. Proc., Vol. 364, p. 1273.

Stiger, M. J., Yanar, N. M., Topping, M. G., Pettit, F. S., and Meier, G. H. (1999) Thermal barrier Coatings for the 21st Century, *Z. für Metallkunde*, **90**, 1069-1078.

Vasinonta, A. and Beuth, J.L., "Measurement of Interfacial Toughness in Thermal Barrier Coating Systems by Indentation," accepted to *Engineering Fracture Mechanics*.

Long Term Durability of Structural Materials
P.J.M. Monteiro, K.P. Chong, J. Larsen-Basse, K. Komvopoulos (Eds)

ELECTROMECHANICAL DEVICES FOR MICROSCALE FATIGUE TESTING

K. Komvopoulos
Department of Mechanical Engineering, University of California, Berkeley, CA 94720, USA

ABSTRACT

Significant advances in micromachine technology have increased the demand for basic understanding of material behavior at the microscale. Understanding of the effect of cyclic loading on the accumulation of damage in micro-structured thin-film systems is of great significance to the long-term durability and performance of microscopic components. The development of novel experimental techniques enabling probing of the material response at scales and conditions similar to those of the actual microdevices is therefore of paramount importance to micromachine technology. To increase the reliability and longevity of microscopic devices, it is essential to accurately determine the material response under both static and dynamic (cyclic) loadings. Most of the previous work has been performed with bulk polysilicon and with very simple devices that cannot resemble the loading conditions encountered during micromachine operation. In this article, the design and basic characteristics of special thin-film polysilicon microstructures suitable for testing under controlled cyclic loading are discussed. Preliminary results demonstrating the novelty and high potential of the proposed experimental procedures for microscale fatigue testing are presented to provide some new insight into the evolution of damage in polysilicon microstructures.

KEYWORDS

Fatigue damage, Cyclic loading, Micromachines, Polysilicon, Microscale mechanical properties

INTRODUCTION

Monitoring of the movement, position, actuation, and signaling of miniaturized devices, known as microelectromechanical systems (MEMS), has been increasingly used to sense chemical, electrical, mechanical, and thermal processes. Rapid developments in this emerging field have been mainly based on a technological basis derived from the integrated circuitry (IC) industry and have recently led to the design, fabrication, and implementation of microsystems spanning a wide range of industrial and medical applications (Wise (1991), Fujita (1997)). With an estimated compound annual growth rate of 50%, it is predicted that in the near future MEMS device revenues will exceed approximately US$10 billion. Due to increasing demands for versatile microsystems, dramatic changes in fabrication techniques, materials, and device dimensions have recently occurred for the purpose of increasing micromachine sensitivity and endurance. While early indications suggest that significant opportunities exist for MEMS, a large number of challenging issues presently prevent the evolution of micromachines from the laboratory to the application world. For such small devices, a number of physical effects have a different significance on the micrometer scale than macroscopic scales. Consequently, the identification of the effects of process parameters on the microstructure and material behavior at micromachine scales and the development of testing methods yielding information about long-term micromachine performance over a wide range of application conditions are of significant scientific and industrial importance. Examples of MEMS where degradation of the material properties may severely limit the operation life include rotary and linear stepper motors, high-speed (>100,000 rpm) geared polysilicon electrostatic micro-engines, electromagnetic motors, gear wheel systems, micro-fluidic devices such as pumps and valves, and mechanical/optical micro-components for communication technology.

The mechanical behavior of polycrystalline silicon has attracted significant attention because it is the most commonly used material in the IC technology. However, most of the research has been concentrated on the static fracture strength of polysilicon. Sharpe et al. (1997) used laser interferometry to measure the strain in microfabricated tensile specimens and determined a tensile strength of about 1.2 GPa. However, the tested specimens exhibited anisotropy and consisted of a two-layer sandwiched structure with columnar grains. Tsuchiya et al. (1997) reported that the fracture strength of polysilicon crystallized from amorphous silicon is in the range of 2-2.7 GPa and decreases with increasing specimen length, while the effect of the specimen width is insignificant. The decrease of the fracture strength was attributed to the larger surface area of the specimen sidewalls, where cracking commenced from defects produced during HF wet etching. Kahn et al. (1996) used a probe tip to separate the end tips of a deeply notched cantilever specimen, and based on the stress at the crack tip (determined from the probe tip position at the instant of fracture) the fracture toughness was found to be equal to 2.3 $MPa\cdot m^{1/2}$, independent of specimen thickness and doping. Tsuchiya et al. (1997) observed that fatigue crack initiation occurred mostly from surface defects along the specimen length and argued that fracture depends on the defect density of exposed surfaces and not on the volumetric defect density. However, a comparison of the data given in the previous studies shows that the two lowest tensile strength values correspond to the specimens with the largest widths. Fracture of flexure elements due to the combined effects of tensile and bending cyclic stresses may exhibit a dependence on specimen width, and thus the conclusion drawn by Tsuchiya et al. (1997) cannot be applied to different types of loading. Furthermore, even though in-plane tensile loading may be applicable to devices having a thin membrane in tension (e.g., pressure sensors and strain gages), load bearing elements in most MEMS devices (e.g., accelerometers and actuators) are flexure systems usually subjected to in-plane bending, similar to the test technique used by Jones at al. (1996) and Connally and Brown (1992). The dynamic characteristics of MEMS devices may change due to fatigue damage accumulation without necessarily leading to fracture. Consequently, quantification of the changes of the dynamic response is critical to the operation and sensitivity of these microdevices.

In addition to the specimen size and type of loading, the effect of fabrication conditions on the fracture strength of polysilicon can be significant. The deposition conditions, such as temperature, working pressure, doping, and annealing, affect the film microstructure and magnitude of residual stress. At low temperatures (~650 °C) silicon films initially deposited in an amorphous state recrystallize to form microstructures comprising equiaxed grains and the residual stress is tensile, whereas at higher deposition temperatures (e.g., ~700 °C) the films possess a columnar microstructure and are under a compressive residual stress (Krulevitch, 1994). Tensile residual stresses may enhance dislocation motion leading to cracking, while excessive compressive stresses may cause buckling. A moderate compressive residual stress is desirable for suppressing dislocation activity. Annealing after deposition may reduce the tensile residual stress, thereby providing additional strength to the film.

Fatigue cracks may also arise by processes quite different from slip band roughening at persistent slip bands, such as pile up of planar dislocations. Kramer (1974) suggested that a surface layer of high dislocation density could be produced from cyclic loading, and eventually become sufficiently strong to support a dislocation pile-up. The stress concentration associated with this pile-up was thought to promote cracking of the hardened layer. If the surface of a component is strengthened or dislocations are blocked by strain aging, the fatigue strength can be increased significantly. Alternatively, if the surface is weakened (e.g., by the formation of corrosion pits or by roughening, i.e., micro-pits produced by wet etching in polysilicon micromachining) the fatigue strength degrades. The reduction of the fatigue life is significant in the low stress/high cycle fatigue region, where the crack initiation process always takes place at the surface, and consumes almost 90% of the total life. Initiation to a single site leads to a large scatter since microplastic flow is a random and microstructure-sensitive material property. In the high stress/low cycle fatigue region (typically below 10^4 cycles), where the crack initiation phase represents ~50% of the total life and many cracks nucleate in the interior, the effect of the surface condition is less pronounced and the scatter in the data decreases considerably with increasing strain amplitude, suggesting a weaker dependence on microstructure.

Fatigue failure is caused by cyclic loading at load levels below those of failure under static loading and consists of the following sequential processes: accumulation of plastic deformation, microcrack initiation, propagation of the most favorable microcrack, and sudden failure of the device (fracture). Compared to continuum plasticity, understanding of plastic deformation at the microscale is still at its infancy. In fact, strain hardening of thin polysilicon structures subjected to cyclic loads has not been studied at the microscale. This is of particular importance because plastic deformation affects the onset of microcrack initiation. Crack initiation mechanisms may involve localized plastic flow in the presence of structural discontinuities, such as grain boundaries, voids, and inclusions, which do not necessarily produce a macroscopic measurable overall plastic strain. Thus, sensitive force microprobe techniques are essential for obtaining such information.

From the above discussion, it is apparent that specialized devices and new testing methods are essential for determining the micromechanical properties and endurance of thin-film microstructures. Therefore, the principal objective of this research was the establishment of a standard methodology for mechanical property testing at the microscale using thin-film microstructures fabricated by standard surface micromachining. Preliminary results are presented to demonstrate the evolution of fatigue damage in polysilicon microstructures subjected to cyclic loading.

EXPERIMENTAL PROCEDURES

Typical MEMS devices, such as rotating rings of gyroscopes, oscillating proof masses of accelerometers, and high-resolution micro-displays, are subjected to millions of mixed-load cycles. For example, assuming a 5 kHz bandwidth, 10 years life, and device operation at 8 h/day for 6 days/week, the cycles to failure for a 20% contingency is $\sim 5.8 \times 10^{11}$. The cyclic thermomechanical behavior of components of microdynamic structures is therefore of paramount importance. However, the use of probe tips or load cells to apply forces on microfabricated test specimens introduces significant errors in force measurements, mainly due to the difficulty to accurately determine the microscopic displacement of a macroscopic device. Hence, it is more advantageous to use on-chip electrostatic actuation.

Basic Features of Fatigue Microstructures

A schematic of the fabricated microstructures for fatigue testing is shown in Fig. 1. The main features of this design are as following.

(a) On-chip actuation is used to generate the applied force using comb drives common to MEMS. This approach is superior to the charged parallel-plate method because the force generated is independent of position. It is obvious that this has important implications since changes in the compliance of the test specimen due to the accumulation of fatigue damage would not otherwise cause a change in the force and, thus, the stress amplitude will increase with time.

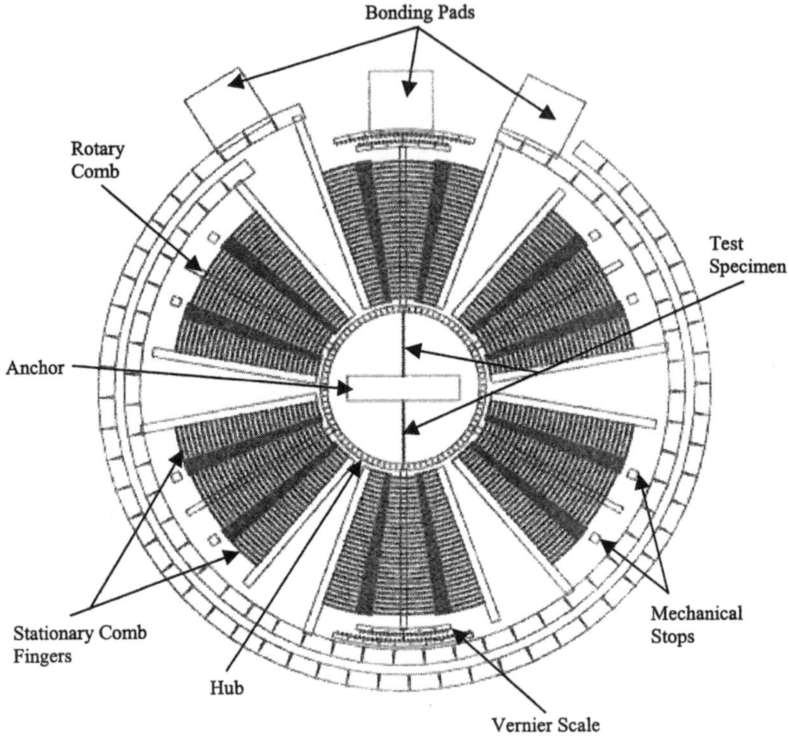

Figure 1: Schematic configuration of a fatigue microstructure.

(b) Cyclic loading is generated by interdigitized comb drives (Tang et al. (1989)). The comb drives are arranged along radial "spokes" extending from a suspended ring held in place by two thin flexures attached to the outside ring surface. These flexures are the *fatigue test specimens*. (Two test specimens have been included in the test microstructure shown in Fig. 1.) This kind of setup is more desirable than a single flexure design because the symmetry of the arrangement causes the comb teeth to follow a very consistent path during actuation. On-chip actuation generates the force required to fatigue the test specimen in a much more controlled manner compared to macroscale loading of microstructures used in previous fatigue studies of MEMS.

(c) Movement is obtained by applying a potential difference between the suspended and stationary comb teeth using an electrical connection made to the test specimen. Upon breakage of the specimen, the comb teeth come into contact and the circuit is shorted out causing the current flow to increase significantly. Since this is easily detectable, the testing can be stopped immediately. If the charged specimen fractures first, the circuit brakes and current flow diminishes. Since it is very difficult to detect such failure, it is desirable to charge the test specimens such that current flows no matter which specimen fractures.

(d) To enhance measurement of the rotation of the ring during actuation, a vernier was fabricated at the end of one of the radial spokes. A matching set of stationary indicator marks was placed near the end to allow measurement of angular displacements as small as 0.2°, to accurately correlate the movement and, in turn, the stress generated by the applied voltage, and to determine whether the compliance of the test specimens changed during testing.

(e) The mask design and process flow were specifically designed to allow control over the dimensions of the test specimens. The height of the beams is controlled by the amount of a single polysilicon deposition. The width of the beams is defined by the anisotropic etch of the polysilicon with an oxide mask.

(f) Because accurate measurement of the residual stress and elastic modulus is necessary in order to accurately calculate the maximum stress on the test specimens, special resonant and residual strain microstructures (Fig. 2) were fabricated together with each set of fatigue test structures. The elastic modulus and residual stress measurements were also used to check whether any significant variations in the mechanical properties occurred across the wafer surface.

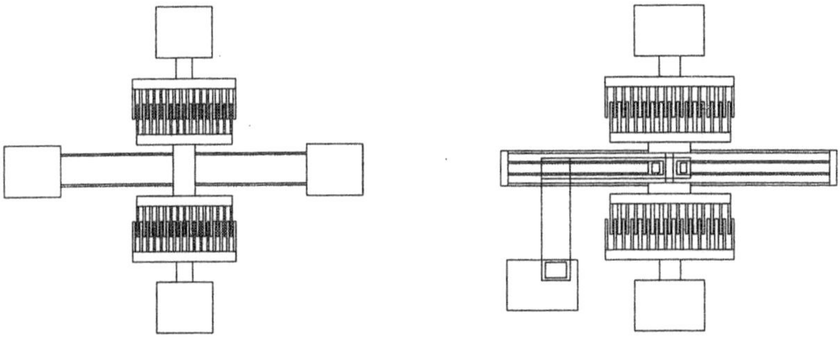

(a) (b)

Figure 2: Schematics of complementary microstructures for determining (a) the residual stress and (b) the elastic modulus of fabricated microstructures.

The first set of comb teeth are used to drive the microstructure to resonance whereas the second set of comb teeth is used to measure the displacement of the structures. Based on a Fourier analysis of the output data, the resonant frequency of the structures was calculated. The elastic modulus was then obtained as a function of the natural frequency of the microstructure shown in Fig. 2(b). Using the modulus of elasticity obtained form the previous experiment the residual stress was calculated from the resonant frequency of the microstructure shown in Fig. 2(a). For accurate fatigue data, the residual stress must be superimposed to the effective stress amplitude applied to the fatigue test specimen.

Fatigue Microstructures

Fatigue testing microstructures were fabricated utilizing a surface with polycrystalline silicon as the structural layer. The surface of a p-type silicon wafer of 1-2 Ω•cm resistivity was first heavily doped with phosphorus in a standard diffusion furnace using $POCl_3$ as the dopant source. Subsequently, a 0.3 μm thick thermal oxide layer produced at 1050 °C and a 0.6 μm thick low-stress LPCVD silicon nitride layer (electrical isolation layer) synthesized at 835 °C were deposited on the wafer. This was followed by a 2 μm thick phosphosilicate glass (PSG) sacrificial layer deposited at 450 °C by LPCVD and annealed at 1050 °C for 1 h in Ar that was subsequently photolithocraphically patterned. Then, the anchor mask was transferred into the sacrificial PSG layer by plasma etching. This step provided anchor holes to be filled with polysilicon. The first structural layer of polysilicon was 2 μm thick and was deposited at 600 °C. A thin layer of 0.3 μm PSG was deposited at 450 °C over the polysilicon layer and the wafer was annealed at 1050 °C for 1 h. The annealing process doped the polysilicon with phosphorus from the PSG layers both above and below it. The polysilicon layer was lithographically patterned using a mask designed to form the first structural polysilicon structural layer. The PSG layer was etched to produce a hard mask for the subsequent polysilicon etch. The hard mask is more resistant to the polysilicon etch chemistry than the photoresist and ensures better transfer of the pattern into the polysilicon. After etching the polysilicon, the photoresist was stripped off and the remaining oxide hard mask was removed by plasma etch. Then, the polysilicon layer was lithographically patterned and etched by RIE and the photoresist was striped off. Finally, the wafer was diced and shipped for sacrificial release and test. The release was performed by immersing the chip into a bath of 49% (or 5:1) BHF at room temperature. This was followed by a 10 min CO_2 supercritical drying, which avoids meniscus formation, thus, reducing the likelihood for stiction due to the high adhesion forces encountered in evaporation drying. Figure 3 shows schematically the process flow of the fabrication of the test specimens, and Fig. 4 shows the basic layout of microdevices with two and three fatigue specimens fixed to an anchor at the center of the devices and a suspended inner ring that can be rotated by the attached comb drives.

A total of about 1040 test microstructures were fabricated on 4-inch Si(100) wafers. The fatigue specimens were tested at resonance by the electrostatic force generated by the rotary actuators. Fatigue damage evolution was studied by observing the time it takes for the stiffness to decrease or equivalently the resonant frequency to change. The Rayleigh method was used to determine the spring constant. This involved equating the potential energy of the test beams at their extreme positions to the sum of the kinetic energies of the beams and the rotational kinetic energy of the actuator. Thus, the natural frequency of the system was monitored, and any deterioration was correlated to fatigue damage accumulation in the test specimens. The strain amplitude was determined from the specimen rotation (between 0.5° and 3.5°).

EXPERIMENTAL RESULTS

Microdevices with fatigue specimens of different dimensions were actuated by a sinusoidal wave generated by a 10 V DC voltage and a 25-40 V bias voltage applied at the resonant frequency of each device. The desired resonance frequency for in-plane rotation was found to be in the range of 2-20 kHz, depending on the number of flexure beams (two or three per microdevice), beam length (30-100 μm), and beam width (2 and 4 μm).

Finite element analysis revealed that the desired vibration mode was not the fundamental mode of the fabricated microstructures. However, even though the in-plane rotation mode of the microdevices with two and three flexure beams was the second and third modes, respectively, the desired modes were well separated from all other modes. Thus, it was fairly straightforward to isolate the desired mode during actuation. To maintain the fatigue microstructures at resonance, the desired natural frequency and amplitude were controlled and tracked continuously over the range of anticipated frequency shifts. A phase detector was used to measure the difference in phase between the driving voltage applied to the forcing electrode and the output voltage of the sensor. Implementation of such control system uses circuitry that can sense motion, determine if the device is at resonance, and then correct the driving frequency accordingly (Sun et al. (2001)).

Figure 5 shows results from short-term fatigue tests performed with two different microdevices having fatigue specimens with cross-section areas equal to 2 μm x 2 μm and lengths of 50 and 60 μm. It is noted that the resonance frequency of both microdevices decreases with time in a similar fashion, indicating that fatigue damage due cyclic loading occurs in the beam specimens. The fatigue mechanism(s) responsible for such change of the dynamic behavior of these microdevices are currently under investigation.

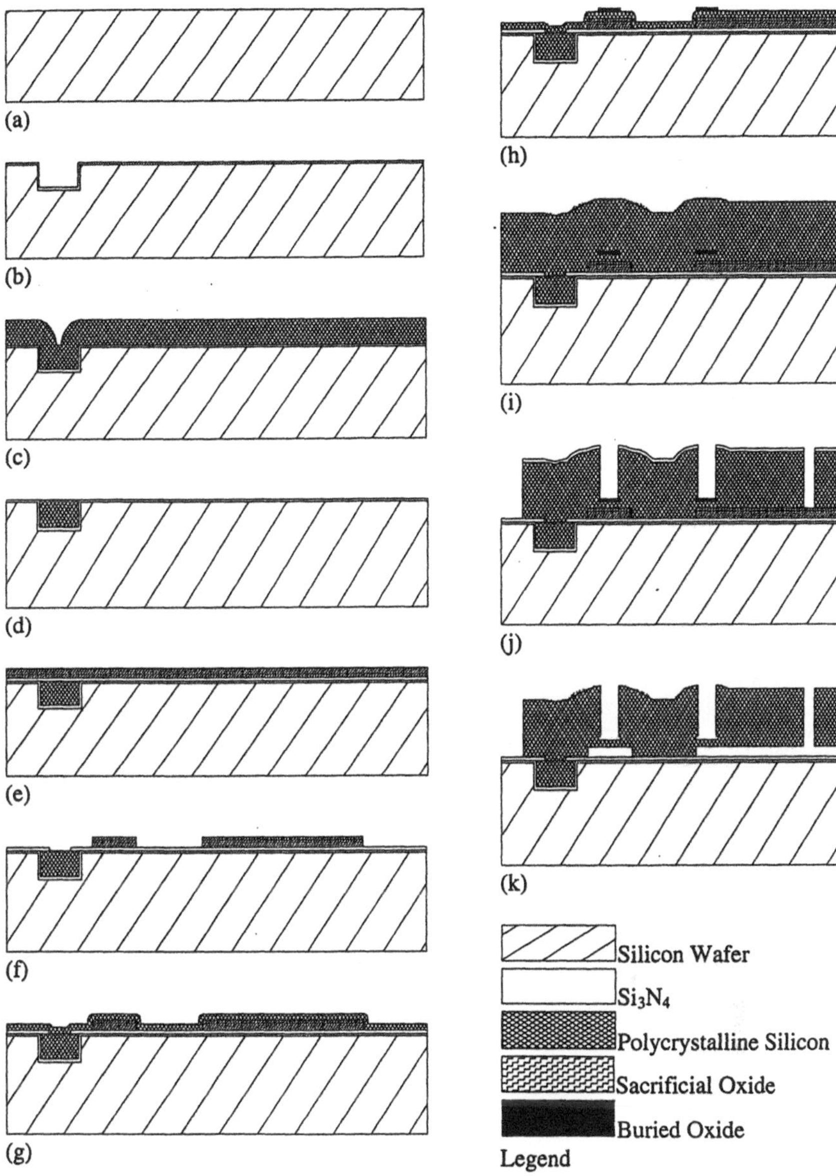

Figure 3: Process flow of fatigue microdevices.

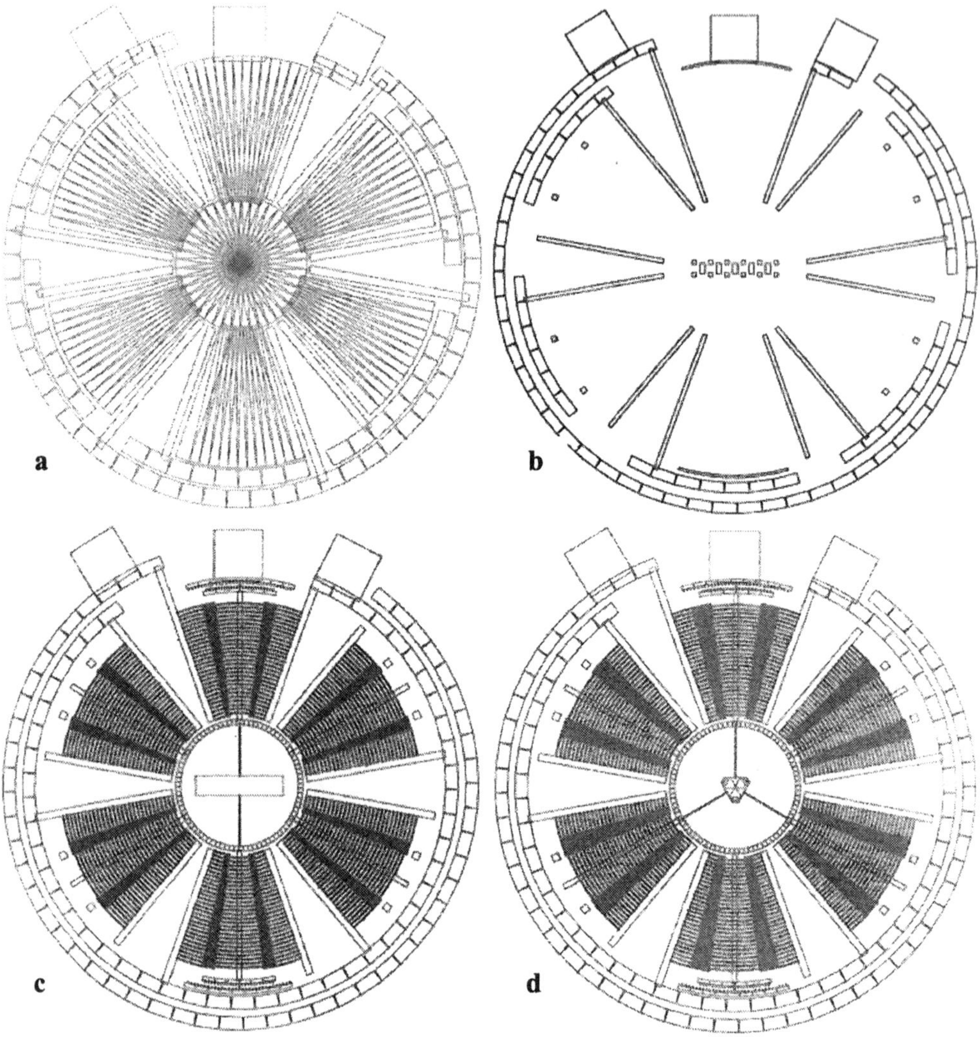

Figure 4: Fabrication layouts of fatigue microstructures: (a) polysilicon ground plane, (b) anchor mask for sacrificial oxide layer, (c) polysilicon structural layer of two-beam fatigue specimen, and (d) variation to polysilicon structural layer for three-beam fatigue specimen.

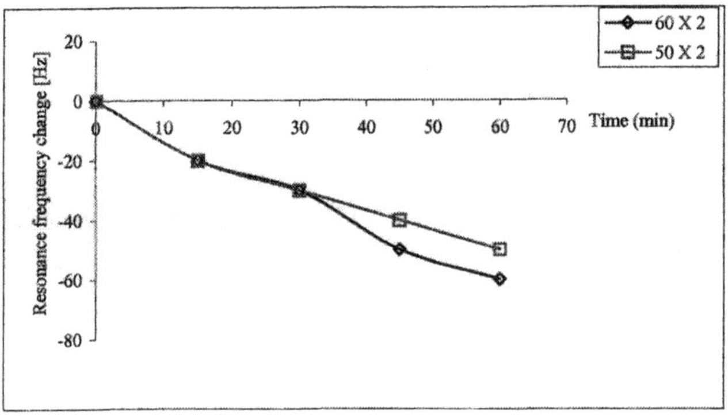

Figure 5: Change of natural frequency of resonating microdevices due to the evolution of fatigue damage in the flexure beams.

SUMMARY

Knowledge of the evolution of fatigue damage due to cyclic loading in microscopic components is of paramount importance to the durability and performance of MEMS. The development of novel experimental techniques enabling monitoring of the material dynamic response at submicron scales is of great importance to micromachine technology. The majority of previous reliability studies in this emerging field have been mostly based on simple test structures that cannot simulate the cyclic loading conditions encountered in MEMS devices. The objective of this work was to bridge this cap by introducing microdevices fabricated by standard surface micromachining that are suitable for fatigue testing under conditions typical of MEMS devices. The microfabrication and basic characteristics of thin-film polysilicon microdevices for fatigue testing under controlled cyclic loading conditions were presented. The microdevices were excited at resonance using on-chip actuation. Cyclic loading was generated by interdigitized comb drives arranged along radial "spokes" extending from a suspended ring held in place by thin flexure beams (fatigue specimens) attached to an anchor at the center of the device. Preliminary results demonstrated the potential of the developed experimental procedures to provide insight into the evolution of fatigue damage in polysilicon microstructures.

ACKNOWLEDGMENTS

This research was supported by the National Science Foundation under Grant No. DMI-9872324. The author gratefully acknowledges fruitful discussions on the design of the fatigue test microstructures with C. Belu and P. Stupar, and experimental assistance by A. Choy, N. Jamali, and X. Sun.

REFERENCES

Connally, J. A., and Brown, S. B., 1992, "Slow Crack Growth in Single-Crystal Silicon," *Science*, Vol. 256, pp. 1537-1538.

Fujita, H., 1997, "A Decade of MEMS and Its Future," *Proc. IEEE Micro Electro Mechanical Systems*, Nagoya, Japan, Jan. 26-30, 1997, pp. 1-8.

Jones, P. T., Johnson, G. C., and Howe, R. T., 1996, "Micromechanical Structures for Fracture Testing of Brittle Thin Films," Micro Electo Mechanical Systems, ASME, DSC-Vol. 59, pp. 325-330.

Kahn, H., Stemmer, S., Nandahumar, K., Heuer, A. H., Mullen, R. L., Ballarini, R., and Huff, M. A., 1996, "Mechanical Properties of Thick, Surface Micromachined Polysilicon Films," *Proc. IEEE Micro Electro Mechanical Systems*, San Diego, Ca, Jan. 1996, pp. 343-348.

Kramer, I. R., 1974, "A Mechanism of Fatigue Failure," *Metall. Trans. A*, Vol. 5, pp. 1735-1742.

Krulevitch, P. A., 1994, Micromechanical Investigations of Silicon and Ni-Ti-Cu Thin Films, Doctoral Dissertation, Department of Mechanical Engineering, University of California, Berkeley, CA.

Sharpe, W. N., Yuan, B., Vaidyanathan, R., and Edwards, B. L., 1997, "Measurements of Young's Modulus, Poisson's Ratio, and Tensile Strength of Polysilicon," *Proc. IEEE Micro Electro Mechanical Systems*, Nagoya, Japan, Jan. 26-30, 1997, pp. 424-429.

Sun, X., Horowitz, R., and Komvopoulos, K., 2001, "Analysis of a Phase-Locked Loop Natural Frequency Tracking System Using the Averaging Method," *J. Microelectromechanical Systems*, submitted.

Tang, W. C., Nguyen, T.-C. H., and Howe, R. T., 1989, "Laterally Driven Polysilicon Resonant Microstructures," *Sensors and Actuators*, Vol. 20, pp. 25-32.

Tsuchiya, T., Tabata, O., Sakata, J., and Taga, Y., 1997, "Specimen Size Effect on tensile Strength of Surface Micromachined Polycrystalline Silicon Thin Films," *Proc. IEEE Micro Electro Mechanical Systems*, Nagoya, Japan, Jan. 26-30, 1997, pp. 529-534.

Wise, K. D., 1991, "Integrated Microelectromechanical Systems: A Perspective on MEMS in the 90s," *Proc. IEEE Micro Electro Mechanical Systems*, Nara, Japan, Jan. 30- Feb. 2, 1991, pp. 33-38.

Long Term Durability of Structural Materials
P.J.M. Monteiro, K.P. Chong, J. Larsen-Basse, K. Komvopoulos (Eds)

FRACTURE AND FATIGUE OF PIEZOCERAMICS UNDER MECHANICAL AND ELECTRICAL LOADS

C. T. Sun

School of Aeronautics and Astronautics, Purdue University
West Lafayette, IN 47907-1282, USA

ABSTRACT

Fracture and fatigue behavior of PZT-4 piezoceramic was studied. Fatigue tests using compact tension specimen under various combinations of electric and mechanical load were conducted to develop a crack growth model. Experimental results indicated that crack growth could be significantly influenced by electric fields. It, however, could not be accounted for by the apparent stress intensity factor alone. In this study, the mechanical strain energy release rate was shown to be a single parameter that could account for both mechanical and electrical loads. Based on the fatigue test results, a fatigue crack growth model in terms of a single parameter was established. To explore the possibility of using the stress intensity factor to predict crack growth in piezoceramics under combined electrical and mechanical loading, the effect of domain switching on the near tip stress field was calculated with a domain switching criterion. The result indicated that the near tip stresses were not dominated by the usual inverse square root singularity. In general, the stress intensity decreases under a negative electric field and decreases under a positive electric field.

KEYWORDS

crack, piezoceramics, fracture, fatigue, electric field, domain switching

INTRODUCTION

Piezoelectric materials have found a wide range of applications in smart structures due to their high modulus and strong electro-mechanical coupling. The use of piezoceramics as actuators in smart structures demands that these materials perform under increasingly high electric and mechanical loads. Durability and reliability of actuators become important issues. Therefore, fracture (failure under monotonic mechanical and electrical loads) and fatigue (failure under cyclic loads) behavior of piezoceramics must be understood and accurately modeled.

In the presence of mechanical and electrical loads, a crack in piezoelectric materials gives rise to singular stress and electric fields near its tip. These strong mechanical and electrical fields produce crack driving forces that extend the crack in a catastrophic manner (fracture) or cumulative manner (fatigue). In either case, the presence of cracks would degenerate the mechanical and piezoelectric performance of the actuator as well.

The phenomenon of crack growth has been studied extensively in many ceramic materials but has only recently been investigated in piezoelectric ceramics. Caldwell and Bradt (1977) studied slow crack growth in PZT using fatigue tests. White et al (1995) studied the effects of cyclic stresses on crack extension at resonant frequency by including thermal dissipation. Cao and Evans (1995) and Lynch et al (1995) performed fatigue tests on Vickers indented specimens under cyclic electric fields above the coercive field and found that electric fatigue is characterized by step-by-step cleavage. Nishikawa et al (1992) performed three-point bending tests on poled and unpoled PZT under cyclic mechanical loading and discussed R ratio effect on the fatigue life. They found that the testing piece at R=0.1 much more easily fails than that at R = 1.0. The mechanism of acceleration effect was attempted in terms of microstructures. Tobin and Pak (1994) showed that fatigue crack growth took place even at field amplitude as low as 5% of the poling field. All test results indicated that there are electric effects on crack growth. However, no fatigue crack growth law including both electro-mechanical effects has been established. This motivated us to conduct a systematic investigation and determine which parameter can properly characterize the fatigue behavior.

In metals, fracture and fatigue crack growth can be described well by using classical fracture mechanics. The stress intensity factor is used to characterize these properties. However, this parameter is not suitable for piezoceramics under combined mechanical and electric loading since the stress intensity factor is independent of the electric field and is unable to account for the effect of the electric field. In recent years, experimental efforts have been made to observe fracture behavior under both mechanical and electrical loading. McHenry and Koepke (1978) measured crack propagation velocities under electric fields and noted that electric fields increased crack speed and that crack propagation direction deviated from its original direction under strong electric fields. Tobin and Pak (1993) performed Vickers indentation tests and found that the apparent fracture toughness of the material was reduced or increased depending on the direction of the applied electric field. Park and Sun (1995) conducted fracture tests on PZT-4 using compact tension specimens and showed that a positive electric field tends to reduce the fracture load, while a negative electric field does the opposite. These experimental results indicate that the apparent fracture toughness of piezoelectric materials vary with the applied electric field.

Pak (1990) attempted to use the total potential energy release rate as a fracture criterion by analogy to the idea of strain energy release rate in elastic solids. However, the presence of applied electric field was found to always reduce the total potential energy release rate, implying that the applied electric field would always increase the fracture toughness. This conclusion contradicts the available experimental data. Based on the experimental observation, Park and Sun (1995) proposed the use of a mechanical strain energy release rate to measure the apparent fracture toughness of piezoceramics. They found that this new parameter was able to account for the effect of the electric field on fracture toughness.

In the first part of this paper, the mechanical strain energy release rate is employed to characterize fatigue crack growth in piezoceramics under various combined mechanical-electrical loads. The model of fatigue crack propagation is derived based on the test data obtained for PZT-4 compact tension specimens. In order to understand the dependency of the crack behavior on electric field, domain switching near the crack tip is investigated. The size of the domain switching zone as well as the stress field are calculated using the finite element analysis in conjunction with a domain switching criterion.

MECHANICAL ENERGY RELEASE RATE

Energy released from the cracked body which creates new crack surfaces during crack extension has been used in classical fracture mechanics. In piezoelectric materials, the total potential energy release rate can be expressed as a path-independent integral, by Pak and Herrmann (1986). For a cracked body shown in Figure 1, this integral is given by

$$J = \int_{\Gamma} (Hn_2 - \sigma_{ij} n_j u_{i,2} + D_i E_i n_i) d\Gamma, \quad i,j=2,3 \tag{1}$$

where $H = \frac{1}{2} C_{ijkl} s_{ij} s_{kl} - \frac{1}{2} \varepsilon_{ij} E_i E_j - e_{ikl} s_{kl} E_i$ is the electric enthalpy, σ_{ij}, s_{ij} and E_i are stresses, strains and electric fields, respectively; C_{ijkl}, e_{ikl}, and ε_{ij} are elastic constants, piezoelectric constants, and electric permitivities, respectively; and n_i is the unit normal vector to the contour Γ. The J-integral when used as a fracture criterion, would indicate that the presence of electric fields always impedes crack propagation. However, all available experimental observations indicate otherwise.

The mechanical strain energy release rate proposed by Park and Sun (1995) includes only mechanical energy released as the crack extends. For Mode I loading in the poling direction x_3, the mechanical strain energy release rate can be obtained using Irwin's crack closure method. For plane strain, we have

$$G_I^M = \lim_{\delta \to 0} \int_0^{\delta} \sigma_{33}(x_2) \Delta u_3 (\delta - x_2) dx_2 \tag{2}$$

where Δu_3 is the crack opening displacement near the crack tip. For a PZT-4 piezoceramic with a center crack, we have

$$G_I^M = \frac{\pi a}{2} (2.28 \times 10^{-11} \sigma_{33}^{\infty\,2} + 2.21 \times 10^{-10} \sigma_{33}^{\infty} E_3^{\infty}) \quad \text{(N/m)} \tag{3}$$

On the other hand, the total potential energy release rate obtained from Eq. (1) is

$$J = \frac{\pi a}{2} (2.28 \times 10^{-11} \sigma_{33}^{\infty\,2} + 1.21 \times 10^{-10} \sigma_{33}^{\infty} E_3^{\infty} - 8.74 \times 10^{-9} E_3^{\infty\,2}) \tag{4}$$

in which a is the half length of the crack, σ_{33}^{∞} and E_3^{∞} are the remotely applied stress and electric field, respectively. Unlike the total energy release rate J, G_I^M may decrease or increase depending on the direction of electric field. Accordingly, the presence of electric field may enhance or impede the crack growth. By adopting G_I^M as a fracture criterion, Park and Sun (1995) were able to predict fracture loads in PZT-4 piezoceramics under combined mechanical and electrical loading fairly accurately. It suggests that it may be a proper parameter to characterize the fatigue crack growth.

FATIGUE TEST USING COMPACT TENSION SPECIMEN

Experimental Procedure

The dimensions of the compact tension specimen and setup of the experiment are shown in Figures 1 and 2, respectively. The poling direction was along x_3-axis. A conductive epoxy was used to make electrodes which were 13.2 mm apart. The side faces were polished in several steps starting with a 600 grit silicon carbide polishing wheel and finishing with a 0.5 μm grain sized alumina polishing pad. The crack was created by cutting the specimen with a 0.46 mm thick diamond wheel perpendicular to the poling direction resulting in a crack length of 6 mm. Subsequently, the crack tip was further sharpened by using a razor blade with diamond abrasive. It is noted that the cutting and polishing were done in water in order to prevent the depoling from excessive heating. Two types of combined loading were used. The first one was tension-tension cyclic mechanical load at a constant electric field; and the second type was a cyclic electric field with a constant tensile load. Table 1 lists the loading conditions. To prevent arcing between electrodes through the air, the specimen was submerged in silicon oil contained in a translucent plexiglass tub. A traveling microscope was used to measure the crack length during the fatigue test.

TABLE 1
LOADING CONDITIONS

	Mechanical	Electrical
Type 1 Loading	P_{max}=67 N, P_{min}=11 N	E=0.53 MV/m
	P_{max}=67 N, P_{min}=11 N	E=0.23MV/m
	P_{max}=67 N, P_{min}=11 N	E=0
	P_{max}=67 N, P_{min}=11 N	E=-0.23MV/m
	P_{max}=67 N, P_{min}=11 N	E=-0.53MV/m
Type 2 Loading	P=75N	E_{max}=0.15MV/m, E_{min}=0
	P=75N	E_{max}=0.08MV/m, E_{min}=0

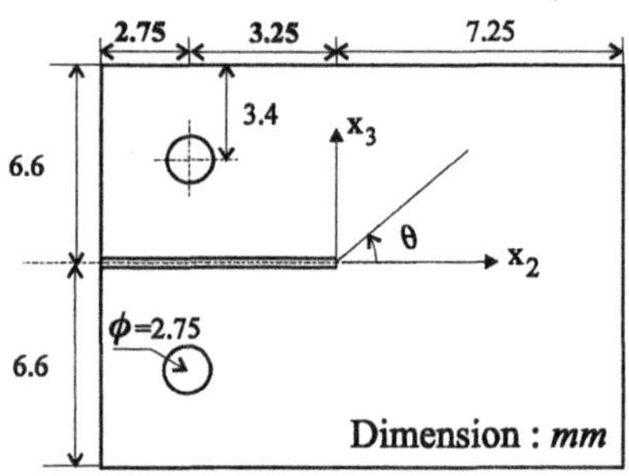

Figure 1: Compact tension specimen, PZT-4

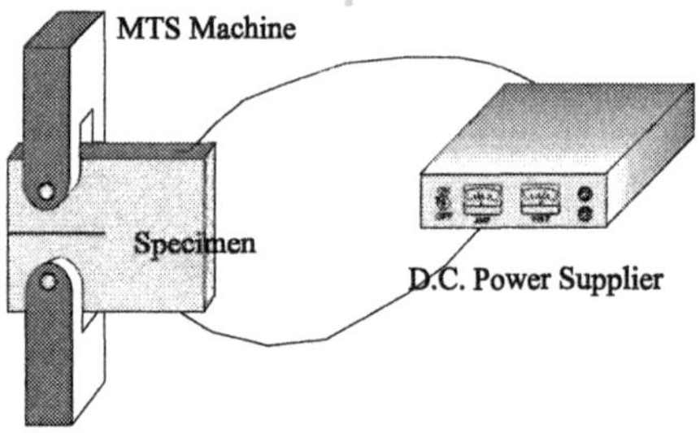

Figure 2: Test setup for fatigue test using compact tension specimen

Finite Element Analysis

The mechanical strain energy release rate corresponding to the crack growth history for each test was calculated using the finite element analysis in conjunction with the crack closure technique. The eight-node plane strain element for piezoelectric materials in ABAQUS was used. Due to symmetry, the half specimen was adopted for the finite element analysis. To illustrate the modified crack closure technique, Figure 3 schematically shows a number of elements near the crack tip. The strain energy released during a virtual crack extension Δa would be the same as the work done in using the crack tip nodal force to close the crack opening displacement if the crack tip node were released. Since the virtual crack extension Δa is taken to be very small as compared with the crack length, the virtual crack opening can be approximated by the crack opening at the node right behind the crack tip. Using this modified crack closure technique, the mechanical strain energy release rate can be calculated as

$$G_I^M = \frac{1}{2\Delta a}\{F_3^e(u_3^{(a)} - u_3^{(b)}) + F_3^f(u_3^{(c)} - u_3^{(d)})\} \tag{5}$$

where $F_3^{(i)}$ is the nodal force in the x_3 direction at the node (i) and $u_3^{(i)}$ is the nodal displacement in the x_3 direction at the node i. For symmetric condition, Eq.(5) becomes

$$G_I^M = \frac{1}{\Delta a}(F_3^e u_3^{(a)} + F_3^f u_3^{(c)}) \tag{6}$$

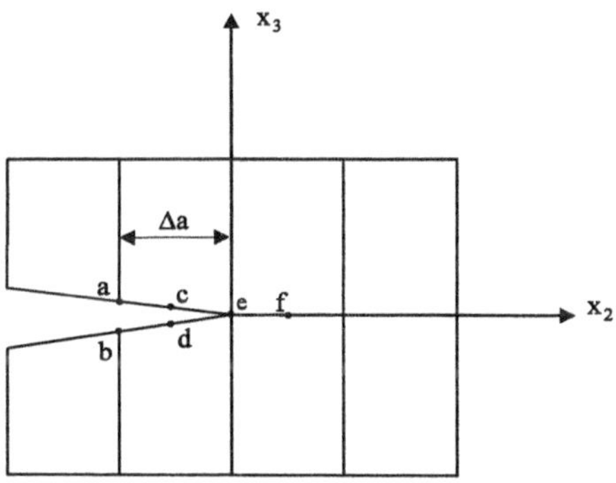

Figure 3: Finite element mesh for illustration of the crack closure technique

Results and Discussion

The crack growth histories corresponding to the two types of loading are presented in Figures 4 and 5, respectively. The test results clearly display the effect of electric field on fatigue crack growth. Specifically, it is noted that a negative electric field tends to slow the crack growth while a positive electric field tends to increase the crack growth.

In metallic structures, the Paris fatigue model has been commonly used to characterize the fatigue crack growth rate da/dN (crack growth per load cycle). The Paris model is expressed in the form

$$\frac{da}{dN} = A(\Delta G_I)^n \tag{7}$$

where A and n are coefficients to be determined from fatigue test data, and the range of strain energy release rate, ΔG_I, is given as

$$\Delta G_I = (G_I)_{max} - (G_I)_{min} \tag{8}$$

which is the range of G_I in the fatigue load cycle.

To establish a fatigue crack growth model for the piezoceramic similar to the Paris model, we obtained first the da/dN data from the results of Figures 4 and 5. Subsequently, the da/dN data were plotted against the range of mechanical strain energy release rate ΔG_I^M. The result is shown in Figure 6. It is interesting to note that in such a plot, all the crack growth curves generated under different loading conditions collapse almost into a single curve. This master curve can be expressed in the form as

$$\frac{da}{dN} = A(\Delta G_I^M)^n \quad \text{(cm/cycle)} \tag{9}$$

with $A = 4.5\times10^{-16}$, n=22.3 and ΔG_I^M is in Nm/m^2.

The above finding indicates strongly that the mechanical strain energy release rate can be the single parameter characterizing the crack growth driving force under combined mechanical and electrical loads. It is evident that the use of such parameter will greatly reduce the need to conduct many fatigue tests under various loading conditions.

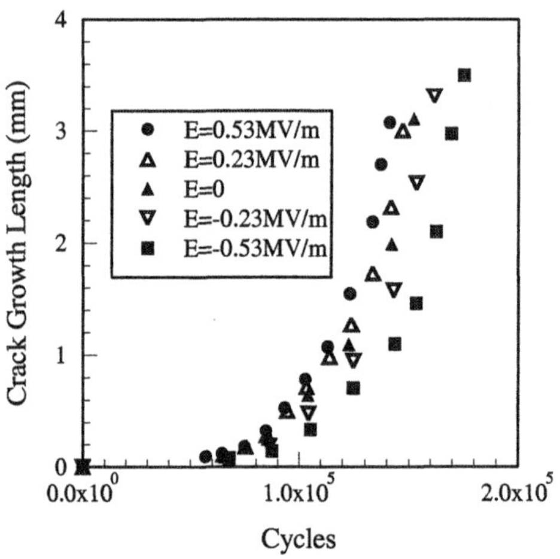

Figure 4 Crack growth vs. number of cycles at various constant electric field for P_{max} = 67N and P_{min} = 11N

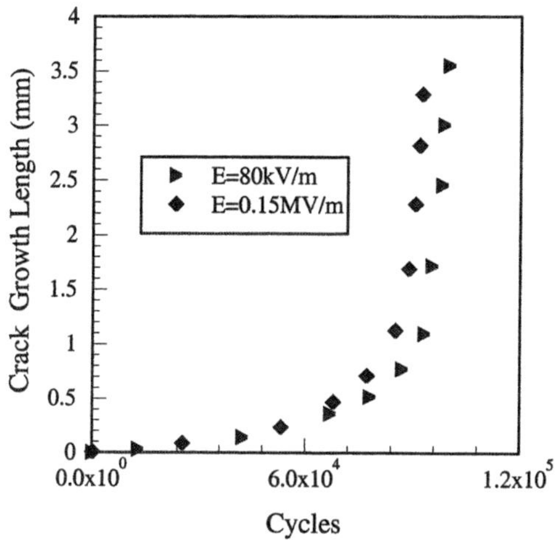

Figure 5: Crack growth vs. number of cycles for cyclic electrical load at a constant mechanical load P = 75N

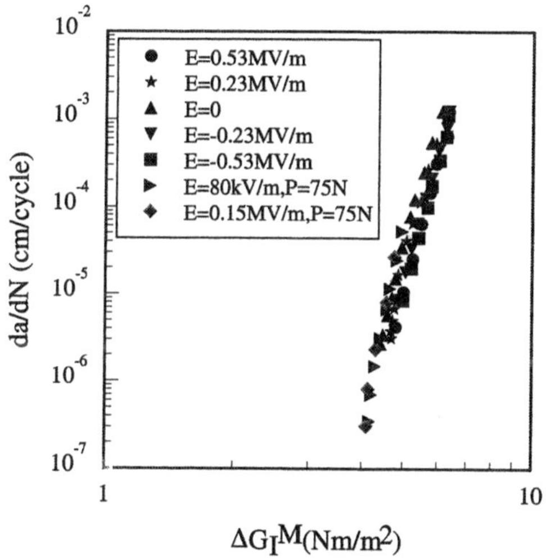

Figure 6: Crack growth rate da/dN vs. ΔG_I^M for all loading cases

STRESS FIELD NEAR CRACK TIP

Under intense electrical and mechanical loads, piezoceramics could undergo polarization switching resulting from microstructural domain wall switching (Lines and Glass, 1977). Many researchers have attempted to investigate the effect of domain switching and the constitutive relations on fracture behavior numerically and experimentally. It is believed that the stresses around the crack tip induced by the polarization switching may play a crucial role in the variation of fracture toughness with respect to electrical loading.. In this study, the finite element analysis in conjunction with a domain switching criterion is used to analyze the domain switching zone and its effect on near tip stresses.

Domain Switching Criterion

The constitutive and governing equations for a piezoelectric material can be written as

$$\begin{aligned} &\sigma_{ij} = c_{ijkp}\gamma_{kp} - e_{pij}E_p \\ &D_i = e_{ikp}\gamma_{kp} + \varepsilon_{ip}E_p \\ &\sigma_{ij,j} = 0 \qquad D_{i,i} = 0 \end{aligned} \tag{10}$$

where σ_{ij} and γ_{kp}, are stress and strain tensors, respectively; D_i, and E_p stand for electric displacement and electric field, respectively; and c_{ijkp}, e_{ikp} and ε_{ip} are tensors of material constants for elastic, piezoelectric, and dielectric properties, respectively.

A criterion for domain switching was proposed by Hwang et al. (1995) which was modified later by Sun and Jiang (1998). This criterion states that domain switching occurs when the change of internal energy density G during switching exceeds a critical value G^c, i.e.,

$$G = \sigma_{ij}d\gamma_{ij} + E_i dD_i \geq G_k^c, \quad k = 90°, 180° \tag{11}$$

where $G^c_{90°}$ and $G^c_{180°}$ are the critical energy values for the 90° and 180° domain switchings, respectively. In Eq. (11), the increments $d\gamma_{ij}$ and dD_i are instantaneous changes of strain and electric displacement, respectively, occurring during domain switching. Thus, the changes of strain $d\gamma_{ij}^o$ and electric displacement dD_i include two parts: the spontaneous strains γ_{ij}^s and the spontaneous polarization P_i^s resulting from the crystal structure changes, the strains $d\gamma_{ij}^o$, and the polarizations dD_i^o produced by piezoelectricity after domain switching.

The domain switching criterion given in Eq. (11) assumes constant stress and electric fields. If the stress and electric fields are affected by the domain switching process, the criterion must be modified to reflect the variations of the stress and electric fields before and after domain switching. We propose using the average values of the stress and electric field and modifying the criterion of Eq. (11) into

$$\frac{\sigma_{ij}^b + \sigma_{ij}^a}{2} d\gamma_{ij} + \frac{E_i^b + E_i^a}{2} dD_i \geq G_k^c, \quad k = 90°, 180° \tag{12}$$

where σ_{ij}^b (E_i^b), and $\sigma_{ij}^a(E_i^a)$ stand for the stresses (electric fields) before and after switching, respectively.

The constitutive relations after domain switching are given by

$$\gamma_{ij} = s_{ijkl}\sigma_{kl} - d'_{kij} E_k + \gamma^s_{ij}$$
$$D_i = d'_{ijk} \sigma_{jk} + \varepsilon_{ij} E_j \tag{13}$$

where s_{ijkl}, d_{ijk} and ε_{ij} are the compliances, piezoelectric and dielectric constants, respectively. The primed quantities denote quantities after domain switching. It is assumed that the elastic properties and dielectric constants of the piezoceramics remain unchanged after switching occurs. Thus, only piezoelectric constants vary with domain switching. After domain switching, the new piezoelectric constants referring to the original coordinate system are obtained as from the coordinate transformation law, i.e.,

$$d'_{ijk} = \beta_{im}\beta_{jn}\beta_{kp}d_{mnp} \tag{14}$$

where β_{ij} is the coordinate transformation tensor associated with rotation of the poling direction (Nye, 1957). If the strains and electric fields are independent quantities, Eq. (13) can be written as

$$\sigma_{ij} = c_{ijkl}\gamma_{kl} - e'_{kij} E_k + \sigma^s_{ij}$$
$$D_i = e'_{ijk} \gamma_{jk} + \varepsilon_{ij} E_j + D^s_i \tag{15}$$

where $\sigma^s_{ij} = -c_{ijkl}\gamma^s_{kl}$ and $D^s_i = -e'_{ijk} \gamma^s_{jk}$. In Eq. (13), γ^s_{ij} and P^s_i associated with domain switching are given here for domain switching in the x_2–x_3 plane with x_3 as the original poling direction. For 180° switching,

$$\gamma^s_{ij} = 0 \qquad i, j = 2, 3$$
$$P^s_2 = 0, \qquad P^s_3 = -2P^s$$

For 90° switching in the x_2–x_3 plane, we have

$$\gamma^s_{22} = \gamma^s , \quad \gamma^s_{33} = -\gamma^s , \quad \gamma^s_{23} = 0$$
$$P^s_2 = -P^s, \qquad P^s_3 = -P^s$$

The spontaneous strain γ^s and polarization P^s are material constants.

An isoparametric plane strain finite element was developed for analysis of piezoelectric materials incorporating the domain switching criterion of Eq. (12). The solution procedure is as follows. For each incremental load level, calculate the stress and electric fields, σ^b_{ij} and E^b_i, respectively, based on the result of the previous switching and loading conditions. The current domain switching zone is first determined by σ^b_{ij} and E^b_i. If switched, the material properties of those switched elements change following the coordinate transformation law, and the spontaneous strains are interpreted into nodal forces and charges. With these updated switching zone and loading, the updated stress and electric fields σ^a_{ij} and E^b_i are then calculated. The subsequent iteration is to use the two sets of solutions, σ^b_{ij} and E^b_i, σ^a_{ij} and E^a_i, to calculate the internal energy change G and use the domain switching criterion of Eq. (12) to update the switching zone as well as the stress and electric fields. In the above incremental solution procedure, the distinction between the switched elements and non-switched elements must be made in order to update the material constants. In addition to material property changes, equivalent mechanical loads and electric charges resulting from switching in each element are added to the finite element nodes.

Numerical Results

The PZT-4 compact tension specimen (see Fig. 1). A pair of vertical forces P are applied through the pin holes, and voltages are applied at the top and the bottom faces. The initial poling direction of the piezoceramic is in the positive x_3 direction. The crack surfaces are assumed to be fully insulated. The eight-node plane strain element for piezoelectric material was used. The entire specimen was modeled since the initial symmetry of the system may be lost after domain switching occurs. The material constants for PZT-4 are listed in Table 2. The values of spontaneous strain γ^s and spontaneous polarization P^s for PZT-4 are 0.39% and 0.34 C/m^2, respectively.

TABLE 2

Material Constants of PZT-4 Piezoceramics

$c_{11} = 13.9 \times 10^{10}$ N/m^2 $c_{13} = 7.43 \times 10^{10}$ $c_{33} = 11.3 \times 10^{10}$ $c_{44} = 2.56 \times 10^{10}$

$\varepsilon_{11} = 6.00 \times 10^{-9}$ C/Vm $\varepsilon_{33} = 5.47 \times 10^{-9}$

$e_{15} = 13.44$ C/m^2 $e_{31} = -6.98$ $e_{33} = 13.84$ –9

The finite element size near the crack tip was about 0.004% of the crack length. Three electric loads in term of nominal electric fields (the applied voltages divided by the specimen height) $E = -0.52, 0, 0.52$MV/m, were considered. The mechanical load P was kept at 94N.

The actual domain wall switching is not a simple on-off process. Instead, it takes a range of loads to initiate and complete. In this study, the coercive values of stress and electric field are chosen to be the values at the end of the total switching process. The critical internal energies for 90° and 180° switching, respectively, are

$$G^c_{180^\circ} = 2E^c_{180^\circ}P^s \quad \text{and} \quad G^c_{90^\circ} = E^c_{90^\circ}P^s = \sigma^c_{33}\gamma^s$$

where $E^c_{180^\circ}$ and $E^c_{90^\circ}$ are, respectively, the coercive electric fields for 180° and 90° ferroelectric switching; σ^c_{33} is the coercive stress for 90° ferroelastic switching. For PZT-4, we have

$$G^c_{180} = 1.36\ \mathrm{MN/m^2}, \quad G^c_{90} = 1.05\ \mathrm{MN/m^2}.$$

The domain switching zones determined using the incremental approach for E = –0.52 and 0.52Mv/m with are plotted in Figs. 7 and 8, respectively. The domain switching zone is very small for the case $E = 0$ with P = 94N. It is noted that the switching zone size is not proportional to the applied electric field. It is evident that the domain switching mode and size are quite different between the positive electric field and the negative electric field. The negative electric field produces a 180° domain switching zone ahead of the crack tip and a 90° domain switching zone behind the crack tip. On the other hand, the positive electric field results in a narrow 90° domain switching zone behind the crack tip

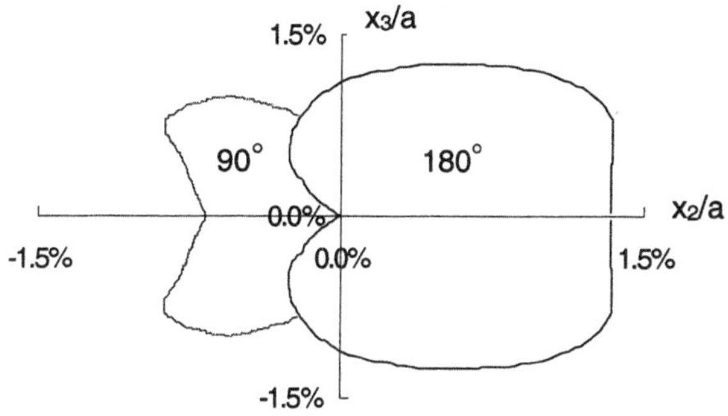

Fig. 7 Domain switching zone for electric load E = - 0.52MV/m and P=94N

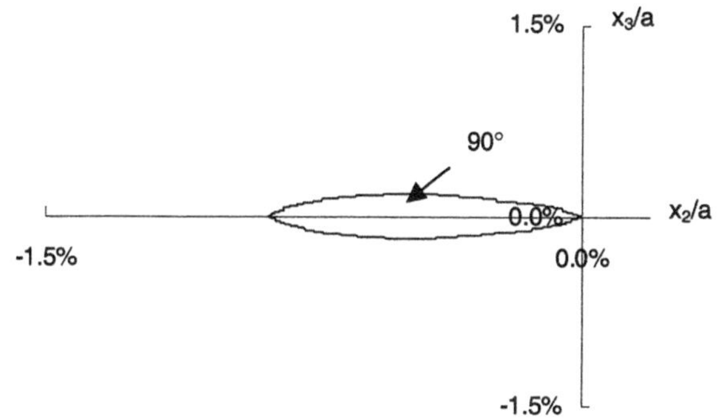

Figure 8 Domain switching zone for E = 0.52MV/m and P = 94N

Distributions of the modified normal stress $\sigma_{33}(2\pi x_2)^{1/2}$ ahead of the crack tip for two loading conditions are shown in Figs. 9 and 10, respectively. It is noted that the near tip stress field is significantly affected by the presence of switching zone, except for the case E = 0 for which the domain switching zone is extremely small. Moreover, the 90° and 180° domain switchings have different effects on the near tip stresses.

Except for the case E = 0, the near tip stress does not seem to be dominated by the K-field with the inverse square root singularity. Hence, a direct application of classical fracture mechanics in using the critical value of the stress intensity factor is not feasible. However, taking the average value of $\sigma_{33}(2\pi x_2)^{1/2}$ over a distance ahead of the crack tip as an effective stress intensity factor, seems capable of explaining the dependence of fracture toughness of PZT-4 on electric field as observed by Park and Sun (1995). Figure 11 shows the predicted fracture loads based on the averaged effective stress intensity factors. The predictions appear to agree fairly well with the experimental data.

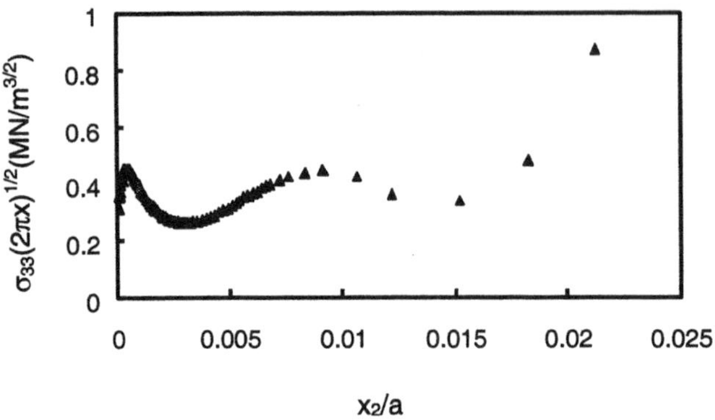

Figure 9 Near tip stress distribution for E = –0.52MV/m and mechanical load P = 94N

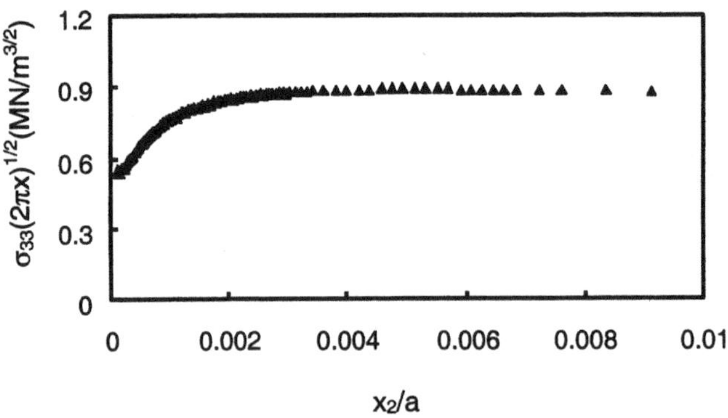

Figure 10 Near tip stress distribution for E = 0.52MV/m and P = 94N

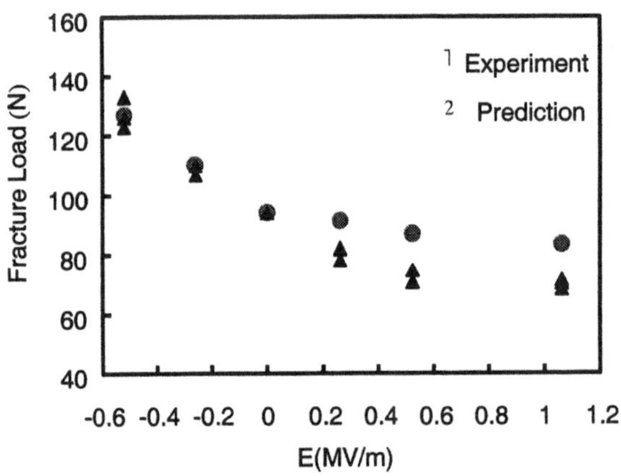

Figure 11 Comparison of predicted fracture loads with experimental data

CONCLUSIONS

Fatigue crack growth in PZT-4 piezoceramics has been investigated. The magnitude as well as the direction of the electric field were found to have a significant influence on the crack growth rate. The mechanical and electrical loads can be combined into a single parameter, i.e., the mechanical strain energy release rate, which can be used to characterize fatigue crack growth for the combined loading. A fatigue crack growth model similar to the Paris fatigue model for metals has been derived and shown to be capable of describing the fatigue crack growth rate in PZT-4 piezoceramics.

From the finite element analysis in conjunction with a domain switching criterion, it was found that domain switching could occur under mechanical and electrical loads, which in turn could affect the stress field near the crack tip. In general, a negative electric field tends to reduce the level of the near tip stresses while a positive electric field tends to increase the level. As a result, the apparent fracture toughness increases under negative electric fields and decreases under positive electric fields.

ACKNOWLEDGMENT

This work was supported by an NSF grant No. 9872330-CMS to Purdue University.

REFERENCES

Caldwell R. F. and Bradt R. C. (1977). Stressing Rate Effects on the Bend and Compressive Strengths of a Piezoelectric Ceramics. *J. Amer. Ceram. Soc.* **60:3-4**, 169-170.

Cao H. C. and A. G. Evans (1995). Electric Field Induced Crack Growth in Piezoelectrics. *J. Amer. Ceram. Soc,* **77**, 1783-1786.

Hwang S. C., Lynch C. S. and McMeeking, R. M. (1995). Ferroelectric/Ferroelastic Interaction and Polarization Switching Model. *Acta Metallica Material* **43**, 2073-2084.

Lines M. E. and Glass A. M. (1977). *Principles and Applications of Ferroelectrics and Related Materials,* Oxford University Press, Oxford, UK.

Lynch C. S., Yang W., Collier L., Suo Z. and McMeeking R. M. (1995). Electric Field Induced Cracking in Ferroelectric Ceramics. *Ferroelectrics* **166**, 11-30.

McHenry K. D. and Koepke B. G. (1978). Electric Field Effects on Subcritical Crack Growth in PZT. In *Fracture Mechanics of Ceramics,* Vol. **5**, R. C. Bradt, A. G. Evans, D. P. H. Hasselman and F. F. Lange, ed., Plenum Publishing Co., New York, NY, 337-352.

Mehta K. and Virkar A. V. (1990). Fracture Mechanisms in Ferroelectric-Ferroelastic Lead Zirconate Titanate (Zr:Ti=0.54:0.46) Ceramics. *Journal of the American Ceramic Society* 73, 567-574.

Nishkawa T. J., Takahashi, A., Hattori and Takatsu M. (1992). Cyclic Fatigue of Electrically Poled Piezoelectric Ceramics. *Fracture Mechanics of Ceramics* **9**. Edited by Bradt, Hasselman, Munz, Sakai and Shevchenko, 493-500.

Nye J. F. (1964). *Physical Properties of Crystals*, Clarendon Press, Oxford.

Pak Y. E. (1990). Crack Extension Force in a Piezoelectric Material. *Journal of Applied Mechanics* **57**, 647-653.

Pak Y. E. and Herrmann G. (1986). Conservation Lawsand the Material Momentum Tensor for the elastic Dielectric. *Int. J. Engrg. Sci.* **24:8**, 1365-1374.

Park S. B. and Sun C. T. (1995). Fracture Criteria for Piezoelectric Ceramics. *Journal of the American Ceramic Society* **78**, 1475-1480.

Sun C. T. and Jiang L. Z. (1998). Domain Switching Induced Stresses at the Tip of a Crack in Piezoceramics. *Proceedings of the 4th European Conference on Smart Structures and Materials,* Harrogate, UK, 6-8 July, 715-722.

Tobin A. G. and Pak Y. E. (1993). Effect of Electric Fields on Fracture Behavior of PZT ceramics. *Proceedings of SPIE, Smart Structures and Materials* **1916**, 78-86.

White G. S., Raynes A. S., Vaudin M. D. and Freiman S. W. (1995). Fracture Behavior of Cyclically Loaded PZT. *J. Am. Ceram. Soc.,* **77:10**, 2603-2608.

Long Term Durability of Structural Materials
P.J.M. Monteiro, K.P. Chong, J. Larsen-Basse, K. Komvopoulos (Eds)

FREQUENCY EFFECT ON THE FATIGUE LIFE OF A CHOPPED FIBER COMPOSITE

B. Regez[1], S. C. Max Yen[1], M. El-Zein[2], and D. C. Wang[3]

[1] Materials Technology Center
Southern Illinois University Carbondale,
Carbondale, IL 62901-6603, USA
[2] Deere Technical Center, Deere & Company,
Moline, IL 61265, USA
[3] Department of Mechanical Engineering
National Yunlin University of Science & Technology
Touliu, Yunlin, Taiwan, ROC

ABSTRACT

In this paper, the fatigue characteristics of chopped fiber composites are presented. Specifically, the effect of stress amplitude and frequency toward the damage accumulation, modulus reduction, and failure characteristics were analyzed whereby an accelerated characterization procedure for fatigue life prediction could be developed.

Two types of chopped fiber composite materials, glass/polyurethane and glass/vinyl ester, were used in this research. Each type of materials was tested under sine wave tensile fatigue loading at 1, 3, or 5 Hertz with constant stress amplitude ranging from 40% to 85% of the static ultimate strength. During each fatigue test, the change in axial modulus, the change in specimen surface temperature, and the development of transverse cracks were measured along with the time to fracture. The development of S-N curves under various stress and frequency conditions allows for the establishment of the empirical shift factors suitable for accelerated characterization. On the other hand, the detailed analysis of the damage evolution in terms of modulus reduction and crack density has revealed some evidence of fatigue failure mechanisms. A normalized modulus reduction equation was introduced to model the fatigue mechanism. This equation represents the fatigue degradation process beyond the instantaneous damage introduced by the applied stress.

In summary, we have found that the application of stress and frequency as the acceleration parameters to forecast fatigue is only possible to a certain extent of stress levels. This, perhaps, is affected by the self-heating mechanism developed during fatigue. On the other hand, the normalized modulus reduction equation gives an indication that the accelerated characterization can be established analytically by obtaining the critical time when the rate of stiffness reduction drastically increases and the limiting value of the normalized modulus. This implies that the normalized modulus at fatigue fracture could be considered as a material property.

KEYWORDS

Composite materials, fatigue, accelerated characterization, damage, modulus reduction, life prediction.

INTRODUCTION

This project intends to develop a design process for service life prediction of machines and structures. Since the design fatigue life of materials and machines is usually far beyond that of the duration of laboratory tests, a life prediction theory based on an accelerated characterization technique must be developed. The accelerated characterization technique to be developed in this project is based on the observation that the material degradation processes under different fatigue conditions (frequency, stress, and temperature) have the same general characteristics but are shifted in time. This shift in time allows one to predict the long-term fatigue life using the data obtained over shorter time intervals. Based on our past studies, the fatigue failure is governed by the reduction in stiffness or the accumulation of strain energy density toward a critical value. Dealing with the history dependent properties such as deformation, modulus reduction, and strain energy loss allows one to address the effects of loading sequence or any arbitrary combination of loads on the fatigue life. This strain energy, density based failure theory has been proven to be effective in an earlier study on the creep rupture of a composite material.

Damage in Composite Materials

In composite materials, damage (in the form of transverse micro cracks, delamination, etc.) is often introduced even at a small fraction of their ultimate strength (Hasan, 1992; McAuliffe, 1996). This implies that the presence of damage does not necessarily require an immediate replacement of a composite machine part. However, the existence of damage becomes very critical when designing a composite structure under a repeated loading and unloading condition. This is due to the fact that the development of damage in a composite material during the reversal of loading may cause a progressive reduction in elastic modulus. Therefore, to accurately model the behavior of a structural composite, it is necessary to determine the relationship between the state of damage and the material properties such as elastic modulus, residual strength/life, and creep. This is, in fact, one of the key concepts of the damage mechanics (Kachanov, 1986).

The continuum damage mechanics (Kachanov, 1986), in general, deals with the materials or structures that contain natural and/or induced defects. For isotropic materials, the formulation of a continuum damage model is very similar to that of the incremental theory of plasticity (Chow and Wang, 1987; Krajanovic, 1985). Moreover, in some continuum damage models, the coupling of the damage state and plasticity has been assumed (Chow and Wang, 1989). At present, the comprehensive continuum damage models for analyzing composite materials are still under development due to the complexity of the damage formation. Furthermore, the experimental procedure that characterizes the parameters of a continuum damage model is generally lacking. Heimbuch et al. (1978) started to develop the S-N curves for the fatigue life of random fiber composite materials for automobile applications. This was followed by a study of the fatigue crack growth phenomenon (Wang, et al., 1983) and the degradation of shear stiffness under shear fatigue (Wang et al., 1984). Reifsnider and Masters (1978) and Reifsnider et al. (1979) studied the degradation of the elastic stiffness and the development of cracks during a continuous fatigue loading condition. They used a concept called "characteristics damage state" as the criterion for the onset of final failure. Yen and Morris (1992) have illustrated the procedure for the determination of the parameters for the creep response under different damage conditions. Also, Yen (1984) has indicated that the state of damage could be substantiated and characterized by the distribution of cracks. It has been shown that the state of damage can be defined by the crack density (Reifsnider and Masters, 1978; Reifsnider et al., 1979), the percentage of ultimate strength (Reifsnider and Masters, 1978), the reduction in stiffness (El-Zein and Berns, 1992; El-Zein et al., 1995), and the change in attenuation of a nondestructive measurement (Reifsnider and Masters, 1978; Reifsnider et al., 1979; Liu, 1992; McAuliffe et al., 1996), etc. The characterization of crack distribution in composite materials has been a very difficult task. The recent development of the optical microscopic image analysis procedure by Yen et al. (1991) has made the characterization of the microscopic features of composite become efficient and accurate. It has been shown that the distribution of fiber/matrix clusters may affect the fatigue resistance of a composite material (Yen et al., 1993).

Under a repeated loading and unloading condition, a polymer usually exhibits the phenomenon of rising surface temperature. This is often attributed to the hysteresis loop developed during loading and unloading. Likewise, a thermoplastic composite material is anticipated to develop hysteresis heating during fatigue conditions (Oldirev, 1977). It is believed that the hysteresis-heating phenomenon is an energy dissipating mechanism for physical changes (or damage development) and must be accounted for in the damage assessment. Therefore, any damage theory formulated by considering the balance in strain energy needs to consider energy dissipation due to the new damage formation.

In summary, a complete characterization of damage state leading to the residual strength assessment must consider three inter-related factors, namely, the loading history, the crack distribution, and the thermal history due to hysteresis.

Accelerated Characterization

The accelerated characterization technique, in general, is a semi-empirical procedure to predict the long-time response (including time-to-failure) of materials using the short-time data. Different accelerated characterization procedures have been used in the past to predict the long-term creep response of composite materials. The Time-Temperature-Stress-Superposition Principle (TTSSP) (Yen, 1984; Yen and Williamson, 1990; Griffith et al., 1980; Yen et al., 1979; Yen et al., 1985; Schapery, 1969; Schapery, 1974) is one of these procedures. The TTSSP is based on creep deformation curves for different thermal-mechanical conditions of the same shape. In addition, increases in temperature and/or stress will shift creep deformation curves to the left on a log-time scale, indicating that increase in temperature and/or stress causes the acceleration of creep deformation. Therefore, by collecting short-time creep deformation data at elevated temperatures (or stresses), long time creep behavior at lower temperatures (or stresses) can be predicted. In general, the TTSSP principle was found to be applicable to the long-term creep response of various kinds of fiber-reinforced polymeric composite materials. More recently, an energy based failure criterion was developed by Yen and Morris (1989) to predict creep rupture phenomenon of a chopped fiber composite material. This failure criterion combines the TTSSP and the free energy failure theory developed by Reiner and Weisenberg (1939). In addition, the stress rupture equation derived by Yen and Morris (1989) resembles the delayed failure theory developed by Christensen, 1981) based on a fracture mechanics approach.

The application of an accelerated characterization procedure to predict fatigue strength is somewhat more complicated than that used in characterizing the viscoelastic response such as creep. The traditional accumulative damage equation, Miner's rule (Miner, 1945), based on the data obtained from the S-N curves does not provide any explicit parameters for accelerated prediction. Based on the time temperature superposition principle, Miyano (1995) proposed an accelerated characterization equation to predict long-term fatigue response of a carbon fiber reinforced composite laminate. Miyano (1995) indicated that the fatigue failure ultimately was contributed by the viscoelastic behavior, thereby, the shift factors for predicting the long time creep could be equally applicable to the fatigue data.

EXPERIMENT

The ultimate goal of this project is to establish an accelerated characterization procedure to predict the life of a material under different fatigue loads. The specific tasks conducted during this study were:

- Determine the basic stress-strain relation
- Establish short-time fatigue life database
- Study the degradation characteristics (modulus and cracks) during fatigue
- Develop a deformation based fatigue life equation
- Determine the parameters for the basis of accelerated characterization

Materials and Specimens

The chopped fiber composite material used in this project is glass/vinyl ester. The fiber volume ratio for glass/vinyl ester was approximately 60% to 63%. The material was fabricated in the form of a rectangular sheet through the compression molding process. The fibers are randomly distributed in the plane of the molded sheet.

Throughout this study, the specimens in the form of flat dog-bone shaped coupons were used. A router with a diamond-coated blade was used to cut the specimens from the composite panel. Each specimen is 152 mm long with a 12.7 mm by 25.4 mm gage section. The grip section is 38.1 mm long and 25.4 mm wide. The radius of curvature from the grip section to the gage section is 62.7 mm. The thickness of the specimen is 3.18 mm. For each test condition, several specimens were used in order to obtain the results of statistical significance.

Instrumentation and Experimental Procedures

The stress-strain relationship and the ultimate strength of the materials investigated in this research were obtained using a MTS machine at a constant rate of 0.504 mm/min.

The fatigue experiments were conducted using sinusoidal function at 1, 3, or 5 Hertz with constant tensile stress amplitude ranging from 40% to 85% of the static ultimate strength. All the fatigue tests were conducted using a PC-controlled servo-hydraulic MTS machine equipped with a National Instruments data acquisition system. The strain data was obtained through a clip gage (12.5 mm gage length) attached to the gage section of the specimen.

During each fatigue test, the axial modulus (referred to as the reduced modulus) and number of transverse (matrix) cracks were measured periodically. The temperature rises due to the self-heating during fatigue were also recorded. It is not certain, however, to what extent the self-heating affects the damage evolution or if it has already been accounted for with other damage parameters. During the fatigue test, a thermal couple was attached to the surface of the specimen at the gage section. The temperature data was recorded periodically using a data acquisition system made by Measurement Group. The use of reduced modulus as the damage parameter has been discussed extensively. The reduced modulus was acquired during an interruption (i.e. the pause) of the fatigue test. During each interruption, the specimen was subjected to a quasi-static tensile test up to the applied stress being used in the fatigue test. A stress-strain was then obtained for each interruption. Upon the completion of the quasi-static test, the fatigue experiment was resumed. Typically, a fatigue test was interrupted 10 times prior to the ultimate fatigue fracture.

Reifsnider et al. (1978, 1979), Yen (1984), and Yen et al. (1992) have demonstrated that the matrix cracks can be used to characterize damage states. The data collection of transverse (matrix) cracks and their distribution over a representative area along the edge of the specimen were conducted during the interruption of the fatigue test. The characterization of matrix cracks was conducted using an image analysis system (Yen et al. 1991; Yen et al., 1993). The image analysis system consists of an optical microscope, a video camera, an image frame grabber, and a computer. An image analysis system works like an optical microscope except all images are interpreted electronically and stored in digital format, namely an image file. The optical images of the edge of the specimen were recorded for analysis and as the permanent records. The stored images allow for additional data analyses at a later time. It should be pointed out that an extensive polishing must be applied to the edge of the specimen in order to obtain an image with detailed features of fibers, matrix, and cracks. The polishing of the specimen was accomplished using microscopic metallic powder of different sizes prior to the fatigue experiment.

RESULTS AND DISCUSSIONS

The typical tensile stress-strain relationship of glass/vinyl ester composite material is given in Figure 1. In the figure, the characterization of the initial elastic modulus is illustrated by a dot line. The last data point reflects the ultimate fracture of the specimen tested. A total of 10 specimens were used to determine the average ultimate strength and the initial elastic modulus. It was found that the initial modulus is approximately 21$\pm$1 GPa and the static ultimate tensile strength is approximately 320$\pm$20 MPa. It should be pointed out that the reduced modulus is approximately 14$\pm$1 GPa if the specimen (Figure 1) was to be unloaded prior to the fracture.

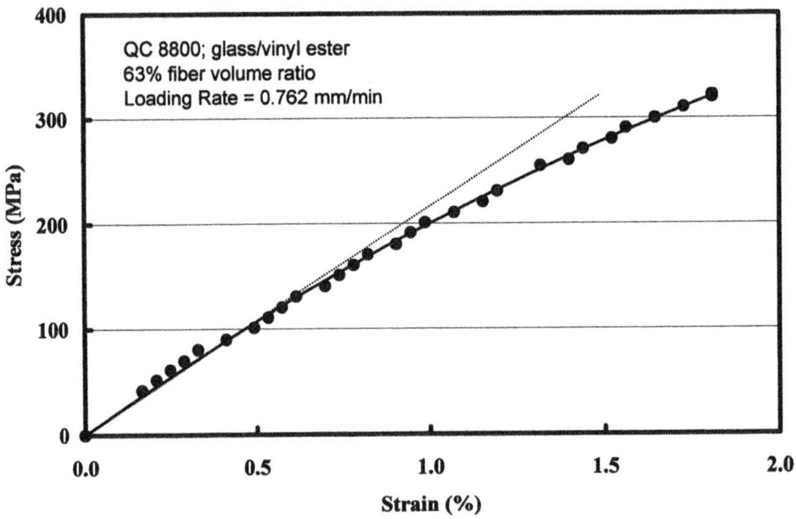

Figure 1: Typical stress-strain relation for glass/vinyl ester composite.

Figure 2 gives the S-N curve of glass/vinyl ester composite due to various stresses at 3 Hz. Three fatigue life data are presented for each stress level. These data reflect the upper bound, medium, and lower bound of the fatigue life data of the specimens tested for a given stress condition. At least five fatigue tests were conducted for each testing condition. The straight line shown in Figure 2 represents the best fit of the data presented. The data scattering shown in Figure 2 falls within the general acceptable range of fatigue life data. Figures 3 and 4 show the comparison of fatigue life data due to different frequencies. The data presented in Figure 3 were obtained through glass/vinyl ester composite with 60% fiber volume ratio while the data shown in Figure 4 were obtained from the same material with 63% fiber volume ratio. In Figure 4, only the average fatigue life data are shown. Both figures show a similar trend in fatigue life versus stress. It should be pointed out that the fatigue life data in Figure 3 and 4 have suggested that the application of accelerated characterization for fatigue life prediction using high stress and high frequency is somehow limited. When comparing to the fatigue life data from a higher frequency to a lower frequency fatigue one would realize that the higher the applied stress the lesser acceleration effect. The empirical shift factor that one could predict the long-term fatigue using short-term fatigue data is illustrated in Figure 3. In fact, the data has indicated that at the stress level above approximately 60% of the ultimate strength, the fatigue life appears to be longer at a higher frequency for the same applied stress. This indeed deviates from the assumption regarding the effect of combined frequency and stress in accelerated life prediction. However, one must also recognize that under the same frequency the fatigue life at a high stress is much shorter (in many decades) than that at a low stress. Therefore, it is still possible to conduct life prediction (or accelerated characterization) using the stress fatigue data under the same frequency. On the other hand, one must keep in mind that with the absence of an analytical equation based on the materials properties and failure mechanisms, the accelerated characterization using the S-N data alone will be purely empirical.

Figure 2: The fatigue life data of glass/vinyl ester at 3 Hz.

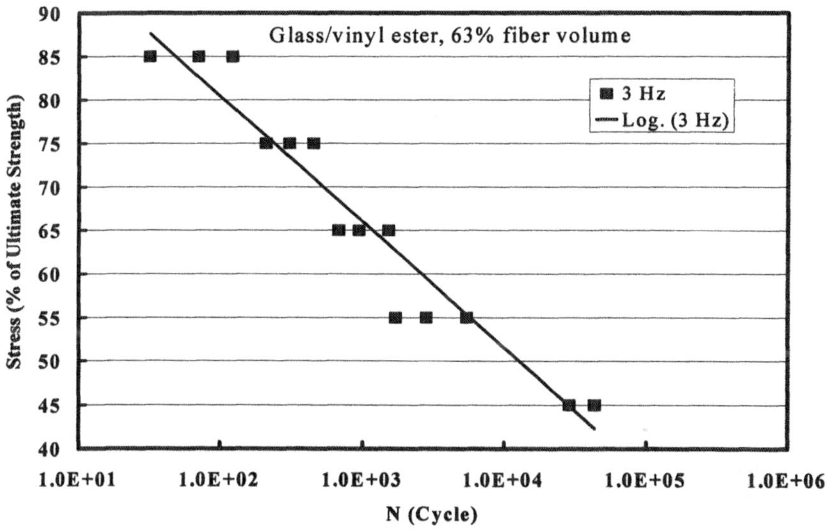

In a later section of this paper, a diagnosis is given regarding the reason why the fatigue life data at a high stress (> 60% of ultimate strength) and high frequency is longer than that of the same stress at a lower frequency. This is to recognize that the intercept of fatigue life curves for different frequencies is rather unusual. In particular, the effect of stress rate, self-heating, and damage accumulation toward the change in loading carrying characteristics and failure mechanisms will be outlined.

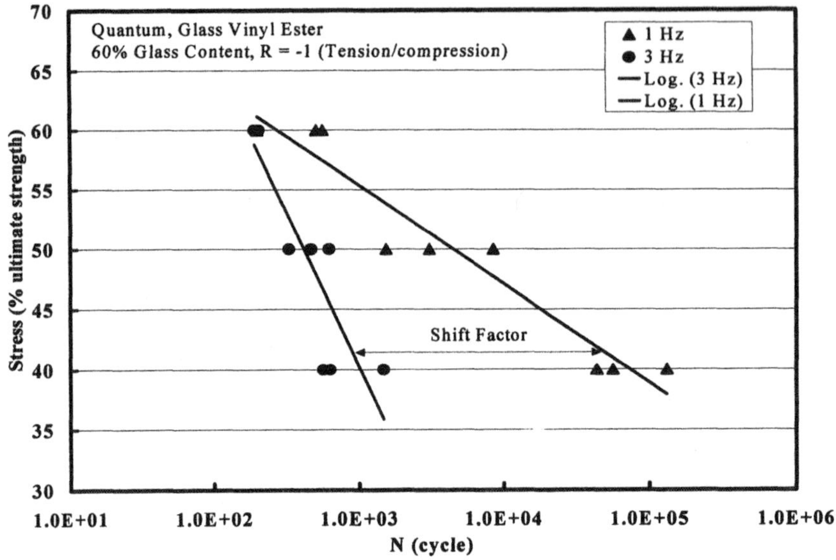

Figure 3: Fatigue life data of glass/vinyl ester under tension/compression fatigue.

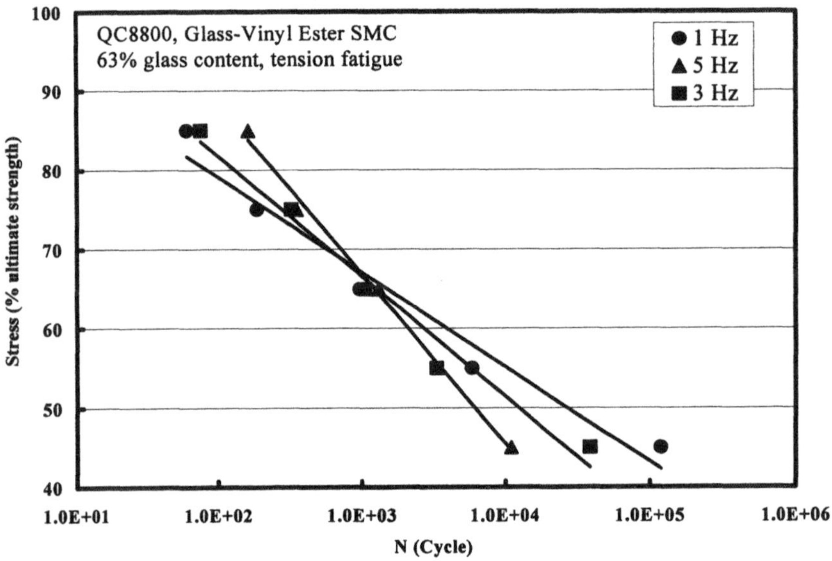

Figure 4: Fatigue life data of glass/vinyl ester under tension fatigue.

During each fatigue experiment, both the surface temperature of the specimen and ambient temperature were recorded until failure. The ambient temperature record was used as a reference to determine whether the self-heating (i.e. the temperature increase) in a specimen was developed. Figure 5 shows the temperature profile for different stresses at 3 Hz. Two distinct thermal characteristics can be deduced from the data shown in Figure 5. It was found that self-heating in a specimen due to fatigue load at low frequency and low stress was rather insignificant. Since the specimens tested at both 20% and 30% of ultimate strength did not fail and were terminated at 1 million cycles, whether the self-heating effect would occur in a later time remained unanswered. On the other hand, a significant self-heating was observed for the fatigue tests under high stress and high frequency. Based on the data shown in Figure 5, for a given fatigue frequency, the higher the stress the faster the heating rate. It was found that at the instant of fatigue fracture the specimen temperature reached approximately 50°C. It was also found that the self-heating characteristics could be fitted by an exponential curve.

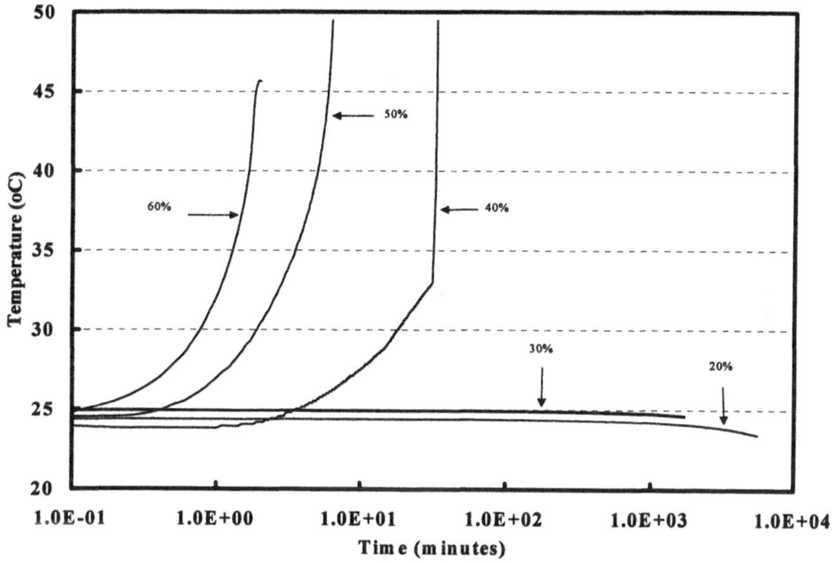

Figure 5: Change of specimen temperature during fatigue test at 3 Hz.

It should be pointed out that the temperature increase in the test specimen during fatigue is rather significant. This is due to the fact that the material properties of glass/vinyl ester such as modulus and strength are temperature dependent. It should be pointed out that the S-N data in Figures 3 and 4 were obtained without taking into account the effect of temperature to the ultimate strength. The adjustment of S-N curves to account for the self-heating effect is not quite simple due to the fact that the specimen temperature is a function of time. Another observation is that the composite material becomes ductile at the elevated temperatures, therefore, it will fail under a different mechanism. In this regard, the increase in ductility could possibly delay the time to failure more then one had anticipated. This has suggested that the time to failure data shown in Figures 3 and 4 at high stress levels have included the effect of ductility induced by the temperature increase of self-heating. Therefore, for the fatigue life data obtained at above the 60% of ultimate under high frequency has included some temperature effect that resulted in a shift of the fatigue life data to the right on the time scale of the S-N plot.

In addition to containing the effect of self-heating, the S-N data shown in Figures 3 and 4 also includes the effect of stress rate. It should be pointed out that the fatigue frequencies used in this study produced high stress rates to the test specimen range from approximately 60 MPa/sec to 500 MPa/sec. On the other hand, the loading rate used in characterization of the ultimate strength is approximately 1 MPa/sec. Knowing that glass/vinyl ester is a rate sensitive material, its ultimate strength would vary with different loading rates. Specifically, a faster stress rate will give a higher ultimate strength. Nevertheless, the static ultimate strength was used to normalize the stress data in the S-N data in Figures 3 and 4. Therefore, the normalized stress data in the S-N plot needs to take into account the effect of stress rate. This would, in general, lower the normalize stress level for all the data shown in Figures 3 and 4. Naturally, the specimens tested at a higher frequency would require a large downward shift in the normalized stress.

During the fatigue test, cracks in the form of transverse crack (perpendicular to the loading axis) and splitting crack (parallel to the loading axis) were developed. Based on the observation from the stress-strain relation, the development of transverse cracks usually occurred at approximately 30% to 40% of ultimate strength. The longitudinal cracks were developed later allowing for the transverse cracks to join together. Likewise, the same sequence of crack development was found during the fatigue load. Figure 6 illustrates the typical crack development process during fatigue at 85% of ultimate strength and 5 Hz. The number in each image gives the number of fatigue cycles when the image was taken. The number of cycles to failure for this specimen was 95 cycles. Each image represents a small portion of a represented area based on which the damage analysis (i.e. crack development) will be conducted. As one can see, the transverse crack tends to develop in the matrix rich region and along the interface between 90-degree fibers cluster and matrix. The fiber cross sectional area in the form of a small circle represents the fiber of 90-degree orientation with respect to the applied load axis. As the fatigue load progressed, new transverse cracks and longitudinal crack developed.

Figure 7 gives an illustration of the number of cracks developed during a fatigue testing. The estimation of the number of cracks was conducted manually using an image analysis system. Since the cracks were developed and distributed in accordance to the random feature of chopped fiber clusters in the specimen, a detailed mechanistic analysis on the effect of cracks has not begun yet. When under a fatigue load, some transverse cracks developed instantaneously. At this instance, the number of cracks (or degree of damage) is analogous to that found in the static test for the same stress level. During the initial stage of fatigue load, the growth of a new crack and the crack propagation appear to be slow. Later, the number of transverse cracks began to increase and the existing ones started to propagate. It is clear that the development of transverse cracks is directly related to the change in modulus of the specimen. In general, the number of crack increased during fatigue can be fitted by a power equation. This suggested that the estimation of modulus reduction with fatigue cycles could also be approximated by a smooth curve. More so, the cracks developed instantaneously can be related to the instant reduction in modulus.

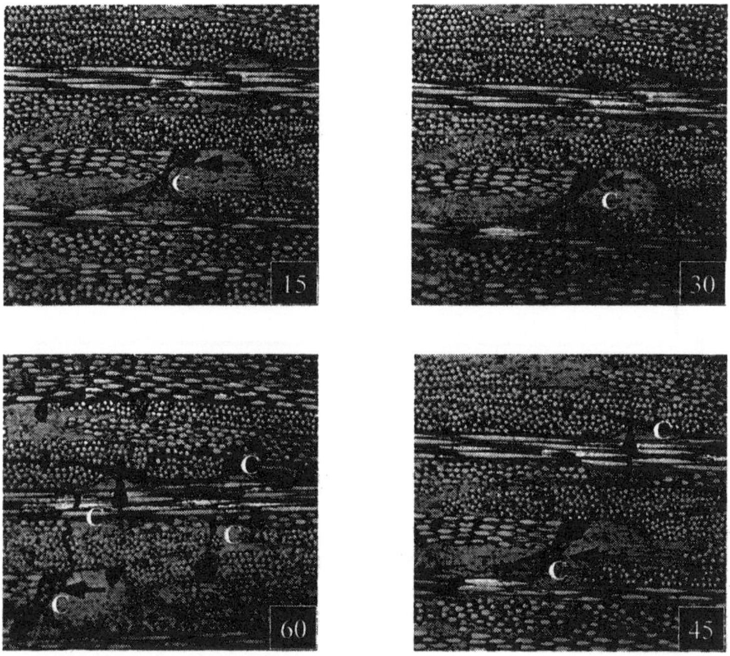

Figure 6: The development of cracks during fatigue (85% at 5 Hz).

The reduced axial modulus of the test specimen was obtained periodically during the fatigue test through the stress-strain experiment. The reduced modulus is the "elastic modulus" of the specimen that has received some duration of fatigue load. This experiment was carried out along with the image acquisition of the edge of the specimen for the analysis of crack development. The first reduced modulus data in a fatigue load was taken immediately after the completion of the first few cycles. The reduced modulus at different stages of fatigue can reveal an insight of damage evolution toward final fracture, thus allowing for the development of a mechanistic based equation for characterizing the kinetics of fatigue. Furthermore, this exercise may lead to the identification of some implicit material characteristics that can be used to generalize the fatigue behavior due to different loading conditions. In this regard, the accelerated characterization of fatigue response and fatigue life can be extended beyond the traditional empirical approach.

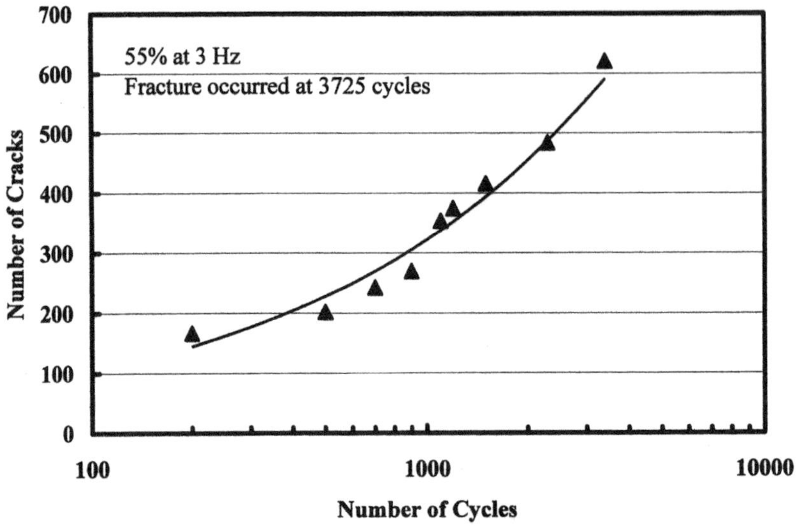

Figure 7: The accumulation of cracks during fatigue (55% of ultimate strength at 3 Hz).

Figure 8 shows the variation of reduced modulus with respect to fatigue cycle for different stress levels at 3 Hz. In the figure, the solid lines represent the curve fit of the data and the symbols are the experimental data. As can be seen from Figure 8, the reduced modulus data shows some degree of scattering but in general, they are within an acceptable range with the best fitting curve. It should be pointed out, however, the characteristics of the reduced modulus curves do not show a consistent trend in analogous to the stresses used in the fatigue experiment. This is primarily attributed to the random variation of microscopic features from one specimen to another. In other words, we have found that the initial number of cracks (or degree of damage) at the beginning of fatigue load is not necessarily proportional to the applied load. This explains why the initial reduced modulus (shown as the first data point in Figure 8) is not proportional to the applied load. One should recognize that the effect of microscopic feature toward the state of damage at different stress levels may be predicted through the concept of damage theory. The data obtained in this research in regard to the initial reduced modulus and damage evolution should serve as the database for such development. More over, it is also possible to use the information on damage evolution and microscopic features as the mechanism for quality evaluation of a manufacturing process.

The finding derived from the figures presented in the previous section has led to the conclusion that a more rigorous prediction or characterization of fatigue behavior needs to be incorporated into the effect of loading history. Specifically, the effect of damage evolution and the deformation history are the key parametric functions to be considered. To this extent, we consider the kinetics of thermal history is embedded within the deformation. It is also believed that the deformation approach to fatigue phenomena can be easily converted into the formulation based on strain energy. In general, it is believed that the deformation history and the reduction of modulus during fatigue at the stress level above the elastic limit (maximum damage free state) must be separated into two components, an instantaneous response and a time-dependent component. The instantaneous component corresponds to the reaction to the magnitude of applied load. This is directly related to the instantaneous development of cracks and/or the instantaneous reduction in modulus as soon as the loading is applied. It is also believed that the instantaneous response is significantly affected by the microscopic features such as the distribution of fiber clusters, the size of matrix rich region, etc. Therefore, to accurately analyze the instantaneous response, a good knowledge on the materials distribution characteristics in relating to manufacturing is needed. The result of this work will be reported in a future publication. The time-dependent component reflects to the process of incremental damage (or damage evolution) beyond the instantaneous response. With the absence of the first data point, the reduced modulus curve shown in Figure 8 is the time-dependent response. The time dependent response is more sensitive to the loading condition and it is less sensitive to the microscopic feature. In other words, the damage evolution will be a progressive process with the absence of abrupt changes. More so, the deformation process or degradation process will be proportional to the loading condition. The rate (or the slope) of modulus degradation shown in Figure 8 has shown the proportionality to the applied stress condition under the same fatigue frequency.

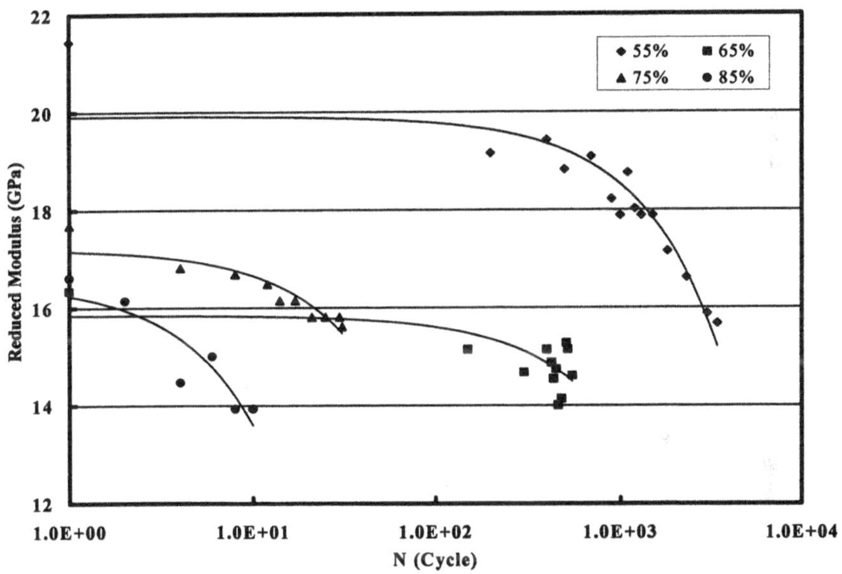

Figure 8: Variation of reduced modulus during fatigue (at 3Hz).

It should be pointed out the characteristics of the time-dependent reduced modulus during fatigue can also be viewed as the frequency response to a driving force, i.e. the fatigue condition. This is analogous to a mechanical system receiving a force excitation. It is then appropriate to model the fatigue data based on the frequency response equation commonly used in the control theory (Kuo, 1995). To this extent, the reduced modulus data were modeled using the following equations developed by Nyquist (1932):

$$\frac{E^*}{E_i} = \frac{1}{1 + J\dfrac{\omega}{C}} \tag{1}$$

and

$$f\left(\frac{E^*}{E_i}\right) = 20 \log \left(\frac{E^*}{E_i}\right) = 20 \log \left(\frac{1}{1 + J\dfrac{\omega}{C}}\right) \tag{2}$$

where $\dfrac{E^*}{E_i}$ = Normalized reduced modulus

E^* = Reduced modulus (as shown in Figure 8)

E_i = Instantaneous reduced modulus (as shown in Figure 8)

w = fatigue frequency in cycles

C = corner frequency where the value of $f(E^*/E_i)$ is equal to –3

J = $\sqrt{-1}$

The unit for $f(E^*/E_i)$ is defined to be dB used in the vibration or noise analysis. The corner frequency, C, can also be defined by the frequency corresponding to the intercept between two lines tangent to the $f(E^*/E_i)$ curve; the initial tangent line and the line with a slope of 20 dB/decade. The corner frequency was used to indicate and define the time when the rate of change increased drastically.

In a way, a plot of the function f with respect to fatigue cycle (ω) reflects a refinement of the data shown in Figure 8. Since each E_i has carried the effect of microscopic feature that can be interpreted as the effect of stress or strain concentration factor. Therefore, dividing E_i with E^* not only normalizes the reduced modulus data to a common reference state of damage (or strain concentration) but also brings all the $f(E^*/E_i)$ curves to a common starting origin (a value of zero) on a chart. The corner frequency was obtained through the graphical fitting of $f(E^*/E_i)$ versus frequency derived from the experimental data. Figure 9 shows the variation of corner frequency with different stresses and frequencies. A least square linear fit was found suitable to model the variation of corner frequency versus stress for a given fatigue frequency. The authors believe that the characterization of the corner frequency is a rather important task in fatigue analysis. In particular, if the specific function expression of corner frequency is obtained, then the variation of $f(E^*/E_i)$ versus fatigue cycles can be established for all conditions. In this regard, the corner frequency may be used to predict the deformation characteristics of a given fatigue condition, thereby achieving a certain degree of accelerated characterization.

Figure 10 gives a comparison between the values of $f(E^*/E_i)$ as calculated from the experimental data and that generated through the use of corner frequency. It should be pointed out once again that the experimental data shown in Figure 10 were used to determine the corner frequency. In general, the analytical prediction of normalized reduced modulus, $f(E^*/E_i)$, based on the equation (2) appears to be in good agreement with experimental data. Moreover, it was found that the value of the normalized reduced modulus under the same stress condition appeared to be proportional to the applied fatigue frequency. It is also possible to shift to the right on the time scale the normalized reduced modulus curve of a high fatigue frequency to predict that of a lower fatigue frequency. It should also be pointed out that in Figure 10, the last data in each experimental curve represents the point at the ultimate fracture. The corresponding value for the normalized reduced modulus at the fracture is given in the figure. It was further found that the value of normalized reduced modulus at the incident of ultimate fracture could be considered as a constant and with a value around –9.85. The validity of the limiting value of the normal reduced stiffness modulus for the composite material used in this study will be further investigated. However, the implication of this finding is that the limiting value of the normalized reduced stiffness modulus can be treated a failure criterion and/or can be considered a material property specifically for fatigue. Therefore, whenever such a value developed in a given material the fatigue fracture occurs.

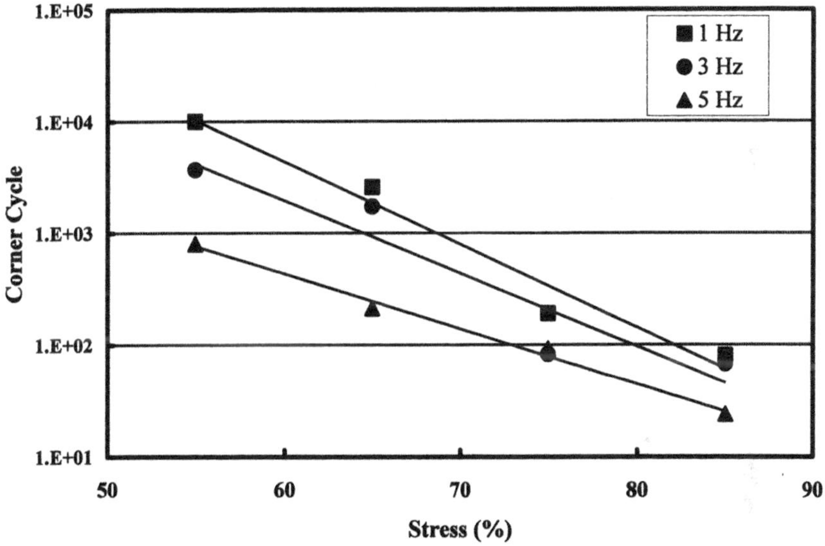

Figure 9: The variation of corner frequency with stress and frequency.

The results obtained in Figures 9 and 10 have suggested that the complete evolution of the normalized reduced modulus under a fatigue load can be predicted up to the ultimate fracture with three key parameters, namely the instantaneous reduced modulus, the corner frequency, and the normalized reduced stiffness at fracture. More so, if the parametric expression of these three parameters is available, then it is possible to predict the fatigue behavior as well as the final fracture. Thus, the concept of the accelerated characterization can be applied.

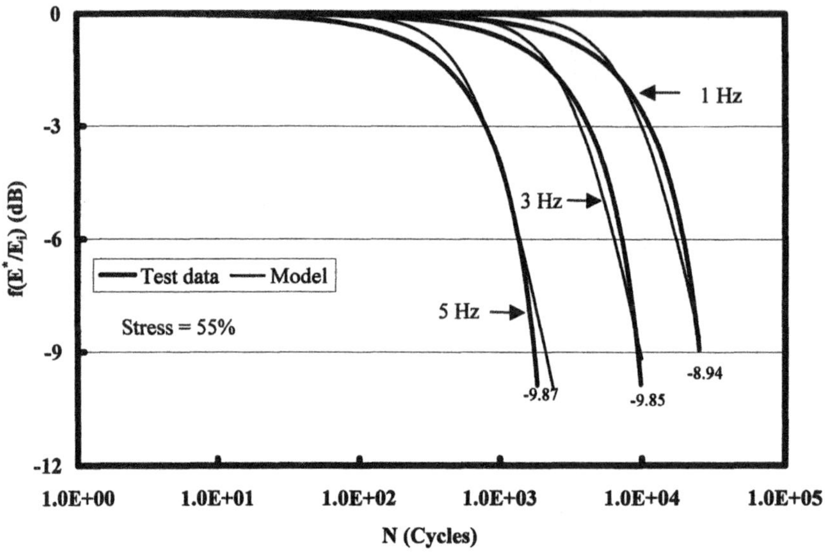

Figure 10: The variation of the normalized reduced modulus with fatigue cycle.

CONCLUSIONS

This research project is aimed at the development of an accelerated characterization procedure based on the fatigue data obtained at different frequencies and stresses. In particular, the authors believed that the deformation approach to characterize the fatigue response could accurately represent the effect of loading history. What follows is a summary of the findings obtained in this project:

- Characterization of fatigue life must address the shift in failure mechanisms due to the effect of stress/frequency effect, self-heating (temperature and heating rate), and damage.
- Empirical shift factors for the accelerated characterization of fatigue life data with the similar failure mechanism may be obtained.
- Self-heating phenomenon during fatigue is significant at high stress and at high frequency.
- The self-heating and crack growth phenomena can be modeled by an exponential equation at the logarithm time scale.
- Fatigue damage can be characterized into an instantaneous damage (cracks or reduced modulus) and a time-dependent damage (cracks or reduced modulus) growth.
- The characteristics of the normalized reduced modulus can be modeled into a frequency

response equation originated from the control theory. This equation is characterized by three parameters, the instantaneous reduced modulus, the corner equation, and the value of normalized reduced modulus at fracture. In particular, the value of the reduced modulus at fracture is a constant and can be considered as a material property for fatigue condition.

- This frequency response equation expressed by the normalized reduced modulus can be extended to conduct the accelerated characterization and fatigue life prediction.

REFERENCES

Chow C. L. and Wang J. (1987). An Anisotropic Theory of Elasticity for Continuum Damage Mechanics. *International Journal of Fracture*, 27, pp. 3-16.

Chow C. L. and Wang J. (1989). An Anisotropic Theory of Continuum Damage Mechanics for Ductile Fracture, *Engineering Fracture Mechanics*. 32, 1989, pp. 791-797.

Christensen R. M. (1981). Life Time Prediction for Polymers and Composite under Constant Load. *Journal of Rheology*, 25(5):517-528.

El-Zein M. S. and Berns D. H. (1992). Life Prediction and Cumulative Damage Analysis in Random Fibers Composites. Proceedings, 6th U.S. Conference on Composite Materials, Orlando, June 22-25.

El-Zein M. S., Yen S. C., and Teh K. T., (1995) Fatigue Life of a Chopped Fiber Composite Under Spectrum Loading. DURACOSY 95 Brussels, Belgium,

Griffith W. I., Morris D. H., and Brinson H. F. (1980). The Accelerated Characterization of Viscoelastic Composite Materials. VPI & SU, Blacksburg, VA, VPI-E-80-15.

Hasan I. N. (1992). Continuous Degradation Process of a Chopped Fibers Composite (SMC-R50) under Axial Tension. MS Thesis, Department of Civil Engineering and Mechanics, Southern Illinois University at Carbondale, Carbondale, Illinois.

Heimbuch R. A. and Sanders B. A. (1978). Mechanical Properties of Chopped Reinforced Plastics. *Composite Materials in Automotive Industry*, S. V. Kulkarni, and R. B. Pipes, Eds., ASME, New York, December, pp. 111-139.

Kachanov L. M. (1986). *Introduction to Continuum Damage Mechanics*. Martinus Nijhoff Publishers, Boston.

Krajcinovic. M. (1985). Constitutive Equations for Damaging Materials. *Journal of Applied Mechanics,* 50, pp. 355-360.

Kuo B. C. 1995. *Automatic Control Systems*. 7th ed. Prentice-Hall, Englewood Cliffs, NJ.

Liu C. T. (1992). Acoustic Evaluation of Damage Characteristics in Composite Solid Propellant. *Journal of Spacecraft and Rockets*, Vol. 29, No. 5, pp. 533-537.

McAuliffe P. (1996). Post Impact Tension/Compression Fatigue of Graphite/PEEK Laminates. MS Thesis, Department of Mechanical Engineering and Energy Processes, Southern Illinois University at Carbondale, Carbondale, Illinois.

McAuliffe P., Yen S. C., Teh K. T., and Huang C. Y. (1996). Post Impact

Tension/Compression Fatigue of Graphite/PEEK Laminates. Presented at 1996 SEM Spring Conference on Experimental Mechanics, Nashville, TN.

Miyano Y. (1995). Long Term Prediction Method for Static, Creep, and Fatigue Strengths of CFRP Composites. Progress in Durability Analysis on Composite Systems, Edited by A. Cardon, H. Fukuda, and K. Reifsnider, A. A. Balkema Publishers, pp. 177-188.

Nyquist H. 1932. Regeneration Theory. *Bell Syst. Tech. J.*, January, pp. 126-147.

Oldirev P. P., Parfeev V. M., and Komar V. I. (1977). Refined Method of Fatigue Life Determination for Polymeric Materials Based on Self-Heating Temperature. *Mechanics of Polymers*, No.5, pp. 906-913.

Reifsnider K. L. and Masters J. E. (1978). Investigation of Characteristics Damage States in Composite Laminates. ASME Publication No. 78-WA/Aero-4.

Reifsnider K. L., Henneke E. G. II, and Stinchcomb W. W. (1979). Defect-Property Relationships in Composite Materials. Air Force Materials Laboratory Report No. AFML-TR-76-81, Part IV, Wright Patterson Air Force Base, OH.

Reiner M and Weisenberg K. (1939). A Thermodynamic Theory of the Strength of Materials. Rheology Leaflet, No. 10, pp. 12-19. Christensen, R. M., (1980), Report No. UCRL-84532, Lawrence Livermore Laboratory, Livermore, California.

Schapery R. A. (1969). On the Characterization of Nonlinear Viscoelastic Materials. *Polymer Engineering Science*, 9:295-310.

Schapery R. A. (1974). Viscoelastic Behavior and Analysis of Composite Materials. *Mechanics of Composite Materials*, Edited by G. P. Sendeckyj. Academic Press, pp. 85-168.

Wang S. S., Chim E. S. M., and Zahlan N. M. (1983). Fatigue Crack Growth in Random Short-Fiber SMC Composite. *Journal of Composite Materials*, Vol. 17, May, pp. 250-266.

Wang S. S., Goetz D. P., and Corten H. T. (1984). Shear Fatigue Degradation and Fracture of Random Short-Fiber SMC Composite. *Journal of Composite Materials*, January, pp. 2-20.

Yen S. C. (1984). Creep and Creep Rupture of SMC-R50 Under Different Thermomechanical Conditions. Ph.D. dissertation, VPI & SU, Blacksburg, Virginia.

Yen S. C., Hiel C., and Morris D. H. (1985). Viscoelastic Response of SMC-R50 Under Different Thermomechanical Conditions. High Modulus Fiber Composites in Ground Transportation and High Volume Applications, Ed. by D. W. Wilson, ASTM STP 876, American Society for Testing Materials, pp. 131-143.

Yen S. C. and Morris D. H. (1989). Accelerated Characterization of a Chopped Fiber Composite Using a Strain Energy Failure Criterion. *Polymer Composites*, Vol. 10, No. 4, pp. 249-255.

Yen S. C. and Williamson F. L. (1990). Accelerated Characterization of Creep Response of An Off-axis Composite Material. *Composites Science and Technology*, Vol. 38, No. 2, pp. 103-118.

Yen S. C., Chu T. C. P. and Jao H. (1991). Applications of Optical Microscopic Image Analysis to Composite Materials. *Experimental Techniques*, Vol. 15, No. 5, pp. 22-25.

Yen S. C. and Morris D. H., (1992). Effect of Damage on the Creep Response of a Chopped

Fiber Composite Material. *International Journal of Damage Mechanics*, Vol. 1, No. 3, pp. 367-382.

Yen S. C., Jao H., Chiang J. H., and Huang C. Y., (1993). Correlation Between Fiber Angle Distribution and Mechanical Properties of Chopped Fiber Composite Using Image Analysis. Proceedings, Conference on Advanced Technology in Experimental Mechanics, Kanazawa, Japan, pp. 73-76.

Yeow Y. T., Morris D. H., Brinson H. F. (1979). Time Temperature Behavior of a Unidirectional Graphite/Epoxy Composite. Composite Materials: Testing and Design (5th Conference), Edited by S. W. Tsai, *ASTM STP 674*, 1979, pp. 263-281.

Long Term Durability of Structural Materials
P.J.M. Monteiro, K.P. Chong, J. Larsen-Basse, K. Komvopoulos (Eds)

ACCELERATED TESTING FOR THE DURABILITY OF COMPOSITE MATERIALS AND STRUCTURES

Yasushi Miyano[1], Stephen W. Tsai[2], Richard M. Christensen[2], and Akira Kuraishi[2]

[1] Materials System Research Laboratory, Kanazawa Institute of Technology, Matto, Ishikawa 924-0838, Japan
[2] Department of Aeronautics and Astronautics, Stanford University, Stanford, CA 94305-4035, USA

ABSTRACT

Developing a testing procedure to establish the lifetimes of materials in extreme service environments is becoming a high priority. With service lifetimes measured in years, it is almost unthinkable to do real time testing under a variety of conditions. Existing accelerated testing methodologies for metals cannot be simply applied to composite materials, since these methodologies are not intended for viscoelastic materials such as polymeric composite materials, which exhibits strong time and temperature dependencies. Our accelerated testing methodology is based on the time-temperature superposition principle for polymeric materials. This principle was originally developed for non-destructive material properties, but recent studies have shown that it can also be applied to failure properties of composite materials. Using this principle as the building block, we have developed a methodology to predict the long-term life of composite materials, such as the static (constant strain rate) strength, creep life, and fatigue life. The details of the methodology and its four required conditions are shown in this paper. The range of applicability of the methodology has been studied for various material systems and test configurations, and was found applicable to most typical polymeric composite materials.

KEYWORDS

durability, fatigue, creep, viscoelastic, time-temperature superposition principle

ACCELERATED TESTING METHODOLOGY

Accelerated testing methodologies for metals have been studied for some time, but not enough studies have been performed on composite materials. One of the most popular tools used to predict the fatigue life of metals is the S-N curve, which is based on the assumption that fatigue life depends on cycles, but not on time. Accelerating the test is an easy task using the S-N curves, since the cyclic loads can be applied at much higher frequencies than the actual loading.

Unlike metals, polymeric composite materials are viscoelastic and their properties exhibit strong time and temperature dependency. For this reason, simply applying the S-N curve to composite materials will not provide accurate prediction of the fatigue life. Accelerating the test becomes a difficult task, since time plays an important role in the fatigue and creep of composite materials. Therefore, we need to develop a new accelerated testing methodology more suited for composite materials.

Our accelerated testing methodology is based on the time-temperature superposition principle of polymeric materials. This principle has been widely employed to characterize the nondestructive properties, and recently, has shown remarkable success in characterizing the failure properties of polymeric composite materials. In this case, elevated temperature states are used to accelerate the mechanical degradations, which occur under loads over long period of time. Using this as our building block, we have developed a methodology to predict the long-term life of polymeric composite materials under various temperatures and loading conditions, such as static (constant strain rate, CSR), creep, and fatigue loading.

Figure 1 shows the outline of our accelerated testing methodology. The four hypotheses shown at the middle of Figure 1 are the conditions that must be met in order for the methodology to work properly. These hypotheses will be tested in order to validate and possibly provide limit to the applicability of the methodology.

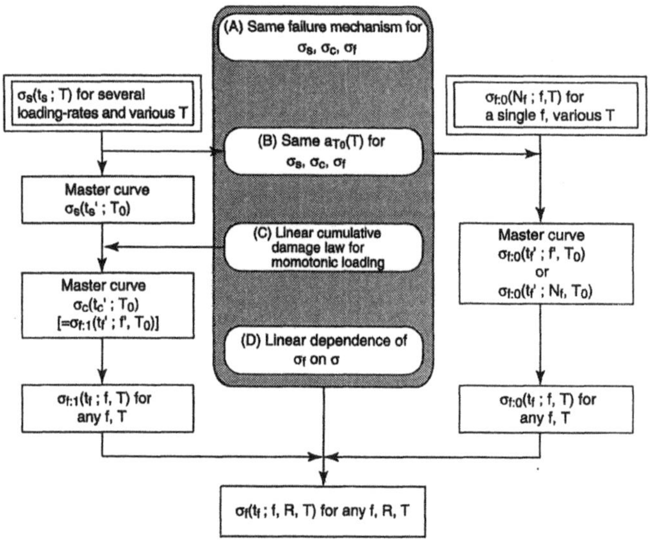

T, T_0 : temperature, reference temperature
f, f' : frequency, reduced frequency
t_s, t_c, t_f : time to failure under CSR(Constant Strain Rate), creep and fatigue loadings
t_s', t_c', t_f' : reduced time to failure
$a_{T_0}(T)$: time-temperature shift factor ($a_{T_0}(T)=t_s/t_s'=t_c/t'_c=t_f/t_f'=f'/f$)
R : stress ratio ($R=\sigma_{min}/\sigma_{max}$)
N_f : number of cycles to failure ($N_f=f\cdot t_f$)
$\sigma_s, \sigma_c, \sigma_f$: CSR, creep and fatigue strength
$\sigma_{f:0}, \sigma_{f:1}$: σ_f for R=0 and R=1

Figure 1: Accelerated testing methodology

Hypothesis A states and assumes that the failure mechanisms under static, creep, and fatigue loading are identical. This allows us to relate these three different loading types as shown in Figure 2. Here, a creep loading is considered as a fatigue loading with stress ratio of $R=\sigma_{min}/\sigma_{max}=1$, and a static (CSR) loading is considered to be equivalent to a half cycle of fatigue loading with R=0.

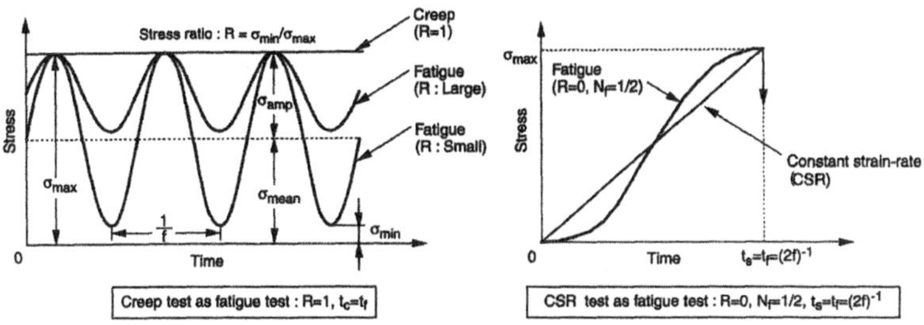

Hypothesis A: Same failure mechanisms for static, creep, and fatigue failure

Figure 2: Relationship among static (CSR), creep and fatigue loading

Figure 3 shows how the static test data for different temperatures are shifted to form a smooth curve called the master curve. The amounts of these shifts are called time-temperature shift factors. In the plot of the master curve, the vertical axis is the strength, and the horizontal axis is the combination of temperature and time to failure, which we call the reduced time to failure. The significance of this master curve is that it can be used to predict the strength under any combinations of temperature and time to failure.

Hypothesis B states that the same time-temperature superposition principle applies for all three types of strengths (static, creep, and fatigue). In other words, we can use the same values of shift factors for all three types of strengths.

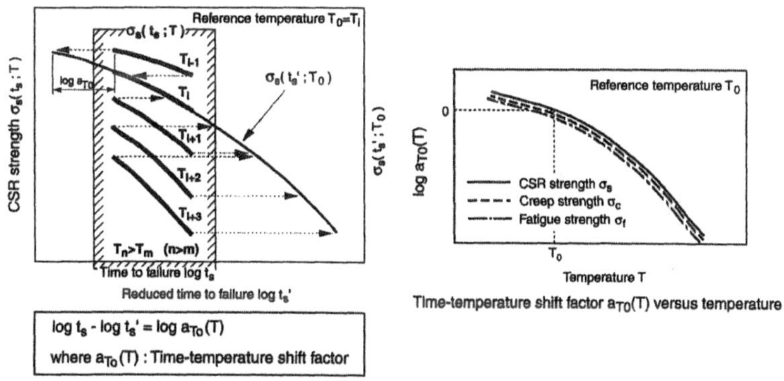

Hypothesis B: Same time-temperature superposition principle for all strengths

Figure 3: Master curves of static (CSR) strength and time-temperature shift factor

Hypothesis C states that the linear cumulative damage law can be used for monotonic loading. This allows us to estimate the life under complex loading, by summing up the damages for individual load steps. Figure 4 shows how the creep life is calculated from the static test results using this law.

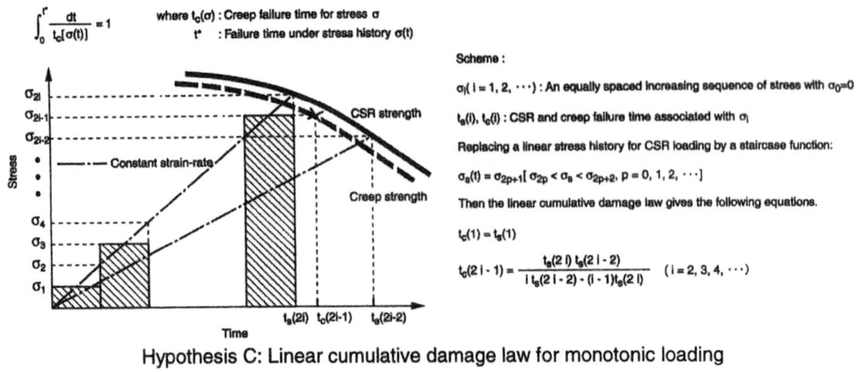

Figure 4: Calculation of creep strength from static (CSR) strength

Figure 5 shows how the prediction curves of the fatigue life are constructed from the static master curve and an S-N curve, according to the Hypotheses A and B. The lower right figure shows the fatigue master curves, which are plotted with the applied load amplitude as the vertical axis, and the reduced time to failure as the horizontal axis. Different curves represent different cycles to failure. These fatigue master curves can be used to predict the fatigue life under any combinations of temperature, time to failure, and cycles to failure. Note that the results are applicable only to the case of stress ratio R=0.

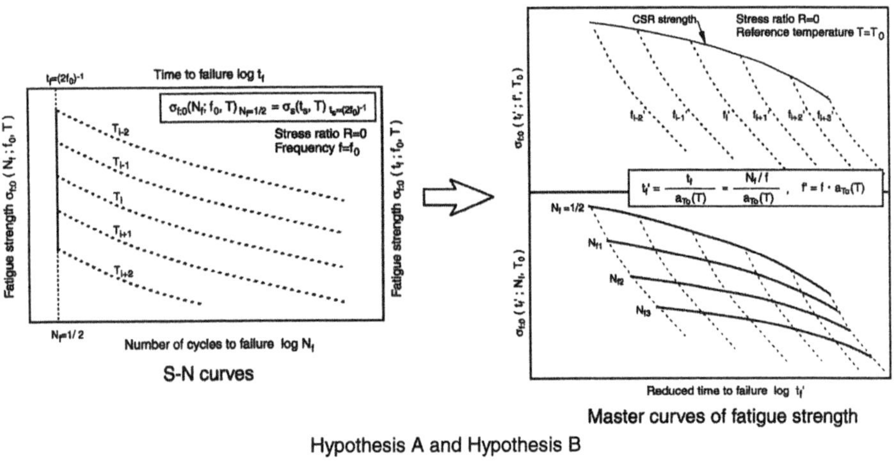

Figure 5: Master curves of fatigue strength

Our last hypothesis, Hypothesis D, states the linear dependence of fatigue strength upon the stress ratio R. Using this hypothesis, we can plot fatigue master curves for various stress ratio values, as shown in Figure 6.

<u>Information Available at This Stage</u>

(a) The fatigue strength $\sigma_{f:1}(t_f'; T_0)$ for stress ratio R = 1
where t_f' : reduced time to failure at reference temperature T_0

(b) The fatigue strength $\sigma_{f:0}(t_f', N_f, T_0)$ for stress ratio R = 0

Fatigue strength, $\sigma_f(t_f; R, f, T)$ at an arbitrary stress ratio R, frequency f, and temperature T

$$\sigma_f(t_f';R,f',T_0) = \sigma_{f:1}(t_f';f',T_0)R + \sigma_{f:0}(t_f';f',T_0)(1 - R)$$

or

$$\sigma_f(t_f;R,f,T) = \sigma_{f:1}(t_f;f,T)R + \sigma_{f:0}(t_f;f,T)(1 - R)$$

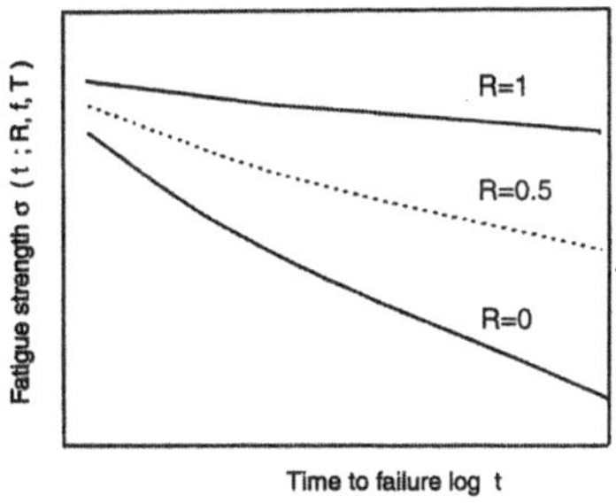

Hypothesis D: Linear Dependence of fatigue strength upon stress ratio

Figure 6: Fatigue strength at arbitrary stress ratio

Most commonly used methods of fatigue life prediction are "point-design", which can only predict fatigue life under certain loading condition and temperature. The result can not be used for different loading conditions or temperatures, making it less useful for the real design. Building a useful database for design can be very expensive and time-consuming, if not impossible.

When all four hypotheses hold, our methodology can predict life under any loading conditions, temperature, time to failure, and cycles to failure. By this method, building a useful database for design can be done in relatively short period of time. The methodology can also be used for material screening and selections.

Table 1 shows an example of the test plan to obtain the fatigue master curves for the bending of a CFRP laminate. The static test, which is used to obtain the shift factor, requires 4 weeks, and the fatigue test, which is used to obtain the S-N curve that we shift, requires 12 more weeks.

TABLE 1

TEST PLAN FOR BENDING TESTS OF CFRP LAMINATES

Static Test	Loading rate:	3 steps (0.01, 1, 100mm/min)
	Temperature:	5 steps (RT~120C)
	Specimens at each step:	5 specimens
	Total number of specimens:	75 specimens
	Testing period:	4 weeks
Fatigue Test	Stress ratio:	0.05
	Maxium cyles:	10^6
	Frequency:	1 step (f=5Hz)
	Temperature:	4 steps (RT~100C)
	Specimens at each step:	20 specimens
	Total number of specimens:	80 specimens
	Testing period:	12 weeks

EXAMPLE

The methodology has been tested for various test configurations, ranging from basic properties to complex systems such as bonded and bolted joints. Let us look closely at the recently obtained results of our bolted joint system.

Figure 7 shows the configuration of the bolted joint system. The bolts connect a GFRP tube and steel rods. Two bolts of different sizes were used so that the lower bolt will always fail first. The tube configuration was chosen, since this configuration allows us to apply compressive loading without any buckling.

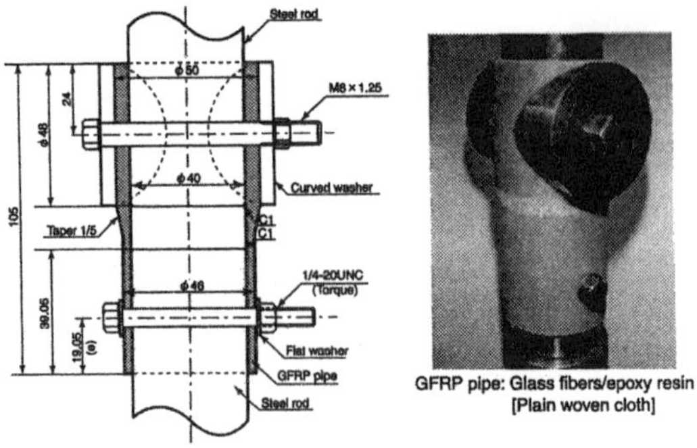

Figure 7: GFRP/metal bolted joint system

Three types of tests were performed, static (constant elongation rate, CER) tests, creep tests, and fatigue tests. Figure 8 and 9 show the fracture appearances of the bolted joints, tested at 25°C and 120°C, respectively. We can observe that the failure mechanisms are identical for the three loading conditions. This shows that Hypothesis A holds for this material and configuration.

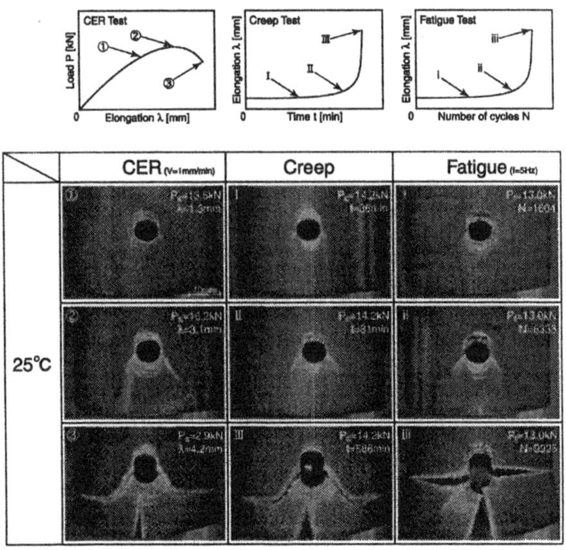

Figure 8: Fracture appearances of GFRP/metal bolted joint during tests at 25°C

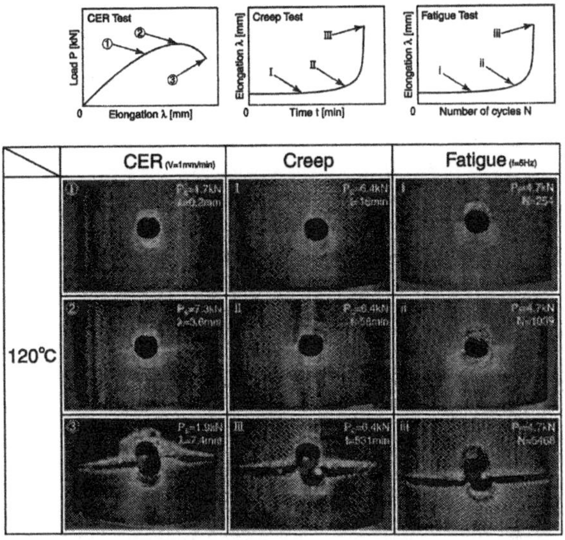

Figure 9: Fracture appearances of GFRP/metal bolted joint during tests at 120°C

Figure 10 is the master curve of static failure load for the bolted joints. The superposition of the individual segments forming a smooth master curve indicates that the time-temperature superposition is applicable to this case. Figure 11 shows the shift factors used to create this master curve. Note that the shift factors for the strength of the bolted joints are almost identical to the shift factors for the storage modulus of the same GFRP material. This indicate a relationship between the viscoelastic effect on the stiffness and that on the strength.

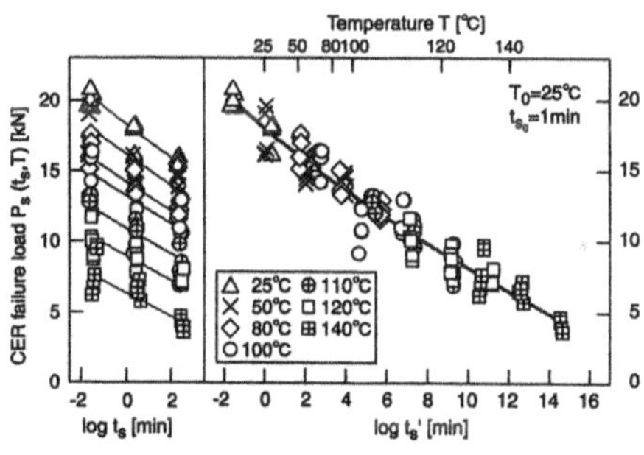

Figure 10: Master curve of static (CER) failure load for GFRP/metal bolted joint

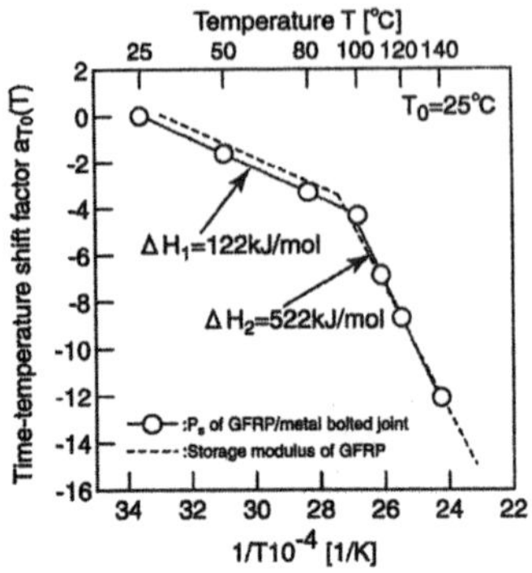

Figure 11: Time-temperature shift factors for static (CER) failure load for GFRP/metal bolted joint

Figure 12 shows the creep test results and the calculated creep master curve using the cumulative damage law. The prediction agrees well with the test results, which shows that Hypothesis C is applicable to this case.

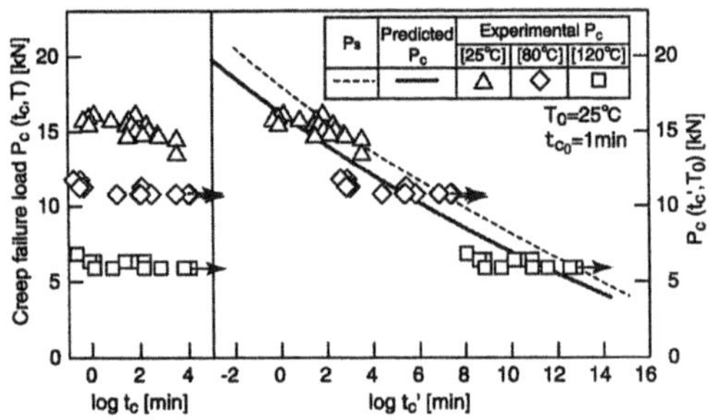

Figure 12: Master curves of creep failure load for GFRP/metal bolted joint

Figure 13 shows the S-N curves of the bolted joints under fatigue loading with frequencies of f=5Hz and stress ratio of R=0. Note that the applicability of this results is limited to this particular loading condition, and temperatures.

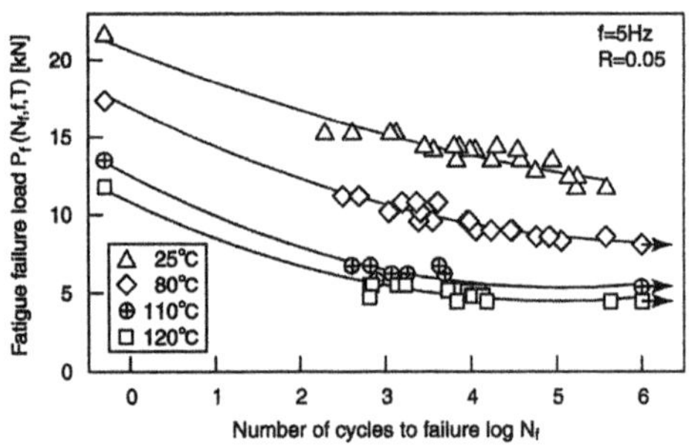

Figure 13: S-N curves of GFRP/metal bolted joint at f=5Hz

By combining the previous static master curve and the fatigue S-N curve, we can create the fatigue master curves shown in Figure 14. Although limited to the stress ratio of R=0, this curve can be used to predict fatigue life under any temperature, time to failure, and cycles to failure. Note that the curves for different cycles to failure overlap on top of each other. This indicate that for this bolted joint system, fatigue life do not depend on cycles to failure. We observed similar trend in several different cases, some of which are shown in the next section.

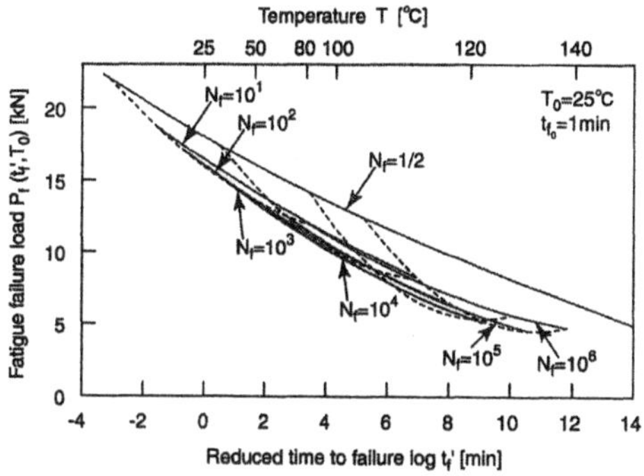

Figure 14: Master curves of fatigue failure load for GFRP/metal bolted joint

Figure 15 shows the S-N curves for a different loading frequencies of f=0.05Hz. When the cycles to failure are the same, the time to failure at this frequencies will be 100 times that for f=5Hz. The prediction agrees well with the actual test results, which shows that Hypothsis B is applicable to this case.

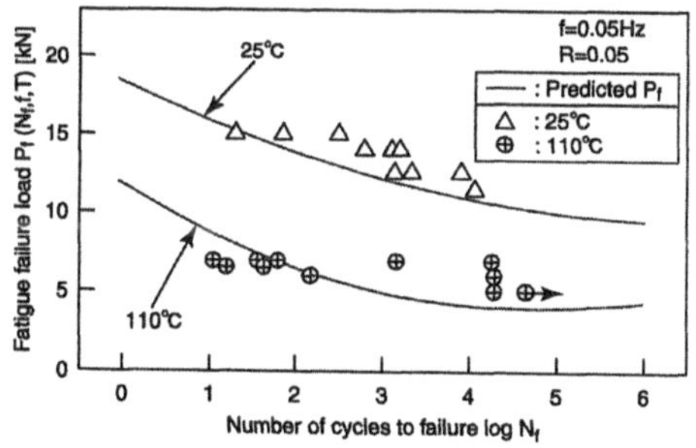

Figure 15: S-N curves of GFRP/metal bolted joint at f=0.05Hz

Figure 16 shows the prediction and the actual test results for the stress ratio of R=0.5. The prediction curves were plotted by interpolating the fatigue master curves for R=0, and creep master curves which are considered as R=1. The predictions and the test results agree well, showing that Hypothesis D is applicable to this case.

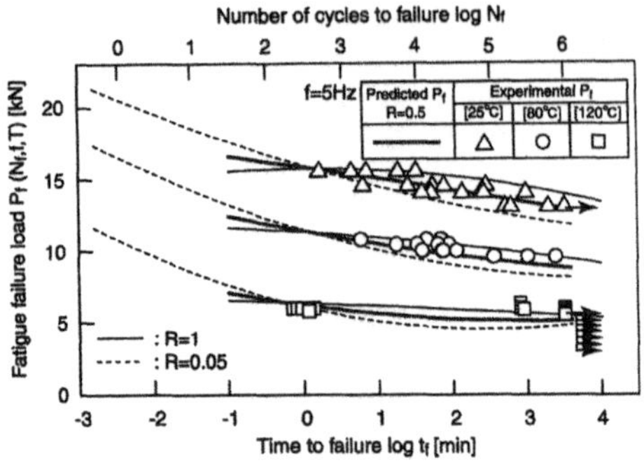

Figure 16: S-N curves of GFRP/metal bolted joint at load ratio R=0.5

RANGE OF APPLICABILITY

The methodology has been tested for various materials systems and test configurations. The methodology works well for most typical polymeric composites, and for most test configurations. The studied cases and the range of applicability are summarized in Table 2. Not all the hypotheses are met in some cases, which does not completely eliminate the usefulness, but merely limits the applicability of the methodology.

TABLE 2
APPLICABILITY OF PREDICTION METHOD

Fiber		Matrix	Type	Fiber/matrix	Loading direction	Hypothesis (A)	(B)	(C)	(D)
Carbon	PAN	Epoxy	UD	T400/828	LT	○	○	○	○
				T300/2500	LB	○	○	○	○
					TB	○	○	○	○
			SW	T400/3601	LB	○	○	○	○
		PEEK	UD	T300/PEEK	LB	○	×	×	△
					TB	△	×	×	×
	Pitch	Epoxy	UD	XN40/25C	LB	○	△	×	○
Glass		Epoxy	SW	E-Glass/Epoxy	LB	○	△	○	×

Notice UD : Unidirectional SW : Stain Woven LT : Longitudinal Tension
LB : Longitudinal Bending TB : Transverse Bending

FRP Joint System	Hypothesis (A)	(B)	(C)	(D)
Conical Shaped Joint of GFRP/Metal	○	○	○	○
Brittle Adhesive Joint of GFRP/Metal	○	○	○	○
Ductile Adhesive Joint of GFRP/Metal	○	○	○	○
Bolted Joint of GFRP/Metal	○	○	○	○

Figures 17 and 18 show some examples of fatigue life predictions of the cases that we have studied. Figure 17 shows the results for the three basic test configurations of CFRP specimens. Horizontal curves indicate weak time dependence, and overlapping curves indicate weak dependence on number of cycles. The important thing to note is that even in the fiber direction, composite materials exhibit time and temperature effect, since the viscoelastic behavior of the matrix affects the overall behavior of composites.

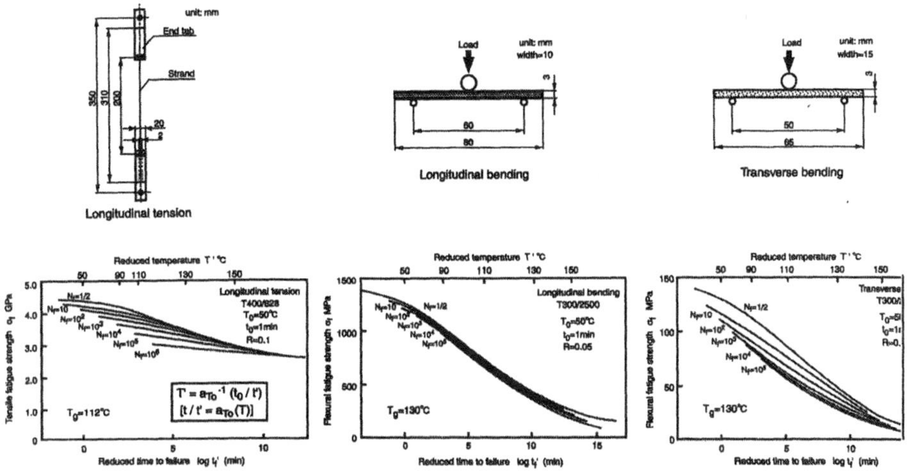

Figure 17: Master curves of fatigue strength of various CFRP specimens

Figure 18 shows the results for the three different joint configurations of GFRP/metal bonded joints. The first example is the conical shaped joint for composite rods, and the other two are the tubular bonded joints, one using a brittle adhesive and the other using a ductile adhesive. It is interesting to note the large difference in the results.

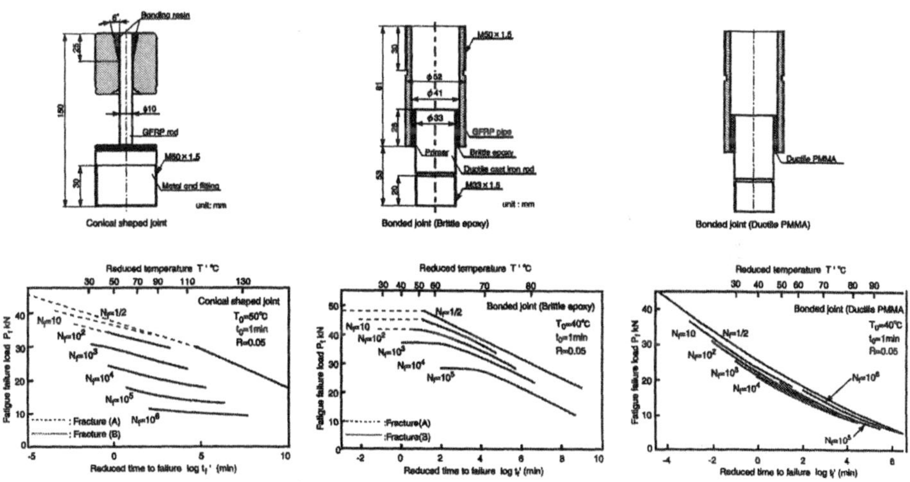

Figure 18: Master curves of fatigue failure load of various GFRP/metal joints

CONCLUSION

A prediction methodology is given for the long-term fatigue strength of polymer composites at an arbitrary stress ratio, frequency, and temperature.

From our experimental findings:
(1) PAN-based CFRP and GFRP/metal joint meet the four hypotheses on which the method is based, regardless of the structural configuration and loading style. The long-term fatigue strength of these composites and structures can be predicted based on the methodology.
(2) The master curves of fatigue strength based on time temperature superposition principle well describe the influence of time, temperature and number of cycles to failure.
(3) The fatigue strength of GFRP bolted joints depends strongly on time and temperature, but less so on the number of cycles to failure and stress ratio.

FUTURE WORK

The proposed methodology has effectively combined the effects of time and temperature on the strength and life of composite materials. Our next goal is to find similar relationship using moisture or other agents. A methodology with time-temperature-moisture superposition will enable us to predict life under any temperature and moisture condition, and also allows us to use moisture and other agents to accelerate the tests.

Our methodology is an extrapolation process in terms of time, and the accuracy and the reliability of the prediction are our major concerns. We are reviewing our methodology from statistical point of view, to maximize the accuracy and the reliability of the prediction, and to create confidence interval of our master curves, which will be important information for the designers.

REFERENCES

Miyano, Y., Nakada, M., Kudo, H. and Muki, R. (1999). Prediction of Tensile Fatigue Life under Temperature Environment for Unidirectional CFRP, *Advanced Composite Materials* **8,** 235~246.

Miyano, Y., Nakada, M., Kudo, H. and Muki, R. (2000). Prediction of Tensile Fatigue Life for Unidirectional CFRP, *Journal of Composite Materials* **34,** 538-550.

Miyano, Y., Nakada, M., McMurray, M. K. and Muki, R. (1997). Prediction of Flexural Fatigue Strength of CFRP Composites under Arbitrary Frequency, Stress Ratio and Temperature, *Journal of Composite Materials* **31,** 619~638.

Miyano, Y., Nakada, M. and Muki, R. (1999). Applicability of Fatigue Life Prediction Method to Polymer Composites, *Mechanics of Time-Dependent Materials* **3**, 141-157.

Miyano, Y., Nakada, M., Yonemori, Y., Sekine, N., and Tsai, S. W. (2000). Time-Temperature Dependence of Tensile Fatigue Strength for GFRP Joint Systems, *Proceedings of 3rd International Conference on Mechanics of Time Dependent Materials*, 194~196.

Yonemori, T., Nakada, M., Miyano, Y. and Tsai, S. W. (1999). Time and Temperature Dependence on Failure load of GFRP/Metal Bolted Joints, *Proceedings of The 6th Japan International SAMPE Symposium at Tokyo, Japan, October 1999* **2**, 1201~1204.

Long Term Durability of Structural Materials
P.J.M. Monteiro, K.P. Chong, J. Larsen-Basse, K. Komvopoulos (Eds)

A UNIFIED APPROACH TO PREDICTING LONG TERM PERFORMANCE OF ASPHALT-AGGREGATE MIXTURES

Y. Richard Kim[1], Roy H. Borden[1], and Murthy Guddati[1]

[1]Department of Civil Engineering, North Carolina State University, Raleigh, NC

ABSTRACT

An asphalt-aggregate mixture is a viscoelastic particulate composite whose deformation behavior is one of the most significant factors contributing to common load-related distresses, such as rutting (permanent deformation) and fatigue cracking, in asphalt concrete pavement systems. The principal objective of this research is to develop test methods and models for predicting long term performance of asphalt concrete that can account for viscoelasticity, damage, volumetric/deviatoric coupling, temperature, and aging under realistic cyclic loading conditions. Preliminary results from constant crosshead rate axial tests at varying temperatures indicate that the time-temperature superposition is valid in asphalt concrete even with a significant level of damage. In support of the continuum damage modeling, discrete element code is developed with an objective of establishing a methodology for modeling asphalt concrete behavior on the basis of composition, microstructure, and interfacial properties at the particle level. Preliminary results from the third scale Model Mobile Load Simulator (Mk. 3) and the surface wave technique demonstrate the promise of the combination of these techniques as a tool for accelerated pavement testing and evaluation.

KEYWORDS

Asphalt concrete, discrete element model, viscoelasticity, time-temperature superposition, continuum damage model, accelerated pavement testing, surface wave test

INTRODUCTION

Asphalt concrete pavement, one of the largest infrastructure components in our nation, is a complex system that involves multiple layers with different materials and combinations of irregular traffic loading and varying environmental conditions. Deterioration of asphalt concrete pavements is mostly related to two common load-related distresses, rutting and fatigue cracking. Rutting is longitudinal depression in the wheel paths usually accompanied by small upheavals to the sides. Recent studies (SHRP 1994, Kim 1994) have proven that rutting is primarily due to permanent deformation in asphalt concrete layers, which is attributable to volume change and shear flow in the mixture. Fatigue cracking on the pavement surface occurs as a result of microcrack initiation, propagation, and coalescence due to repetitions of traffic loading and/or temperature cycling over extended periods of time. Fatigue cracking is also governed by the deformation behavior of the asphalt concrete layer, among all the layers in the pavement system. Therefore, the performance of asphalt concrete pavements is closely related to the performance of asphalt concrete. In order to predict the performance of asphalt concrete with reasonable accuracy, better understanding of its deformation behavior under realistic conditions is urgently needed.

Asphalt concrete is a viscoelastic particulate composite that consists of aggregate particles and asphalt cement matrix. When the asphalt-aggregate composite is subjected to repeated traffic loading at lower temperatures, distributed microstructural damage occurs primarily in the form of microcrack nucleation and growth due to the embrittled binder and high stress concentrations along the aggregate-binder interfaces. Therefore, for this type of damage the role of the binder and the variables that influence the properties of the binder (e.g., aging, adhesion, etc.) become important. At higher temperatures, the asphalt binder becomes too soft to carry the load and therefore the principal type of damage is permanent deformation due to volume change (i.e., densification) and rearrangement of aggregate particles. Therefore, a reliable performance prediction model should account for the effects of various constitutive factors that affect the aggregate-binder and aggregate-aggregate interactions.

Developing a realistic mathematical model of the mechanical behavior of asphalt concrete with growing damage is a complicated problem. The complexity is attributed to the viscoelastic hereditary effects of the binder, complex nature of describing the damage evolution, and the coupling between these two mechanisms. Even though this type of model is developed, verification of the model in actual pavements is difficult, especially for fatigue cracking, because the fatigue performance in pavements is described by the extent of the cracked area whereas in the model, it is represented by mechanistic parameters, such as stiffness. In a recently completed FHWA project (DTFH 61-92-C-00170), the PI has made some advancements in the area of constitutive modeling and in-situ testing of asphalt concrete. A uniaxial viscoelastic continuum damage model has been developed by applying the elastic-viscoelastic correspondence principle to separate out the effect of viscoelasticity and then employing internal state variables based on work potential theory to account for damage evolution under loading and microdamage healing during rest periods. Through the verification study, it was found that the constitutive model has an ability to predict the hysteretic behavior of the material under both monotonic and cyclic loading up to failure, varying loading rates, random rest durations, multiple stress/strain levels, and different modes of loading (controlled-stress versus controlled-strain).

Another accomplishment in the FHWA project was the development of a nondestructive technique for in-situ characterization of fatigue and healing of asphalt concrete in flexible pavements. This technique is based on the surface wave test method and dispersion analysis techniques developed at NC State University. It was demonstrated, using the SHRP pavements at the Turner-Fairbank Highway Research Center and Mn/ROAD pavements, that asphalt concrete regains a significant portion of its original strength during rest periods, which in turn lengthens the fatigue life of asphalt concrete pavements. Another important finding is that this material loses its structural integrity in the early stages of its fatigue life, long before cracks appear on the pavement surface.

Therefore, the current project is building upon the previous FHWA project, to develop prediction models for fatigue and rutting of asphalt concrete by accounting for the effects of aging, temperature, and multiaxial states of stress and then to verify these models through in-situ testing of pavements. Both laboratory and in-situ characterization of the effects of fundamental damage processes on thermomechanical behavior are being addressed. The principal objective of this research is to develop a comprehensive and unified approach to predicting long term performance of asphalt-aggregate mixtures by expanding the capability of both laboratory and in-situ performance prediction test methods and models for fatigue cracking and rutting of asphalt concrete. The outcome of this research will lead to better understanding of dissipative deformation behavior of asphalt concrete in more realistic environmental conditions and states of stress in relation to permanent deformation and fatigue cracking, and therefore more accurate prediction of performance of asphalt concrete pavements. The resulting laboratory test methods and models will provide engineers a means of developing or selecting mixtures more resistant to permanent deformation and fatigue cracking. Also, the ability of the in-situ test methods will allow engineers to monitor deterioration of the asphalt concrete layer in the pavement system, which will enhance the engineers' understanding of when to rehabilitate or how to prolong the pavement's useful lifetime.

COMPUTATIONAL MODELING OF ASPHALT-AGGREGATE MIXTURES

Considering the microscopic nature of mechanical interaction between asphalt and aggregates, micro-mechanical modeling methods can be utilized to obtain a reliable computational tool to characterize asphalt concrete. In this effort, a discrete element modeling tool has been developed for the analysis of asphalt concrete. In what follows, description of the tool, underlying discrete element method, limitations and future extensions are discussed.

The Discrete Element Method

Discrete element modeling (DEM) involves approximating the media of analysis using discrete granular elements mechanically interacting with each other. DEM is thus ideally suited for the simulation of granular materials such as soils. It is also possible to use DEM to understand micro-mechanical behavior of continuous heterogeneous media such as asphalt concrete by discrete element approximation. Such understanding is expected to shed some light on complex phenomena of damage growth involving fracture, void collapse and other microstructure evolutions.

There are two critical aspects in discrete element modeling: approximation of continuous media with a discrete element mesh, and modeling the interaction between neighboring elements. The discretization process involves the choice of element size, shape and arrangement. Element sizes are determined by the scale of micro-mechanical activity, and in the case of asphalt concrete it is generally at meso-scale (around 10 microns). The choice of element shape depends on internal microstructure – in the case of asphalt concrete it appears sufficient to use circular discs (2-D) or spheres (3-D). The arrangement can be ordered or disordered (random). In our study, ordered arrangements are used with varying material properties to simulate heterogeneous asphalt concrete. Once the discretization is complete, the interaction model is determined based on the mechanical properties of the constituent materials. In the remainder of the section, our approach of micro mechanical analysis of asphalt concrete is outlined.

Discrete Element Modeling Tool for Asphalt Concrete

A software tool has been developed to utilize the discrete element modeling procedure to analyze asphalt concrete. The tool includes not only the numerical discrete element simulator, but also a preprocessor that enables easy simulation of realistic specimens. The structure of the software tool is shown in Figure 1.

The preprocessor, written in Visual Basic, is used to convert a realistic asphalt concrete specimen to discrete element models that are then analyzed using a discrete element simulator written in FORTRAN. The underlying ideas behind these programs are explained in the next two sections.

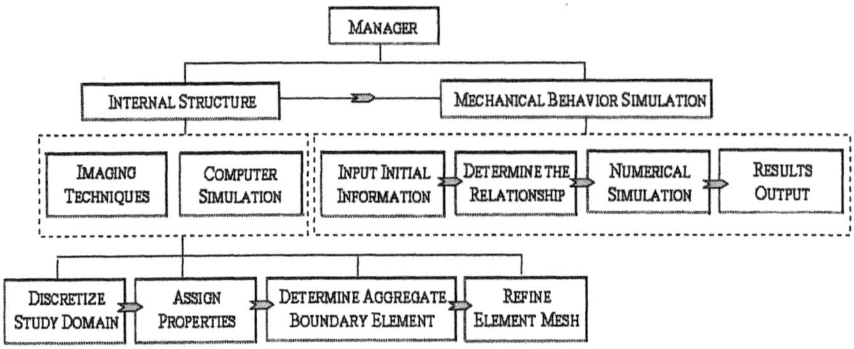

Figure 1: Structure of the discrete element model

Approximation of continuous medium as discrete elements (Preprocessor)

Asphalt concrete is a multi-phase material consisting mainly of three phases – asphalt, aggregate and air voids. A discrete element model of such material essentially involves two types of discrete elements – asphalt elements and aggregate elements. The air voids are simulated by simply not having any elements. For a realistic asphalt concrete specimen, one often requires a rather complex and large discrete element mesh that is extremely time consuming to create manually. The preprocessor contains algorithms that enable automated and realistic generation of discrete element "mesh". A unique aspect of our discrete element representation is that each aggregate particle is represented not by a "clump" of discrete elements, but by "boundary representation", i.e., discrete elements are used only on the boundary of the aggregate particles and not in the interior. This is justified because individual aggregate particles are rigid and will not generally be fractured. Such a boundary representation significantly reduces the number of discrete elements, and thus reduces the computational cost.

The preprocessor includes two different approaches to generate the discrete element mesh. The first approach involves imaging the actual microstructure of asphalt concrete specimens. This method will be useful in verifying/calibrating the DEM against the responses measured from actual tests. The second approach involves statistically generating the microstructure using mix design information, such as aggregate gradation, asphalt content and air void content.

Once the microstructure of asphalt concrete specimen is generated from one of the two approaches described above, a discrete element mesh is generated from the microstructure using the procedure depicted in Figure 2.

Modeling the interaction between elements

Once the discrete element "mesh" is generated, it is necessary to establish models for the interaction between adjacent elements. Proper representation of this interaction is key to accurate simulation of mechanical behavior of asphalt concrete. The following paragraphs outline different types of interaction models relating the inter-element force with distance between the elements.

Aggregate-aggregate interaction: The interaction between aggregate elements can be further divided into two sub-categories – interaction between elements in the same aggregate particle, and between two different aggregate particles. In the former case, a linear elastic spring is assumed to be linking these elements. In the latter case, however, a more complicated nonlinear spring is used to model the separation or contact between two aggregate elements (Figure 3).

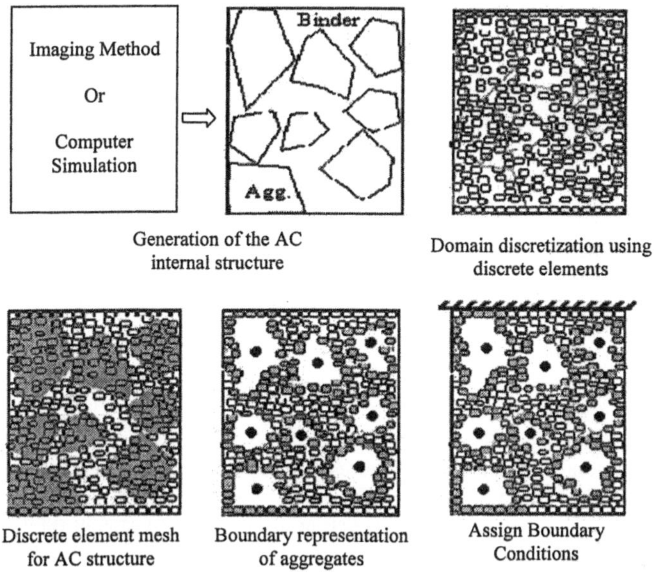

Figure 2: Generation of discrete element mesh

Figure 3: Aggregate-aggregate interaction models:
(a) Linear spring for interaction between elements in the same aggregate;
(b) Spring-gap model for interaction between elements from two different aggregates.

Asphalt-asphalt interaction: The interaction between two asphalt binder elements is more complex in nature due to the deformability of these elements. Noting the viscoelastic nature of asphalt binder, a viscoelastic spring is used for the interaction. Also, noting that two particles cannot indefinitely deform when they are compressed together, a Hertzian contact law is coupled with the viscoelastic spring to more accurately represent the interaction. Figure 4 illustrates the hybrid model used asphalt-asphalt interaction.

Essentially, before two neighbor elements are compressed together, i.e., $d_{ij} \geq 2r$, the interaction is viscoelastic in nature:

$$F_{ij}(t) = \int_0^t E(t-\tau)\frac{d}{d\tau}\left(d_{ij}\right)d\tau \qquad for\ d_{ij} > 2r.$$

In the above, F_{ij} indicates the interaction force between elements i and j, d_{ij} is the distance between the elements, and E is reminiscent of relaxation modulus and is represented by Prony series.

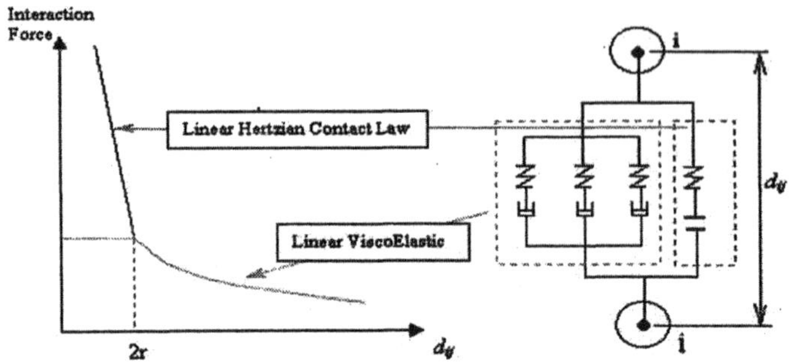

Figure 4: Binder-binder interaction model

When the elements are compressed together ($d_{ij} < 2r$), a penalty function (stiff spring) is used to simulate the impenetrability:

$$F_{ij} = Kd_{ij} \qquad for\ d_{ij} < 2r.$$

Here, K is a large spring stiffness taken as a thousand times the maximum physical spring stiffness encountered in the entire discrete element model.

Asphalt-aggregate interaction: A model similar to asphalt-asphalt interaction model is used for this purpose. This is justified because most of the deformation is in the binder, which indicates that the deformation as a function of the interaction force is half as much as that of asphalt-asphalt model, i.e.,

$$F_{ij}^{Aggregate-Binder}\left(d_{ij}\right) = F_{ij}^{Binder-Binder}\left(2d_{ij}\right).$$

Verification of the model

A series of numerical experiments are conducted to verify the discrete element modeling approach and the software tool. One of such experiments is a typical creep test simulation (Figure 5). A small asphalt concrete element is subjected to uniform step load, and the average displacement is observed. It can be seen that the discrete element model appropriately predicts the expected creep response. Many other tests have been performed and further verification is underway.

Current Limitations and Future Extensions

To date, a major part of the discrete element model, including the simulation of internal structure, has been completed. The Fortran program has been verified using some standard examples in the linear range. However, it appears that the current implementation of DEM can not simulate nonlinear cracking phenomena in a realistic manner. It appears that a better representation of the interaction between elements is necessary. After careful evaluation, it has been decided that a lattice-based micro-mechanical model may be better suited for modeling asphalt concrete. Such an analysis of asphalt concrete is currently underway.

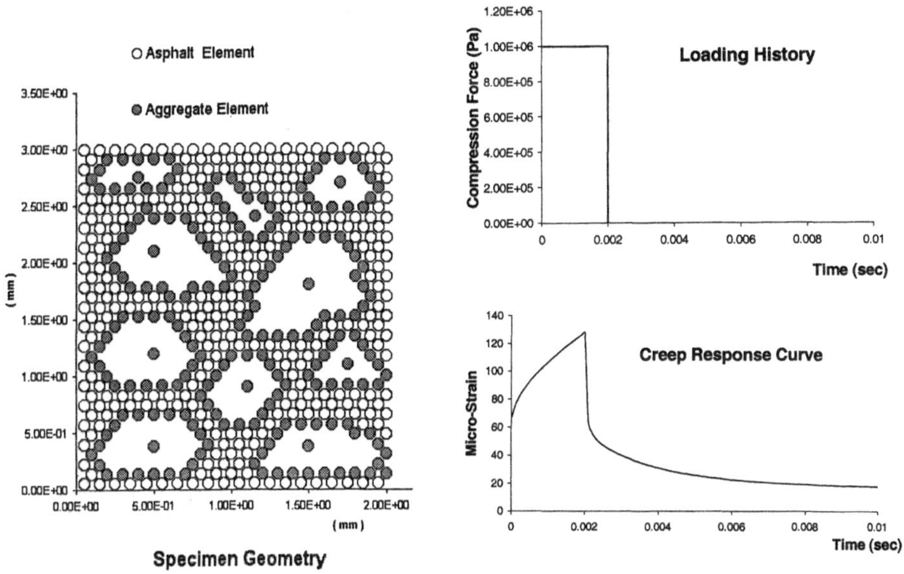

Figure 5: Numerical verification: creep test simulation using discrete element model

VISCOELASTIC, CONTINUUM DAMAGE MODELING

It has been shown that asphalt concrete in its linear viscoelastic state is a thermorheologically simple material. That is, time-temperature superposition can be applied given that the material is undamaged. As such, data from complex modulus testing conducted within linear viscoelastic limits at different frequencies and temperatures could yield a single continuous master curve for the variation of dynamic modulus with frequency at a given reference temperature.

However, for comprehensive material modeling, laboratory testing often extends to the damaged state where macrocracks in the asphalt concrete matrix start to develop. If it can be shown that time-temperature superposition holds for that state, laboratory testing required for comprehensive material characterization can be reduced. For that purpose, the research group conducted preliminary constant crosshead rate tests in uniaxial tension mode at different temperatures and strain rates. Time-temperature superposition is deemed to hold if the shift factors determined for producing a continuous stress versus time master curve for a reference strain level are temperature dependent only and match those for the undamaged state.

Testing Program

Specimens used in this study are fabricated from 12.5 mm Maryland State Highway Administration Superpave mixtures. Specimens are 75 mm in diameter and 140 mm in height, cut and cored from 150 by 170 mm specimens compacted using the Superpave gyratory compactor, ServoPac. The testing machine used is UTM-25, a servo-hydraulic closed loop machine fabricated by IPC. Strains are measured by spring-loaded LVDTs; two with 75 mm gage length and two with 100 mm gage length. Using two gage lengths enables the determination of the onset of macro-cracking. The test protocol for the time-temperature superposition in tension consists of complex modulus (frequency sweep) testing followed by a constant crosshead rate in tension until failure.

Frequency Sweep Testing

Haversine loading in tension and compression sufficient to produce a total strain amplitude of 70 micro-strains was applied at 20, 10, 3, 1, 0.3, and 0.1 Hz and at -10, 5, 25, and 40 C. The shift factor, a_T, used to shift the dynamic modulus vs. frequency curves at -10, 5, and 40 C along the frequency axis to form a continuous master curve at 25 C (Figure 6), is defined as follows:

$$\text{Log}(F_R) = \log(F \times a_T)$$

where F_R = reduced frequency at reference temperature (25 C);
F = frequency before shifting for a given temperature; and
a_T = shift factor for temperature T.

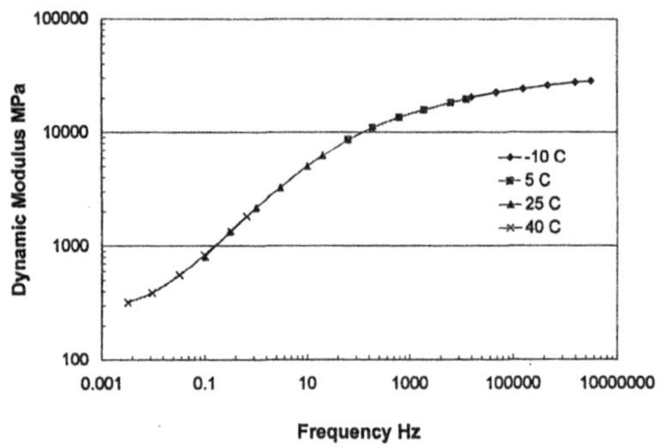

Figure 6: Master curve for dynamic modulus based on a reference temperature of 25 C

Time-temperature superposition of damaged asphalt concrete

After frequency sweep testing and after allowing enough time for any accumulated strain to be recovered, the specimen is pulled at a constant crosshead rate until failure. Testing temperatures are the same as those of frequency sweep, while crosshead rates, as originally proposed, are 0.0005, 0.0015, 0.0045, and 0.0135 strains per second. Several strain levels for which the master curves are to be constructed are selected given they are common to all tests. For each selected strain level and testing temperature, the corresponding stress level and time obtained from the strain rates used are cross-plotted to form a stress versus time cross plot. This is repeated for all testing temperatures and selected strain levels. Then, to construct the master curve at 25 C for a given strain level, the stress-time cross plot for that strain level and for each temperature is shifted along the time axis using the shift factor a_T defined as follows:

$$\text{Log}(t_R) = \log(t/a_T)$$

where t_R = reduced time at 25 C,
t = time before shifting for a given temperature, T, and
a_T = shift factor for temperature T.

If the time-temperature superposition is valid at varying damage states, the shift factors determined from the dynamic modulus tests should be applicable to the stress-strain data with damage.

Since shift factors a_T are not yet known, those from dynamic modulus are used to estimate the LVDT strain rates. Using earlier developed relationships between crosshead and LVDT rates, the corresponding crosshead rates are then determined and applied.

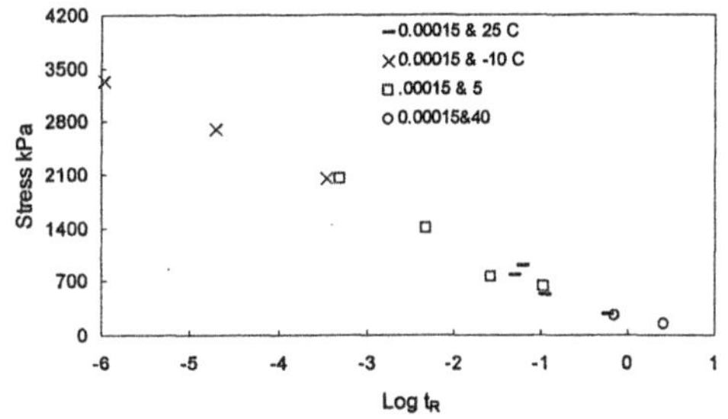

Figure 7: Master curve for a strain level of 0. 00015 and a ref. temp. of 25C

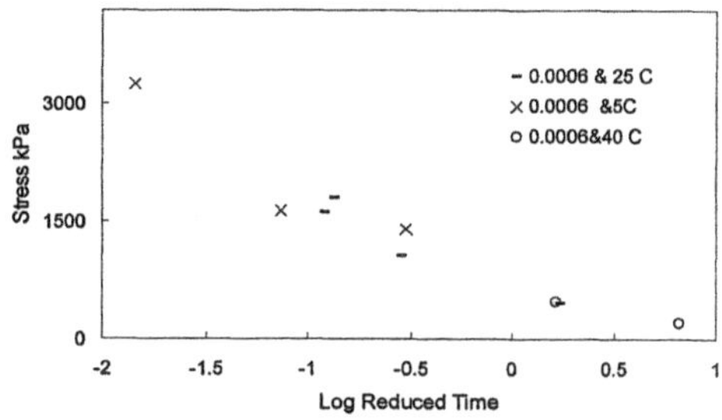

Figure 8: Master curve for a strain level of 0.0006 and a ref. temp. of 25C

Figure 7 is a master curve for a strain level of 0.00015 using tests done at -10, 5, 25, and 40 C. Figure 8 is a master curve for a strain level of 0.0006; since this strain is not achieved at -10 C testing conditions, only those conditions at 5, 25, and 40 C are included in the analysis. The shift factors used for both curves are the same as those from dynamic modulus. Master curves for five strain levels at a reference temperature of 25 C are presented in Figure 9. Individually, all the master curves constructed using shift factors from non-damaged state, complex modulus testing, are continuous and overlap at long reduced time values. For shorter reduced times the curves start to deviate to the right, higher stresses for a given reduced time, as strain level increases.

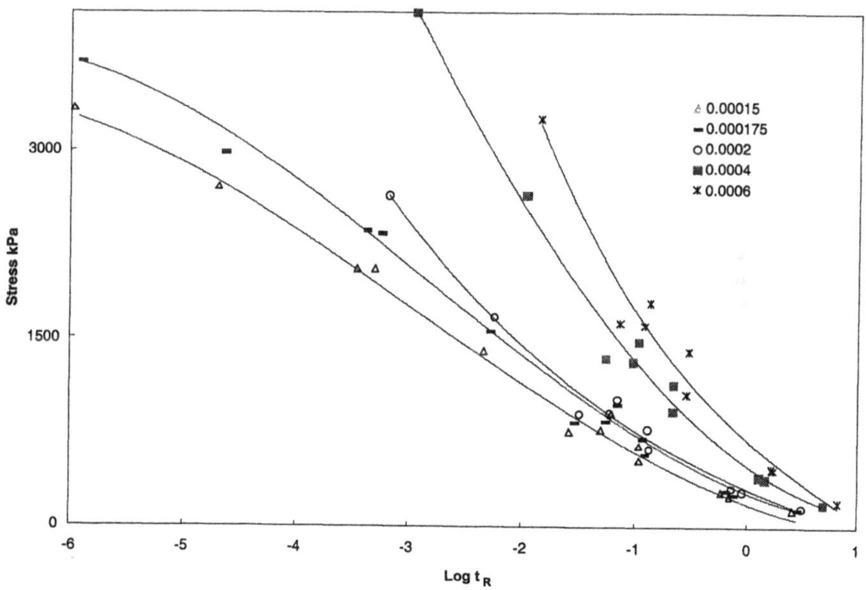

Figure 9: Master curves at 25 C for strain levels of 0.00015, 0.000175, 0.0002, 0.0004, and 0.0006

In spite of the problems encountered in constant crosshead rate tests in tension, preliminary results regarding applicability of the time-temperature superposition principle for damaged state are promising. Shift factors determined for the non-damaged state were used to construct master curves for low strain levels, pre-peak behavior, for the damaged state. More replicates at current and additional strain rates need to be tested to complete the master curves at current and higher strain levels that represent peak and post-peak behavior.

ACCELERATED PAVEMENT TESTING

Using the third scale Model Mobile Load Simulator (Mk.3), fatigue and rutting performance of different asphalt concrete mixtures can be tested in a controlled environment. This testing has the capability of bridging the gap between field and lab testing. Through instrumentation of the pavement, resulting data can be used in verification of the performance prediction models.

Third Scale Model Mobile Load Simulator

The Mk. 3 is a third generation machine. Initially, the Texas Mobile Load Simulator (TxMLS or MLS) was developed by the Texas Department of Transportation (TxDOT), the Center for Transportation Research (University of Texas at Austin), and the Victoria Machine Works (VMW) to simulate full scale loading on pavement in a mode and method better than previous Accelerated Pavement Testing (APT) devices. To accomplish this, standard truck components were implemented to the extent that it was feasible.

The Mk.3 is able to simulate a Super-single tire at a $1/4^{th}$ scale or one tire in a dual tire system at a $1/3^{rd}$ scale with a maximum load of 2700N. The Mk.3 has a tire width of 80mm and is capable of wandering from zero to 80mm following a general load distribution curve. The maximum speed of the Mk.3 is 9 km/hr, which is approximately 7200 wheel applications/hr. Looking at the Mk.3, it is about 270 cm long and 80 cm wide. It has four wheels that are loaded by two springs on either side. The wheels are connected and run in a conveyor belt fashion. Each of the four feet that support the Mk.3 are set into skate boards that run perpendicular to the movement of the wheels allowing for lateral wandering. A rounded guide a couple centimeters off the pavement aids in reducing the reaction as the wheels are introduced to the pavement. Further information on the MLS, Mk.10, and Mk.3 can be found in Hugo (1994), Kim et al. (1998) and McDaniel and Hugo (1998).

Pavement Instrumentation

To give a complete picture of a pavement's behavior, various sensors were used. Strength, temperature, strain due to bending, deformation, and cracking were the areas of interest. A short discussion of each of these follows.

Stress wave testing is used to determine the strength of the pavement. Accelerometers are located at 15 cm spacing to receive the signals produced by the impact of a marble. These signals are displayed on an oscilloscope in their raw form and stored in a computer for analysis. A stress wave analysis program (SWAP) has been developed at NC State to analyze this raw data. First, the dominant frequency is determined through a Fast Fourier Transform (FFT). Using this dominant frequency, the Short Kernel Method is applied to the data. This method filters out all of the other frequencies that are represented in the data taken by the accelerometers. By filtering the raw signal, a pattern can generally be seen between each of the four sensors. This pattern represents the arrival of the wave at each of the accelerometers.

Figure 10 shows the arrival at Channel 1 as peak 1 and the arrival at channel 2 as peak 2. The distance between sensors and time from the peak arrivals allows calculation of the speed of the wave, called the phase velocity. The phase velocity and the mid-depth temperature can be used to determine the apparent modulus of the pavement slab. Further studies on stress wave analysis and SKM can be seen in Katzke (1997) and Kim (1997).

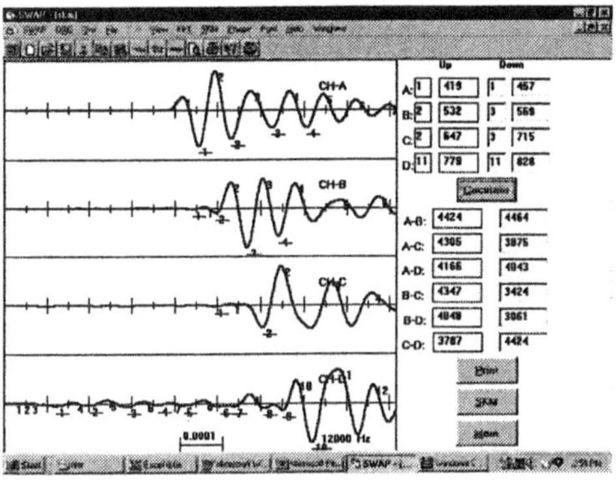

Figure 10: The Short Kernel Method

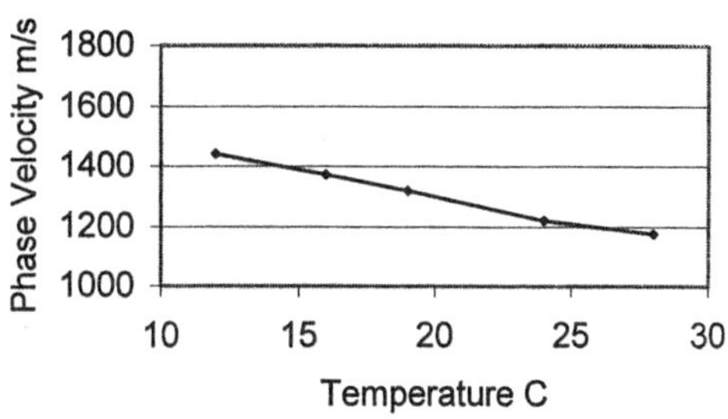

Figure 11: Location 1 pre-loading temperature profile

The strength of an asphalt concrete pavement is very susceptible to temperature. Thermocouples were placed at the top, bottom, and middle of the pavement to keep a record of the pavement temperature, and monitor the effect of the loading on the pavement. Due to the unknown increase in pavement temperature once loading occurred, a pre-loading pavement profile was done at five temperature settings. This profile can be seen in Figure 11. Knowing the effect of temperature on the phase velocity at a testing location, the initial point in the damage profile is determined by finding the phase velocity for the resulting test temperature. Generally, the increase in temperature due to loading is one degree Celsius.

Stripe-type strain gauges measure the strain induced at the bottom of the asphalt layer. Preliminary estimates of the strain expected here were analyzed with the current testing conditions using ELSYM5 a multi-layered elastic analysis program. The expected strain values are around 70 microstrains.

Two methods of deformation determination are being utilized. The first method called Multi-Depth Deflectometer (MDD) uses a dial gauge to measure the surface and the bottom of the asphalt layer. An increase in deformation will occur equally in both readings to represent the bending of the asphalt layer without deformation within the layer. The profilometer profiles only the surface of the pavement. Readings are taken perpendicular to the movement of the wheel at seven locations. A sample profile in Figure 12 shows a 1-mm rut that could not be detected by the naked eye.

To relate the decrease in apparent modulus (or phase velocity) with cracking, the area of cracking is determined. Prior to testing the pavement surface was painted white to help accentuate cracks. Digital pictures of the tested area are taken and traces of the pavement are made on Mylar paper to aid in this area calculation.

Currently, testing is being performed on a 3-inch pavement to look at healing of pavement. Healing is the strengthening of the asphalt concrete due to a rest period and the viscoelastic nature of the asphalt concrete. Microcracks begin, but are able to heal themselves when the pavement is not loaded.

The load being used in this test is 2600 N and the application rate is 6000 wheels/hr. To show that healing occurs it is necessary to maintain a constant cooler temperature to promote fatigue failure. To show the effects of healing, two pavements will be tested. The first, being currently tested, is run with only the necessary rest period to maintain the Mk.3 and perform desired tests. The second specimen will have 24-hour rest periods.

A preview of the second slab can be seen in the first slabs phase velocity vs. wheel loads plot, in Figure 13. Due to the unknown length of testing or maintenance after stopping the Mk.3, stress wave tests are done immediately after loading stops and immediately prior to loading. The increases in phase velocities and length of each rest period vary, but each rest period shows this trend of healing.

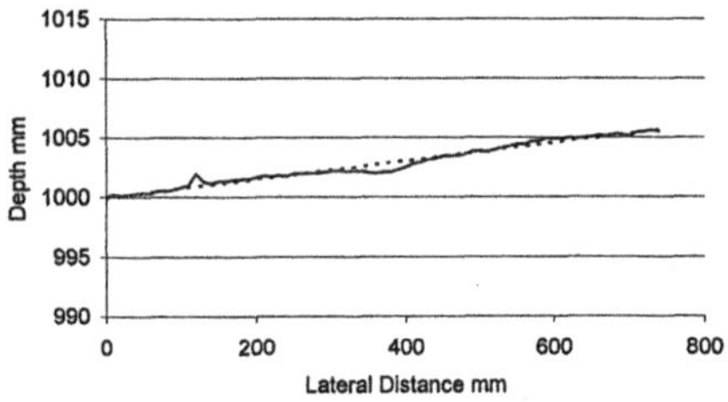

Figure 12: Surface profile

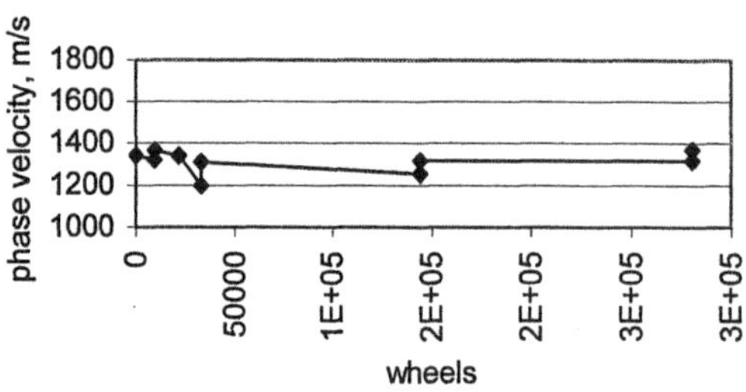

Figure 13: Damage test results as of 10/20/2000

ACKNOWLEDGMENT

This work is supported by the National Science Foundation. The work presented under Viscoelastic, Continuum Damage Modeling section was also supported by the National Cooperative Highway Research Program 9-19 project through the Arizona State University. All the specimens used in demonstrating the time-temperature superposition in damaged materials were fabricated by Arizona State University. The authors greatly appreciate Professor Matt Witczak at ASU, the PI of the NCHRP 9-19 project, for his support in this research.

REFERENCES

Hugo.F. (1994). Some Factors Affecting the Design and Use of the Texas Mobile Load Simulator. *ASTM STP* 1225, 67-88.

Katzke, Evan D. (1997). Nondestructive Evaluation of Asphalt Pavement Surface Layers Using Stress Wave Testing. M.S. Thesis. North Carolina State University, Department of Civil Engineering, Raleigh, NC.

Kim, S. M., Hugo, F, and Roesset. (1998). Small-Scale Accelerated Pavement Testing. *Journal of Transportation Engineering*, March/April, 117-122.

Kim, Yongon. (1997). In-Situ Evaluation of Microcrack Damage and Healing of AC Pavements Using StressWave Tests, Ph.D. Dissertation, N.C. State University, Raleigh, North Carolina.

McDaniel, M., and Hugo, F. (1998). The Use of the MMLS to Investigate the Fatigue of Asphalt Pavements at Low Temperatures. Paper No. 981097 presented at the 77th Annual Transportation Research Board meeting, Washington D.C.

Appendix

FUTURE RESEARCH TOPICS SUGGESTED AT NSF WORKSHOP ON LONG TERM DURABILITY BERKELEY, OCT 26-27, 2000

Broad Research:

1. "General model" for durability prediction for different classes of materials. Extrapolation of durability behavior of conventional material to new materials.
2. Develop graphs that indicate the durability performance of materials exposed to various environmental conditions ("Ashby" map).

Damage Characterization and Techniques:

1. Develop short-term tests, which can truly predict long term behavior under realistic service and environmental conditions.
2. Develop standard measurement methods for damage characterization.
3. Role of "initial" and "damage-induced" anisotropy.
4. Include uncertainty in processes and data to improve life prediction confidence.
5. Characterize damage under complex loading (both static and cyclic).
6. Use of neural network for durability characterization and prediction.
7. Information and model uncertainty.
8. Experimental testing for micro/nanoscopic devices and samples under multi-axial fatigue testing at different temperatures and humidity levels.
9. Central facility in nano-fabrication (NSF sponsored facility already exists - NNUN = National Nanofabrication Users Network).

Linking Scales:

1. Identify source and initiation of damage - link nano/micro scale processes (at nano/microstructural level) to macro-damage behavior at material level.
2. Relate damage behavior to structural performance - link material to structural level.
3. Link up between laboratory durability study and field performance of actual structures.

Durability Enhancement:

1. Science based approach to develop materials/structures which avoids the lack of durability in current materials/structures.
2. New solutions/approaches which avoid deterioration in the first place - eliminate source of damage which evolves into poor durability.
3. Integrate smart design into structural systems to avoid durability problem.
4. Implementation of durability enhancement approach into structural design.

Specific Materials/Structures:

1. Durability of repair materials and repaired structures/systems.
2. Durability of new materials, including smart materials, nano-materials, bio-materials, microstructure tailored materials.
3. Durability of systems requiring extremely long life, such as nuclear waste disposal (over 1000 years).

Others:
1. What to do with current large amount of data on durability?
2. Identify relationship between aging and long term performance.

AUTHOR INDEX

KEYWORD INDEX

FSC
www.fsc.org
MIX
Papier | Fördert
gute Waldnutzung
FSC® C083411

Kolibri 360 GmbH
Ferdinand-Jühlke-Straße 7
99095 Erfurt, Deutschland
produktsicherheit@kolibri360.de